ARITHMETIC,

PRACTICALLY APPLIED,

FOR ADVANCED PUPILS,

AND FOR PRIVATE REFERENCE,

DESIGNED AS A

EQUEL TO ANY OF THE ORDINARY TEXT-BOOKS ON THE SUBJECT.

BY HORACE MANN, LL.D.,
THE FIRST SECRETARY OF THE MASSACHUSETTS BOARD OF EDUCATION,

AND PLINY E. CHASE, A.M.,
AUTHOR OF 'THE COMMON-SCHOOL ARITHMETIC.'

PHILADELPHIA:
PUBLISHED BY E. H. BUTLER & CO.
1857.

JUST PUBLISHED,

THE ELEMENTS OF ARITHMETIC, PART FIRST, for Primary Schools. The SECOND PART, for Grammar Schools, which is in course of preparation, will complete the following Arithmetical Series:—

ELEMENTS OF ARITHMETIC, PART FIRST. By HORACE MANN and PLINY E. CHASE.
ELEMENTS OF ARITHMETIC, PART SECOND. By HORACE MANN and PLINY E. CHASE.
COMMON-SCHOOL ARITHMETIC. By PLINY E. CHASE.
ARITHMETIC PRACTICALLY APPLIED. By HORACE MANN and PLINY E. CHASE.

E. B. MEARS, STEREOTYPER. C. SHERMAN, PRINTER.

MR. MANN'S PREFACE.

The appearance of my name on the title page of this Arithmetic, requires me to state the extent of my connection with its authorship, and of my responsibility for its execution.

Believing the idea of the work to be original, I will attempt its elucidation. In seeking for the elements or materials of its questions, it proposes to take a survey of all the vocations of life, of all the facts of knowledge, and of all the truths of science, and to make a selection from each department of whatever may be most interesting and valuable. It does not confine itself to the playthings of the nursery, or to the commodities of the market place, and to the *money* they will cost, or make, or lose. On the contrary, the present work proposes to carry the student over the wide expanse of domestic and social employments; to introduce him to the various departments of human knowledge so far as that knowledge has been condensed into tables, or exhibited in arithmetical summaries, and to make him acquainted with many of the most wonderful results which mathematical science has revealed. Instead of groping along the mole-path of an irksome routine, with little other change than from dollars and cents, to pounds and pence, or some other familiar currency, and with little other variety than from cloth to corn, or some other common-place commodity, it derives its examples from biography, geography, chronology, and history; from educational, financial, commercial, and civil statistics; from the laws of light and electricity, of sound and motion, of chemistry and astronomy, and others of the exact sciences. Trades, handicrafts, and whatever pertains to the useful arts, so far as they are the subject of numerical statement, and their facts possess arithmetical relations, together with all the ascertained and determinate results of economical or political knowledge, and of scientific discoveries, are laid under contribution, and are made to supply appropriate elements for the questions on which the youthful learner may exercise his arithmetical faculties.

In this way, and without departing from the most rigid rules on which an arithmetical text-book should be constructed, I have supposed that a work may be prepared which shall exemplify in the best manner the science of numbers, and be full of useful knowledge also; and which, while it exercises the student's powers of calculation, shall enlarge his acquaintance with the varied business of the world, and with many of the most interesting results of applied science.

In a universe like this, where every star has been weighed in a mathematical balance, and all inter-stellar spaces have been measured by a mathematical line; where the orbits of all the planets have been traced as by a compass, and their velocities graduated to their distances by an unchanging law; where not only wind and tide, but every particle of dust in a hurricane, and every drop of water in a cataract, know their exact places by an infallible rule; where the gravitation of matter, the radiation of heat, and the diffusion of light, at all times and instantaneously, adjust their force to their distance with unerring precision; where every chemical combination is formed on some fixed principle of proportion, and the atoms of every crystal arrange themselves around their nucleus in geometric lines; and where, whatever other contemplations or volitions occupy the Infinite Mind, it is still true, as was said by the old Greek philosopher, that "God geometrizes;"—I say, in such a universe, built, weighed, measured, compounded, and arranged on mathematical principles, why should not the arithmetical exercises of those minds which have entered it, to dwell in it forever, embrace something more than the market price of commodities, the gain or loss in trade, and the interest and discount of banks?

Why, for instance, cannot a child be taught to count the bones in his hands, as well as the nuts in his pocket; to add together the number of bones in all the different parts of his body, or to subtract the number of those contained in his head or in his hands, from the number of those contained in his feet, as well as to add or subtract the number of apples or of cakes in the possession of James, John, and Joe; and why can he not, by such exercises, be led to enrich his mind with anatomical or physiological facts, instead of stimulating his imagination with the provocatives of appetite? Why cannot a child add together the population of the different States of this Union, or of the different nations of Europe or of the world, and thus learn the sum of the population of the whole earth and of its parts, as well as to add naked columns of abstract figures? In addition, subtraction, multiplication, and division, why cannot the pupil use examples whose elements or data are drawn from the distance between different historical epochs, from the ages of distinguished men, from the date of one discovery or invention to that of another discovery or invention, or from the rise to the fall of dynasties, or from a comparison of the heights of different mountains, or the lengths of different rivers, or of degrees of longitude on different parallels of latitude, or the distance from city to city by land, or from port to port by sea; and thus live in the perpetual presence and company of most important truths pertaining to history, chronology, biography, and geography, and so familiarize himself with these classes of facts, without devoting special time or effort to their acquisition, just as he becomes familiar with the faces and the names of his school-fellows and his townsmen, merely because he has always lived amongst them? If the arithmetical exercises of the pupil direct his attention almost exclusively to the shop of the retailer or the counting-room of the merchant, then he does not enjoy even a pedlar's opportunity to become acquainted with men, events, times, places, and things,—with the great results of business and of civilization, as they now exist in the world. Instead of wearying the learner with endless reiterations about bales, boxes, barrels, and bushels, or dollars, dimes, cents, and mills, why not open before him some of the vast storehouses of truth, and display some specimens of their endless variety and beauty? Let the teacher, taking the learner by the hand, follow the farmer, the craftsman, the architect, the manufacturer, the road-maker, the mill-wright, the ship-wright, the watch-maker, and the long catalogue of others who are employed in the mechanic arts, or other branches of useful industry; or, rising into the sphere of the educated or professional laborer, let him observe the optician, the electrician, the mathematical instrument maker, the astronomical observer, the telegraph operator, and borrow from them all some of the curious facts pertaining to their respective arts and professions, and convert them into the pleasing and instructive elements of arithmetical problems. I can conceive of a work so replete with the facts of technology and science, that it shall be examined with interest and profit by any one who is only seeking after valuable information, and which, at the same time, shall be perfectly adapted to the mere student who only seeks after the best means of arithmetical practice.

Among the advantages of such a work as is here proposed, the two following seem to me unquestionable:

1. The pupil, while studying Arithmetic for its own sake, will acquire some knowledge of many other things. Cicero observes that the face of a man will be tinged by the sun, for whatever purpose he may walk abroad. So, by daily familiarity, the mind of the student will be replenished with useful facts, and imbued with a scientific spirit, although the acquisition of the facts and the spirit be not the direct object of his study. So far as the examples of his text-book have been drawn from the actual business of life, as it goes on around him, the learner cannot but see that his studies have a practical bearing and are connected with obvious realities. A great number of facts, such as dates, sums, quantities, distances, will be impressed on his mind; and thus a species of most valuable knowledge, which, as we all know, it is most difficult to acquire in after-life, will be gratuitously bestowed upon him. Doubtless great differences will exist among pupils, in regard to the amount of information they will obtain from being domiciled, as it were, among the determinate truths of business, art or science; but even the most dull and stupid will be constrained to learn something, not only of the existence of various sciences, and of various kinds of business, of which they would otherwise be forever ignorant, but also of the nature and distinctive characteristics of those sciences or employments. All will be saved from great misconceptions; and doubtless the curiosity of many will be awakened for further information. These advantages embrace not only an increase of positive knowledge, but an enlargement of the mind's scope in regard to the subjects of knowledge.

It is unnecessary here to remind the observing man, how the understanding of any one thing, by the self-activity of the faculties, generates the power of understanding many other things, and each of these, in their turn, of many more, and so on in geometrical progression. *The knowledge of any one truth acts as an introducer and interpreter between us and all its kindred truths.*

2. There are advantages of another kind, which appear to me of not inferior magnitude, though I am fully aware that the views I am about to present, will affect different minds very differently, according to their Theory of Mind. My belief is that Arithmetic, in its strict and technical sense, addresses but one faculty of the mind, or, at most, but a very limited group of the faculties. No

other study pursued in our schools is so restricted, either in regard to the mental powers, which it calls into exercise, or the objects which it brings under their cognizance. Hence, during the hours devoted to Arithmetic in our schools, most of the mental faculties lie dormant, or play the truant by employing themselves upon forbidden objects. Doubtless this intense exercise of a single faculty, or of a limited group of faculties, and the non-exercise of all the rest, is one of the main reasons why Arithmetic, when not taught with great ability, is so often an irksome study. Reading and Geography, for instance, cover the widest field of interesting subjects, and it is impossible for any teaching to be so dull, or any circumstances so repulsive, as wholly to despoil them of their charms. If we would invest Arithmetic with similar attractions, we must draw its examples from as wide and as rich a field. Then will it interest new faculties,—faculties which, otherwise, it never addresses. All metaphysicians know, from the principles of their science, and all laborious students know, from their own experience, that nothing refreshes or re-creates a wearied faculty so certainly and so speedily as the genial exercise of some other faculty. Such exercise is more restorative than absolute quiescence. It is with the faculties of the mind as with the muscles of the body, they should have alternate exercise and rest; and the most healthful and agreeable relaxation for any organ that is tired of exercise, is the exercise of another organ that is tired of repose. The footman who travels over a long and level road, where the same muscles are subjected to a perpetual recurrence of the same strain, rejoices at the sight of a hill in the distance; for he prefers to put a new set of muscles to the hard service of carrying his body up a hill, rather than to compel the fatigued ones to continue their lighter task. Equally cheering and recuperative must the alternation be, when, after addressing one set of faculties with one combination of agreeable truths, we appeal to another set of faculties to determine certain arithmetical relations which exist between those truths.

Should any teacher dissent from this doctrine, that alternate exercise and rest, like the ever-alternating systole and diastole of the heart, is the law of all our powers, both bodily and spiritual; or should any pupil be found, whose senses are such perfect non-conductors of truth, as to exclude all information though the atmosphere by which he is surrounded is saturated with it, still, no loss or harm would be incurred; for the single process of arithmetical training can be carried on by the aid of the examples contained in this book as well as by those in any other.

Such is an outline of the present work. Since its conception first flashed upon my mind, I have pondered upon it much, and have conversed respecting it with many gentlemen not only of great mathematical attainments, but of varied scholarship, and both reflection and conversation have deepened my conviction of its value.

Justice to my associate, Mr. CHASE, requires me to say a word in regard to the shares of responsibility and of merit which attach to us respectively as the joint authors of this work. In communicating my plan to him, I unfolded its whole scope and purpose as it lay in my own mind; I indicated the various sources whence materials for its construction might be drawn, and I have rendered him some aid in the collection of those materials. The residue of the work,—the definitions, the rules, the statement of the questions, and the answers, so far as answers are given, together with the arrangement of the topics, or subjects,—is substantially his. So far as there is skill in their selection, or science in their statement, or accuracy in their results, I shall gladly join with others in awarding the merit to him.

HORACE MANN.

MARCH, 1850.

P. E. CHASE'S PREFACE.

THE following work does not profess to be a mere Arithmetic of the ordinary stamp. It takes ground that has hitherto been unoccupied, and its plan in most respects is entirely new.

In the "Common-School" Arithmetic the rules and principles that are usually taught, have been very fully explained and illustrated, but in that work, as in all of the other similar treatises that are now in general use, the primary object is merely the inculcation of processes,—the practical application of those processes being introduced only incidentally. But every teacher is aware that practical exercises, more numerous than it would be possible to insert in an elementary work, without rendering it of an inconvenient size, are necessary in order to make a thorough arithmetician. This necessity can often be but partially supplied, and consequently most pupils find whenever they enter into active life, that the calculations of their business, whatever that business may be, are all to be learned anew, and that all the Arithmetic they have studied at school is of little value, except for the expertness it may have given them in the simple operations of the fundamental rules.

I have long believed that a Sequel to Arithmetic,—a work not designed to take the place of any of the ordinary text-books, but absolutely requiring a familiarity with some one or more of them before it can be studied at all,—might be so prepared as to supply the want to which I have alluded. By giving numerous examples similar to those which are constantly occurring in the various walks of life, the student may be enabled to prepare himself better at school, for his future employment, and by the incidental introduction of subjects of general interest, the study may be made pleasant, and the thoroughness which is of the first importance in every undertaking may be more readily secured.

There are many difficulties connected with the selection and arrangement of materials for such a treatise, which may be urged in excuse for any deficiencies that exist in the present volume. I trust, however, that there will be few objections brought against the execution of our undertaking, for which the book itself does not afford a remedy. Any fault of arrangement may be easily corrected, by taking up the chapters in a different order from the one I have adopted; any redundancy of examples may be avoided, by omitting such portions as appear of the least importance; any deficiency may be supplied, by framing additional questions from the abundant materials, in the shape of tables and remarks, which are interspersed throughout the book. This latter peculiarity of our plan, we regard as one of its greatest recommendations. Teachers can make new questions for their classes, to any extent they deem expedient, and they may find it a most valuable exercise to allow their pupils to form questions for each other, from the data that are given in the text-book.

In many places, the explanations have been simplified by the use of letters. If their use should seem, at first, too algebraical, I think all objection will be removed upon examination of the manner in which they are employed. Every scholar who has any tolerable degree of familiarity with Arithmetic, should be able to reason as readily with *a*s and *b*s, as with apples and beans; and I flatter myself that the manner in which I have applied the simplest rules of analysis to operations with letters and symbols, will be of great use to all who study the following pages.

The character of the work having rendered it necessary that it should be, like other school-books, principally a compilation, I have endeavored to search for the *best* authorities, and to give due credit to the various writers that I have consulted. I am also indebted to officers of the government, and to teachers of some of our leading institutions of learning, for many valuable suggestions. I have added many rules and illustrations that are entirely original, and I hope that the result of our joint labors will be found a valuable companion in every schoolroom into which it may be introduced.

I consider myself most fortunate in having obtained the valuable assistance of my associate, HORACE MANN. His proposal, that the work should be rendered highly instructive as well as practical, will doubtless commend itself to general favor; and for any merit that may be found in the execution of the plan indicated in his preface, I am greatly indebted to the suggestions and criticisms with which he has favored me, during his revision of the manuscript and proof-sheets.

PLINY E. CHASE.

TABLE OF CONTENTS.

Art. XIII.—The Laboratory.

LIST OF AUTHORITIES.

Allen, W., American Biographical and Historical Dictionary.
American Almanac.
American Temperance Society, Reports.
Babbage, C., Economy of Machinery and Manufactures.
——— Bridgewater Treatise.
Bache, A. D., Reports on Weights, Measures and Balances.
Barlow, P., New Mathematical Tables.
Barnard, H., School Architecture.
Beckman, J., History of Discoveries and Inventions.
Belknap, J., American Biography.
Benjamin, A., Practice of Architecture.
Bigelow's Technology.
Biographia Americana.
Biographie Portative Universelle.
Bonnycastle's Arithmetic.
Boston Custom-House, Table of Currencies.
Bowditch, N., Practical Navigator.
British Husbandry.
British Sanitary Reports.
Budge, J., Practical Miner's Guide.
Buist, R., American Flower Garden Directory.
Carpenter, W. B., Mechanical Philosophy, Horology, and Astronomy.
Carpenters' and Builders' Assistant.
Cavallo, T., Elements of Natural Philosophy.
Chambers's Educational Course.
——— Information for the People.
Colman, H., Agricultural Reports.
——— European Agriculture.
Crossley & Martin, Intellectual Calculator.
Daniell, J. F., Chemical Philosophy.
Deane, S., New England Farmer.
Desaguliers, J. T., Course of Experimental Philosophy.
Draper, J. W., Text Book on Chemistry.
Emerson, G. B., Report on the Forests of Massachusetts.
Encyclopædia Americana.
——— Britannica.
——— Edinburgh.
——— Rees'.
Evans, O., Young Mill-Wright and Miller's Guide.
Ewing, A., Practical Mathematics.
Ferguson, J., Lectures on Select Subjects.
Gillespie, W. M., Manual of Road Making.
Gregory, G., Dictionary of Arts and Sciences.
Griffiths, T., Chemistry of the Four Seasons.
Hare, R., Chemistry.
Hatfield, R. G., American House Carpenter.
Herschel, Sir J., Outlines of Astronomy.
Hunt, F., Merchants' Magazine.
Hutton, C., Recreations in Mathematics and Natural Philosophy.

Ingram, A., Concise System of Mathematics.
Jackson, C. T., Report on the Geology and Mineralogy of New Hampshire.
Jamieson, A., Mechanics of Fluids.
Johnson, C. W., Farmer's Encyclopædia.
Johnson, G. W., Dictionary of Modern Gardening.
Johnson, L. D., Memoria Technica.
Johnson, W. R., Report on American Coals.
Keith, T., Complete Practical Arithmetician.
Lavoisne, C. V., Genealogical, Historical, Chronological, and Geographical Atlas.
Leslie, J., Philosophy of Arithmetic.
Library of Useful Knowledge.
Liebig, J., Animal Chemistry.
Lloyd, H., Lectures on the Wave Theory of Light.
London, Edinburgh, and Dublin Philosophical Magazine.
Loudon, J. C., Encyclopædia of Agriculture.
M'Culloch, J. R., Commercial Dictionary.
M'Gauley, J. W., Lectures on Natural Philosophy.
Mahan, D. H., Elementary Course of Civil Engineering.
Mass. Board of Education, Annual Reports.
Mint, U. S., Manual of Coins.
Moseley, H., Illustrations of Mechanics.
Murray, H., Encyclopædia of Geography.
Nesbit, A., Treatise on Practical Arithmetic.
Newton, I., Universal Arithmetic.
Nicholson, J., Operative Mechanic.
Norie, J. W., Epitome of Practical Navigation.
Owen, R. D., Hints on Public Architecture.
Parnell, E. A., Applied Chemistry.
Partington, C. F., Account of the Steam Engine.
Patent Office Reports.
Pierce, B., Elementary Treatise on Sound.
Pilkington, J., Artists' Guide and Mechanics' Own Book.
Pratt, J. H., Mechanical Philosophy.
Priestley, J., Description of a System of Biography.
Ranlett, W. H., The Architect.
Rogers, J., Vegetable Cultivator.
Scientific American.
Shaw, E., Practical Masonry.
Smeaton, J., Experimental Inquiry concerning the natural powers of water.
Somerville, M., Connexion of the Physical Sciences.
Tredgold, T., Principles of Warming and Ventilating Public Buildings.
Tucker, G., Progress of the United States in fifty years.
United States Almanac.
Ure, A., Dictionary of Arts, Manufactures and Mines.
——— The Cotton Manufacture of Great Britain.
Wade, J., British History, Chronologically Arranged.
Walsh, M., Mercantile Arithmetic.
Wistar, C., System of Anatomy.
Year Book of Facts.

ARITHMETIC.

I. SYMBOLS.

1. THE following characters, or symbols, are frequently employed in Arithmetic to represent the operations that are to be performed upon quantities:

The sign $+$ (*plus* or *and*) shows that the numbers between which it is placed, are to be added together. Ex.: $5+6$; $3+10+27$.

The sign $-$ (*minus* or *less*) shows that the quantity which follows it, is to be subtracted from the one which precedes it. Ex.: $11-8$; $29-16$.

The sign $=$ (*equal*) shows that the quantity which follows it, is equivalent to the quantity which precedes it. Ex.: $4+19=27-18+14$ (read 4 *plus* 19 *equal* 27 *minus* 18 *plus* 14); $65-27+3=41$.

The sign $\times$ (*times*) shows that the numbers between which it is placed, are to be multiplied together. Ex.: $4\times 15=60$; $3\times 10=6\times 5$. Multiplication is also sometimes expressed by a point, thus: $4\cdot 15=60$; $3\cdot 10=6\cdot 5$.

The sign $\div$ or $:$ (*divided by*) shows that the former of two quantities is to be divided by the latter. Ex.: $240\div 15=16$; $108:9=36:3$. Division is also represented by placing the dividend above, and the divisor below a horizontal line. Ex.: $\frac{276-95+15}{2\times 7}=14$.

A *vinculum* $\overline{\quad}$, or a *parenthesis* ( ), shows that several quantities are to be collected into one. Ex.: $(\overline{8-4}+\overline{16\div2})\times3=36$; $(\overline{2\times7}-11)\times3=\overline{14-11}\times3=3\times3=9$.

A small figure placed above any quantity, at the right hand, denotes a power of that quantity. Ex.: $9^2=81$; $(1+3)^3=64$; $((\overline{\overline{2\times5}-6}\times3)-10)^5=(\overline{\overline{10-6}\times3}-10)^5=(\overline{4\times3}-10)^5=(12-10)^5=2^5=32$.

The *radical sign* $\surd$, prefixed to a quantity, is used to denote some root of that quantity. If no figure is written above the sign, the square root is indicated. To express the 3d, 4th, or *n*th root, ^{3}, ^{4}, or ^{n}, is placed at the left hand. Ex: $\sqrt{6-2}=\sqrt{4}=2$; $\sqrt[3]{18+9}=\sqrt[3]{27}=3$; $\sqrt[4]{(8\times3)-(7+1)}=\sqrt[4]{24-8}=\sqrt[4]{16}=2$.

2. Examples to be read and solved by the Pupil.

1. $8+9+7-3+6-5-11=?$ *Ans.* 11.
2. $4-2+6-1+13\times$[a]$5=?$
3. $8\times$[a]$4-3+5+16-27=?$
4. $27\times3\times5\div45=?$ *Ans.* 9.
5. $(6\times8)+(\overline{42:14}\times\frac{10}{2})-(65\div5)=?$
6. $(\frac{27}{9}\times\frac{32}{4}\times3:2)+(\overline{15\times12}-10)=?$ *Ans.* 206.
7. $(4\times11\times3)-(\overline{6\div3}\times\overline{144:36}\times\frac{156}{13})=?$
8. $\sqrt{8+1+5-7+(6\times\overline{4+3})}=?$
9. $\sqrt[3]{27}\times\sqrt{64}\div\sqrt[5]{24+8}=?$
10. $\sqrt[9]{42+22-(37+\overline{3\times5})-\sqrt{87+(2\times3\times4+10)}}=?$

II. TEST QUESTIONS.

Every pupil should be required to give written answers to as many of the questions in this article as he is able to

[a] The pupil should be taught to say, 13 *times* 5, 8 *times* 4, rather than 13 *multiplied by* 5, 8 *multiplied by* 4.

solve. The teacher will thus understand the proficiency of each member of the class, and can give such explanations as may be most desirable. The questions should be reviewed from time to time, until they can all be answered without any mistakes.[a]

3. Questions on the Theory of Arithmetic.

1. What is Arithmetic? What is a number? In what different ways may numbers be expressed? What are the digits, and why are they so called? What different names has the character 0? Can you give any reason for calling naught the figure of place? What are the fundamental operations of Arithmetic? What is given, and what is required in each?

2. What is Numeration? What method of numeration is generally adopted? What is the peculiarity of that method? What is a unit? What is meant by the *simple* value of a figure?—by the *local* value? What is the difference between a *place* and a *period?* What names are given to the places in each period? How do you read units?—tens?—tens and units?—hundreds?—hundreds, tens, and units? Repeat the names of the first eight periods.[b] How do we read whole numbers? Can the Arabic method represent numbers less than a unit? What are such numbers called? How are they distinguished from whole numbers? How do we read decimals?

---

[a] "A teacher ought not only to instruct his pupils, but also to interrogate them frequently, and test their proficiency." *Quintilian.*

[b] The Numeration table may be continued to any extent we please. The periods above Sextillions, to Vigintillions, are, Septillions, Octillions, Nonillions, Decillions, Undecillions, Duodecillions, Tredecillions, Quatuordecillions, Quindecillions, Sexdecillions, Septendecillions, Octodecillions, Novemdecillions, Vigintillions. The following illustrations will show how difficult it is for us to form any distinct idea of the value of very large numbers.

If every man, woman, and child, on the face of the globe, were to count at the rate of four every second, without ceasing day or night, *the whole amount of all that they could count in* 5000 *years, would be less than one sextillion.*

If the sun, all the planets, and all the fixed stars that are visible through the most powerful telescope, were reduced to a powder so fine that 100000 grains would be less in bulk than a drop of water, and if one of these grains were destroyed in every million years, *the whole visible universe would probably be annihilated in less than one vigintillion years.*

3. What is the effect of removing a figure two places to the left?—two places to the right?—one *period* to the left?—to the right?—two periods to the left?—to the right? What is meant by Notation? How do you write whole numbers? In writing five sextillion, thirty billion, and seven thousand, what would you place in each period? If you were to omit the naughts in any one period, what effect would it have on all the figures above it? How do you write decimals? How would you make 5702 represent hundredths?—millionths?—thousands?—thousandths? What is the effect of placing naughts at the right hand of decimals?—at the left hand?—at the right hand of integers?[a]—at the left hand? What is the value of tens multiplied by thousands?—hundreds by hundreds?—ten-thousands by tens?—hundred-thousands by hundreds?—tens by hundreds?—thousands by thousands?—tenths by tens?—thousandths by hundreds?—thousandths by hundredths?—thousands by hundredths?—millions by millionths?—How would you represent 9000 by adopting 9 for the base of the numerical system?—by adopting 5 for the base?

4. What is FEDERAL MONEY? How may it be written? What is the probable origin of the sign that is usually prefixed to dollars? How may dollars be reduced to cents?—to mills? How may cents be reduced to dollars?—to mills? How may mills be reduced to cents?—to dollars? If there are decimals below mills, how may they be read?

5. Since removing the decimal point in any number changes the place of each figure, what is the effect of removing the decimal point two places to the left?—to the right?—three places to the left?—to the right?—seven places to the left?—to the right? Then how can you most readily find 10, 100, or 1000 times any number?—.1, .01, .001, or .0001 of any number?

6. What is ADDITION? What sign is used to denote addition? What is the sign of equality? How do you arrange numbers that are to be added together? Where do you commence the addition? What do you do with the sum of each column? Why do you not always place the whole sum of a column underneath the column? How do you find whether your answer is correct? Can any one of a series of numbers be greater than the sum of the whole series? How can you prove addition by casting out 9s?

7. What is SUBTRACTION? What is the remainder?—the subtrahend?—the minuend? What is the signification of the termi-

---

[a] An *integer* is a whole number: as *seven*, *forty-nine*.

nation *nd* in many arithmetical terms? If you add the difference of two numbers to the less number, what will you obtain? If you take the difference from the greater number, what will you obtain? How do you find the minuend, if the remainder and subtrahend are given?—the subtrahend, if the remainder and minuend are given? What is the sign of subtraction? Of what operation is subtraction the opposite? How do you write numbers in subtraction? Where do you begin to subtract? If any figure is greater than the one above it, what may be done? Explain the reason of this. What may be done when the subtrahend has more decimal places than the minuend? How do you prove subtraction? In finding the difference between two numbers, which must be the minuend? How can you tell which is the larger number? Can you think of any method of proving addition by subtraction? How can you prove subtraction by casting out 9s?

8. What is MULTIPLICATION? Of what is it an abbreviation? What is meant by the multiplicand?—the product?—the multiplier?—the factors?—a composite number? Name some composite numbers. What is the sign of multiplication? Can the multiplier and multiplicand exchange places? Give an illustration in proof of your answer. How do you multiply by a single figure? How may you multiply by a composite number? How do you proceed when the multiplier consists of a number of figures? How many decimals must there be in the product? What do you do if there are not decimals enough? What may be done, if there are naughts at the right hand of either factor? How can you most readily multiply by 10, 100, 1000, &c.? Explain the reason for the several modes of multiplication. How do you prove multiplication? What must you know before you can find the value of any number of things? How can you prove multiplication by casting out 9s?

9. What is DIVISION? Of what is it an abbreviation? What is meant by the dividend?—the quotient?—the divisor?—the remainder? Of what are the divisor and quotient factors? What are the different modes of expressing division? Give, in your own words, a rule for division, and explain the reason for each step of the process. Of what operation is division the opposite? How many decimals must the quotient contain? Why? How do you find the dividend, when the divisor and quotient are given?—the divisor, when the dividend and quotient are given? What is the difference between short and long division? What may be done, if there are naughts at the right hand of the divisor? How

can you most readily divide by 10, 100, 1000, &c.? How do you prove division? How can you avoid frequent trials, in finding the true quotient figure? When the value of any number of things is given, how do you find the value of one? When the value of a number of things is given, how can you find the number of things? Can you think of any method for proving multiplication by division? How can you prove division by casting out 9s?

10. When the SUM OF TWO NUMBERS and one of the numbers are given, how do you find the other? When the greater of two numbers and their difference are given, how do you find the less? When the less of two numbers and their difference are given, how do you find the greater? When the factors are given, how do you find the product? When the product and one of the factors are given, how do you find the other factor? When the value of any number of things is given, how do you find the value of any *other* number? By which of the fundamental rules do you increase a number? By which do you diminish a number? In what cases would you employ addition?—subtraction?—multiplication?—division? What is meant by the 3d,—7th,—15th power of a number?

11. What is a PRIME NUMBER?—a composite number? Name all the prime numbers less than 100. How can you tell whether any number can be divided by 2, 5, 3, 9? How do you find the prime factors of any number? How do you determine whether any number is a prime? What is meant by a multiple?—a common multiple?—the least common multiple? How do we find the least common multiple of two or more numbers? Show that this process will give the least common multiple. What is a common divisor?—the greatest common divisor? How do we find the greatest common divisor? Why do we proceed in this manner? What is cancelling? How do you cancel? When you have cancelled all the numbers in either the dividend or divisor, what must be put in their place? When the divisor is a composite number, how may we obtain the quotient?

12. What are FRACTIONS? In how many different ways may they be read? Show why each of these ways may be adopted. What is meant by the numerator?—the denominator?—the terms of a fraction? What is shown by the numerator?—by the denominator? How is the value of a fraction affected by increasing the denominator?—the numerator?—by diminishing the numerator?—the denominator? What is a mixed number? How would you find what part 324 is of 187? Why?

13. What is REDUCTION? How do we reduce fractions to whole numbers?—to mixed numbers?—to decimals? Explain these reductions. What are proper fractions?—improper fractions? When is a fraction greater than 1?—less than 1?—equal to one? How would you write a decimal in the form of a fraction? Explain the rule for reducing whole or mixed numbers, or decimals, to fractions. Explain the method of reducing compound to simple fractions; fractions to their lowest terms; fractions to a common denominator; complex to simple fractions; fractions to others having any given numerator or denominator. Explain the rules for addition, subtraction, multiplication, and division of fractions. What is meant by an integer?—by the reciprocal of a number? Take examples of your own, and form rules for finding the sum and the difference of two fractions, each of which has 1 for its numerator;—for finding the sum and the difference of two fractions which have the same numerator?

14. What are INFINITE DECIMALS? What is the repetend? How is it distinguished? To what is every repetend equivalent? Show that it is so. How then would you reduce any infinite decimal to a fraction? How may addition and subtraction of infinite decimals be performed without reducing them to fractions?

15. What are COMPOUND NUMBERS? Repeat the table of Federal Money; English Money; Troy Weight; Apothecaries' Weight; Avoirdupois Weight; Long Measure; Cloth Measure; Square Measure; Cubic Measure; Dry Measure; Liquid Measure; Circular Measure; Time Measure. How many shillings in a guinea? How many dollars in a pound sterling? How do we find the area[a] of a rectangular surface? How many cubic feet make a foot of wood, or a cord foot? How many cord feet in a cord? How many cubic feet in a ton of round timber?—a ton of square timber?—a ton of shipping or storage? How do we find the solid contents of any rectangular solid?

16. For what purpose are each of the weights and measures used? How many cubic inches in a half peck?—in a common gallon?—in a gallon of milk or of malt liquor? For what purpose is the hogshead (measure) used? How many degrees in a quadrant?—in a sign of the zodiac? How many days in a leap year? What years are leap years? Repeat the number of days in each month. In mercantile business how many days are considered as making a month?

---

[a] The number of square feet, rods, or acres, &c., in any surface, is called its *area*.

17. Give, in your own language, a rule for reducing higher denominations to lower; lower denominations to higher; for compound addition; compound subtraction; compound multiplication; compound division. Explain each rule. Can fractions and decimals be reduced by the same rules as whole numbers?

18. What is meant by PER CENT.? How do we compute any required percentage? Why? What is meant by commission?—insurance?—premium?—policy?—underwriters?—taxes?—stocks?—the par value?—the dividend on a stock? When is a stock said to be above par?—below par?—at a discount?—at an advance? What part of the original cost is stock worth, that sells at 16⅔ per cent. below par?—at 12½ per cent. advance?—at a discount of 17⅞ per cent.?—at 25 per cent. above par? What is meant by duties?—specific duties?—ad valorem duties?—draft?—tare?—gross weight?—net weight? In custom-house business, what allowance is made for leakage and breakage?—for draft? How do we find what per cent. is gained or lost by any transaction?

19. What is INTEREST?—simple interest?—compound interest? What is the principal?—the rate?—the amount? Give, in your own language, the General Rule for Interest;—the Bank and Business Rule. Show how each rule is obtained. What is a promissory note?—an endorsement? What is the usual mode of computing interest on notes that are settled within a year from their date? What is the *Legal* Rule? How do we find the *amount* of any sum at compound interest? How do we find the *compound interest?*

20. How do you find THE RATE, when the principal, interest, and time are given?—the time, when the principal, interest, and rate are given?—the principal, when the time, rate, and interest are given?—the principal, when the time, rate, and amount are given? Explain the reasons for each of these methods.

21. What is DISCOUNT?—equitable discount?—bank discount? How do you find equitable discount?—bank discount? What are days of grace? How many days of grace are usually allowed?

22. What is ANALYSIS? What is meant by known and unknown quantities? What should we endeavor to do, in solving difficult questions? How may this usually be done? Give a General Rule for analysis.

23. What is RATIO? How is it expressed? How may it be indicated? Which is the usual mode, and how is it read? What is the antecedent?—the consequent? How do you find the ratio between different denominations?

24. What do you understand by REDUCTION OF CURRENCIES? Do the people of the United States ever have occasion to make such a reduction? Why? Name the currencies in common use. How may each be reduced to Federal Money? How may Federal Money be reduced to each currency? How do you reduce English Money to Federal Money, including the premium of exchange? Federal Money to English Money? Give your reasons. What is the cause of the premium?

25. What do you understand by PRACTICE? Repeat the table of parts of a dollar. Give some examples illustrating the use of the table. What is meant by given terms?—by terms of demand? Give the rule for complex analysis, and illustrate it by an example of your own.

26. What are DUODECIMALS? In what are they used? What are the denominations of duodecimals? How are they marked? What are those marks called? How do you add or subtract duodecimals? How do you find the product of any two denominations? Why? Multiply 4 7′ 10″ by 3″ 5‴. How do you find the quotient of any two denominations? Why? Divide 8′ 2″ 6‴ 10⁗ 4′′′′′ by 2′ 8″ 4‴.

27. What is a PROPORTION? How is it usually written, and how read? In what other ways may it be written? Why? What are the extremes of a proportion?—the means? Prove that the product of the extremes is equal to the product of the means. If any three terms of a proportion are given, how may the other be found? Which term usually represents the unknown term? How do you find the fourth term? What class of questions may be stated in the form of a proportion? How would you arrange the first and second terms, if you wished the fourth term to represent the multiplication of the third term by a ratio?—the division of the third term by a ratio? What do you understand by direct, and by inverse ratio?

28. What is placed as the THIRD TERM in a proportion? State the rule for simple proportion. What is this rule sometimes called? Is the value of the third term ever affected by more than one ratio? Illustrate your answer by an example, and state that example in the form of a compound proportion. Why do you invert the several ratios by which the third term is to be multiplied? State the rule for compound proportion.

29. What is the origin of ARBITRATION OF EXCHANGE? Give an example. In what different ways may arbitration be effected? Which is the usual method? Repeat the Chain Rule.

30. What is FELLOWSHIP? How may it be solved by Proportion? What is Compound Fellowship? Give the rule. Explain examples of your own, both by analysis and by proportion, in Simple and Compound Fellowship.

31. What is meant by AVERAGE? Give an example. How do you find the average of a series of quantities? What is Alligation? —alligation alternate? What is the rule for alligation medial? What is Equation of Payments? How does it differ from alligation medial? Give, the rule. On what supposition is this rule founded? What is done with fractions of a day? Give the rule for alligation alternate. Explain the rule by an example of your own.

32. What is INVOLUTION? What is the first power of a number? —the second power?—the fifth power?—the eleventh power? What is the exponent? What is the second power often called, and why?—the third power?

33. What is EVOLUTION? What is the root of a number? What is the radical sign, and how is it used? What is the square root? How may we determine the number of figures that the square root will contain? Explain the method by which the square root is found, by multiplying 39 by 39 and extracting the square root of the product. What must be done when any trial divisor is not contained in the dividend?—when any figure obtained for the root proves too large? How may approximate roots be obtained?

34. What is the CUBE ROOT of a number? How may we determine the number of figures that the cube root will contain? Explain the method by which the cube root is found, by raising 39 to the third power, and extracting the cube root of the result. What must be done when the trial divisor is not contained in the hundreds of the dividend?—when any figure obtained for the root is too large? How may approximate roots be obtained? How may the trial divisors, after the first, be conveniently found?

35. What is a SERIES?—a natural series?—an arithmetical series?—a geometrical series?—a harmonical series? What is meant by Arithmetical Progression? What are the extremes?—the means?—the common difference? Explain the rule for finding one of the extremes and the sum of all the terms; for finding the common difference and sum of the terms; the number of terms and the sum of the terms; the number of terms and common difference; one extreme and the common difference.

36. What is meant by GEOMETRICAL PROGRESSION?—by the ratio?

—the extremes?—the means? Explain the rule for finding the last term, when the first term, ratio, and number of terms are given; for finding the sum of all the terms. How may compound interest be found by series?—by the table? How may we find by the table, the present worth of any sum at compound interest?—the time in which any principal will amount to a given sum?—the rate at which the principal will amount to a given sum in a given time?

37. What is an ANNUITY?—an annuity certain?—a contingent annuity?—a perpetual annuity?—an annuity in possession?—an annuity in reversion?—the present worth of an annuity? How do you find the amount due on an annuity?—the present worth of an annuity certain?—the present worth of a perpetual annuity?—the present worth of an annuity in reversion?—an annuity, the present worth being given? What does permutation show? Explain the rule.

4. PRACTICAL QUESTIONS.

All the information necessary to enable the pupil to answer these questions, will be found in the body of the work.

1. How do you find the price at which goods must be sold, in order to gain any required amount? 2. How do you find the original cost, the selling price and the loss being known? 3. How do you find the whole amount of an invoice, from the original cost and the charges upon the merchandise? 4. How do you find the amount gained by any sale? 5. The loss upon a sale? 6. The net proceeds of a sale? 7. The balance of an account? 8. The average price of several ingredients? 9. The quantity of several ingredients, that will make a mixture of a given average value?

10. How would you determine the number of feet of boards[a] in the floor of a room? 11. The number of bricks in the walls of a house? 12. The number of shingles or slates[b] on a roof? 13. The number of clapboards[c] on a house? 14. The amount of painting and plastering[d] in a house? 15. The quantity of timber in the frame of a house? 16. The cost of glazing[e] a house?

---

[a] Lumber is measured by Board Measure, 1 foot $= \frac{1}{12}$ cub. foot.

[b] 1000 shingles, or 500 slates, are allowed to 100 square feet.

[c] Give the *data* requisite to determine the number.

[d] Painting and plastering are estimated by the square yard.

[e] Glazing is estimated by the square foot. Builders usually furnish windows by the piece.

17. How would you find the amount of earth to be removed in excavating a cellar,[a] when the surface of the ground is level? 18. When the surface is uneven?[b] 19. The quantity of gravel[c] required for making a road? 20. For an embankment? 21. For an embankment on uneven ground, the top of the bank being level? 22. The quantity of stone in a wall?[d] 23. In estimating walls, how would you allow for corners? 24. For windows, doors, gates, or other openings?

25. How would you estimate the number of yards of carpeting required for covering a floor? 26. The number of rolls of paper[e] necessary for papering a room? 27. The quantity of canvass in the sails of a ship? 28. The weight of a stone wall or pillar?[f] 29. The weight of iron pipes or pillars?[f] 30. The weight of lead pipe?[f] 31. The weight of a brick wall?[f] 32. The weight of a wooden bridge?[f] 33. The weight of wire cables?[f] 34. The weight of a ship in the water?

35. How many days from June 3d to August 19th? 36. From May 24th to September 9th? 37. From Jan. 7th, 1848, to March 6th? 38. From Feb. 27th, 1859, to Nov. 11th? 39. Give the shortest method that you know for finding the number of days between any given dates. 40. If you give a note on Monday, payable in 90 days, on what day of the week will it become due?[g] 41. On what day would a note at 60 days become due, if given on Saturday? 42. On Thursday? 43. On Tuesday? 44. On Friday? 45. On Wednesday? 46. A note at 30 days, given on Wednesday? 47. On Tuesday?[h] 48. On what day of the month would a note fall due, if dated June 27th, at 30 days? 49. March 9th, at 60 days?[i] 50. Aug. 1st, at 90 days? 51. May 18th, at 3 months?

---

[a] Excavations are estimated either by the cubic yard, or by the "square", of 216 cubic feet.

[b] In excavating uneven ground, the depth may be measured at several different points, and the average of all the measurements taken.

[c] Gravel is estimated by the "square" (see note [a].)

[d] Stone is measured by the "perch" of $24\frac{3}{4}$ cubic feet.

[e] A roll of paper-hangings contains $4\frac{1}{2}$ square yards.

[f] See Table of Specific Gravities.

[g] 93 days = 13 weeks and 2 days. The note will therefore become due on Wednesday.

[h] If the last day of grace falls on Sunday, or on a holiday, the note must be paid on the day previous.

[i] 9 and 63 are 72. Deducting 31 days for March, 41 remains. Deduct 30 days for April, 11 remains. The note will be due May 11.

52. October 31st, at 4 months?[a] 53. Sept. 25th, at 6 months? 54. If the 1st of June falls on Monday, how many days' interest shall I lose by giving my note at 60 days, dated May 3d? 55. How would you determine the distance of a cannon, or a thunder cloud, by observing the flash and report?[b] 56. Suppose the nearest fixed star to be suddenly destroyed, how long would its light continue to reach us?[c] 57. How long would it take a locomotive, at the rate of 30 miles an hour, to travel the same distance that light goes in a single second? 58. To travel as far as the distance from the earth to the sun?[d] 59. How far does the earth move in its orbit, while a ray of light is coming from the sun?[e]

60. Can you tell why a difference of 15° in longitude, makes an hour's difference in time?[f] 61. In latitudes where a degree of longitude is equivalent to 45 miles, how far must you travel to find 10 seconds difference of time? 62. If you travel east, will you find the time earlier or later than in the place from which you start? 63. Why? 64. When it is noon at Philadelphia, what time is it at New York? 65. Boston? 66. St. Louis? 67. London? 68. Paris? 69. Washington? 70. Portland? 71. St. Petersburg? 72. Canton? 73. Astoria? 74. Suppose the chronometer of a vessel to be regulated by the Boston time, in what longitude is the vessel, if the sun passes the meridian at half-past nine by the chronometer? 75. If an unbroken telegraph wire should be extended from Boston to Oregon City, at what time could the people at the latter place hear of a transaction that occurred in Boston at 20 minutes past 3 P. M.?[g] 76. What place would the same news reach by telegraph, at noon of the same day? 77. If your watch keeps accurate time, and is correct when you leave

---

[a] A note for months, falling due on the 31st, is considered as due on the *last day* of the month.

[b] Sound moves at the rate of a mile in about 5 seconds.

[c] The distance of the nearest fixed star is upwards of 21000000000000 miles. Light moves at the rate of 192000 miles in a second.

[d] The mean distance of the earth from the sun, is about 95 million miles.

[e] The circumference of the earth's orbit is about 600 million miles.

[f] How many degrees does any point on the earth's surface pass over in a day?

[g] No perceptible time will elapse during the passage of electricity over the wire.

Boston, how would you set a watch right that is five minutes slower than your own when you reach Buffalo?

78. How would you determine the area of any piece of land, if you had no measure but a yard-stick? 79. Knowing the width of a rectangular field, how would you find the length of a strip that would contain an acre? 80. How would you estimate the area of a field by pacing?[a] 81. If a road 3 rods wide, runs 380 feet through my land, what amount of damages can I claim, the land being worth $150 an acre?[b] 82. The length of a certain field is four times the breadth, and the area is ten acres;[c] how many rods of wall will enclose it? 83. What must be given, in order to determine the altitude of a triangle? 84. The base? 85. The area?

86. How would you estimate the contents of a load or a pile of wood?[d] 87. Of a pile of boards?[e] 88. Of a pile of bricks?[f] 89. Of a stack of hay?[g] 90. Of a heap of earth? 91. Of a wall? 92. Name some object that you think would measure about 1 foot;—1 yard;—3 yards;—1 rod;—some distance that you think about a furlong. 93. Draw a line about 1 inch long; 3 inches; 6 inches; 9 inches; 1 foot; 4 inches; 8 inches. 94. How much do you suppose a hogshead[h] of water would weigh?[i] 95. If you should wish to weigh 15 pounds, but had mislaid your weights, how could you form an estimate with water? 96. Give an estimate of all the dimensions of the room you are in?

97. Suppose a note to be given for $1000, interest payable semi-annually, to what would it amount in 4 years, at compound interest? 98. At simple interest? 99. Allowing simple interest on the

---

[a] A pace is estimated at 3 feet. An ordinary step is about 2¼ feet.

[b] $64.77.

[c] The field can be divided into four squares, and the side of one of those squares will be the shorter side of the field.

[d] Give the answer for piles of regular and irregular shapes.

[e] A cubic foot = 12 feet Board Measure.

[f] 27 bricks measure 1 cubic foot. The number of bricks required to build a house, may be estimated by dividing the number of cubic feet in the walls by .04.

[g] The area of a circle may be conveniently found by squaring $.\dot{8}$ of the diameter. Hay is sometimes sold in the stack, by the cubic foot. 400 feet of trodden hay, weigh about one ton.

[h] A hogshead of measure, is intended.

[i] A cubic foot of water weighs about 1000 ounces.

principal, and also on the interest after it becomes due?[a] 100. Allowing compound interest when notice is given, supposing the debtor to be notified at the end of each year?[a] 101. How would you determine the face of a note, that would yield $1000 in 3 years, at simple interest?[b] 102. At compound interest?[b] 103. At annual interest, allowing *simple* interest on the interest?[b] 104. How would you determine the face of a note to be discounted at bank, in order to obtain any required sum?[b]

105. How many shillings are equivalent to .1 of a pound? 106. To what decimal of a pound, is 1 shilling equivalent? 107. To what decimal of a pound, are 24 farthings equivalent? 108. Then how many must you add to any given number of farthings, to represent their value in thousandths of a pound? 109. Can you give any rule, deduced from the answers to the preceding questions, for reducing shillings, pence and farthings, to the decimal of a pound, by inspection? 110. Can you reverse the process, and give a rule for reducing any decimal of a pound to shillings, pence and farthings, by inspection?

111. In computing interest at 6 per cent., in how many months will an investment gain .01 of itself? 112. In how many days will it gain .001? 113. Can you give a convenient rule for determining by inspection, the interest on $1 at 5 per cent., 6 per cent., and 7 per cent., for any given time? 114. Give similar rules for finding the amount, and the present worth of $1 for any given time. 115. Can you think of any other abbreviations in computing interest?[c]

---

[a] The laws of different states vary with regard to compound interest. In many places it is collected on all notes; in some cases the note is renewed each year and the interest is included in the new note; by the laws of some states, compound interest may be collected, provided notice is given when the interest becomes due; and in some states, simple interest is allowed on the principal, and also on all the interest from the time each payment becomes due till the final settlement.

[b] After finding how much $1 would yield in the given time, how would you find the *number* of dollars required to yield the desired amount?

[c] The following rule will be found very convenient in computing interest on notes and accounts, when the rate is 6 per cent. *Multiply* 1 *per cent. of the principal by one-half the even number of months, and if there is an odd month, add* 30 *to the number of days. Divide the days by* 6, *and multiply* .001 *of the principal by the quotient. If there are any re-*

116. If you intrust a certain sum to a factor, to cover the whole amount of his purchases and commission, how would you find the amount he can lay out? 117. In selling a certain invoice of merchandise at wholesale, a discount of 15 per cent. was made from the retail price. The clerk, in making out the account, calculated 15 per cent. on the whole cost, and deducted it from the bill. Could you have told him any readier mode of obtaining the result?

118. How would you determine the present value of a widow's dower?[a] 119. The value of a pension, payable during the life of one or more individuals?[a] 120. The amount that should be annually contributed to secure the payment of any desired sum, at a person's decease?[a] 121. The amount of an annual payment, for securing a weekly contribution during illness?[b] 122. The amount of a legacy, sufficient to erect a bridge and provide funds for all the repairs that it will ever probably need? 123. The amount of weekly savings necessary, to make a young man worth $5000 in 20 years? 124. The amount of weekly savings necessary for cancelling a debt, with all the interest, in any given time?

125. How would you reduce Sterling to Federal Money, at 9 per cent. premium? 126. Federal to Sterling Money, at $7\frac{1}{2}$ per cent. premium? 127. How would you find the value of stock at a discount of 27 per cent.? 128. At an advance of $18\frac{3}{4}$ per cent.? 129. How are interest and discount computed in Banks? 130. How do you compute percentage on English Money?

131. If you have the diameter of a circle given, how would you find the diameter of a circle that is 16 times as large? 132. $\frac{1}{4}$ as large? 133. 2.56 times as large? 134. 49 times as large? 135. If you know the area of a field, what would be the area of a similar field, each side of which is $\frac{1}{2}$ as long? 136. 3 times as long? 137. 7 as long? 138. If a ball 2 inches in diameter, weighs $1\frac{1}{2}$ pounds, what would be the weight of a similar ball, 6 inches in diameter? 139. What would be the diameter of a similar ball that would weigh 96 pounds? 140. If a tree 1 foot in diameter, yields 2 cords of wood, how much wood is there in a similar tree that is 3

---

maining days, take as many 60ths of 1 per cent. Add the products thus obtained, and their sum will be the interest at 6 per cent.

[a] All these questions are solved by estimating the probable duration of life. After that is determined, what remains to be done?

[b] The average amount of sickness is supposed to be known.

feet 6 inches in diameter? 141. If a hollow sphere 3 feet in diameter and $2\frac{1}{3}$ inches thick, weighs 12 tons, what are the dimensions of a similar sphere that weighs 324 tons?

142. Knowing the original cost of any article, how would you determine the price at which it must be sold, in order to gain any given per cent.? 143. How do you find the percentage gained or lost in any transaction? 144. The dividend that a bankrupt can pay upon each dollar of his debts? 145. The percentage of increase in the population of a place? 146. The original cost, by knowing how much per cent. is gained or lost by selling at a given rate? 147. The entire value, by knowing the value of any given percentage?

148. How would you determine the time at which a debt could be cancelled by a note for any particular amount? 149. The time at which several debts can be cancelled by a single payment? 150. The amount of interest due on an unsettled account, there being debits and credits embraced in the account? 151. The average time for settling an account, in which there are charges with different times of credit? 152. Can you give more than one method for averaging, or equating an account?

153. Why do we begin at the left hand in division? 154. Do you know of more than one mode of proof for each of the simple rules. 155. What is meant by the Arithmetical Complement of a number? 156. Can you perform a number of additions and subtractions at a single operation, by using the Arithmetical Complement? 157. Can you multiply by three or more figures at a single operation?

158. How do you find the cost of articles sold by the hundred or thousand? 159. Knowing the difference of longitude between two places, how do you find their difference of time? 160. Knowing the difference of time, how do you find the difference of longitude? 161. If d represents the diameter of a circle, c the circumference, and a the area, how would you find d, c and a being known? 162. Knowing c and d, how would you find a? 163. Knowing a and d, how would you find c? 164. How would you find the rate of insurance, knowing the premium and the amount insured? 165. The amount necessary to insure, in order to cover the premium and expenses of collecting, in addition to the loss? 166. The cost and rate per cent. of profit or loss being given, how would you find the amount of profit or loss? 167. The cost given, how would you find the selling price to gain or lose a specified

rate per cent.? 168. Cost and selling price given, how would you find the rate per cent. of profit or loss? 169. Selling price and rate of profit or loss given, how would you find the cost? 170. From the prime cost how would you find the selling price so as to gain any proposed percentage, and allow a discount for ready money?

171. How would you estimate the quantity of grain in a rectangular bin? 172. In a circular bin? 173. In a pile against the side of a building? 174. In a pile in the corner of a building? 175. How would you find the true weight, by a pair of false scales? 176. How would you find the value of any estate, knowing its annual rent? 177. Knowing the cost, rent, and annual outlay for taxes and repairs, how would you find the rate of interest that any estate yields? 178. How would you find what quantity of stock may be purchased for any given sum, allowing for brokerage? 179. How would you find the rate of interest gained by money invested in stocks at any given price? 180. How would you find what sum must be laid out in any kind of stock to produce a given annual income?

181. What must be given, and in what manner would you proceed, to determine the area of a rectangular field? 182. To determine either side? 183. To find either side of a right angled triangle? 184. To find the area of the surface of a sphere? 185. The solidity of a sphere? 186. The solidity of a cone? 187. Of a cylinder? 188. How would you find the area of an irregular field? 189. The solidity of an irregular body? 190. The specific gravity of a body? 191. The tonnage of a ship? 192. The contents of a pan with slanting sides? 193. Of a cylindrical pail? 194. Give the dimensions of a box that would hold a bushel? 195. A peck? 196. A wine gallon? 197. A wine quart? 198. A beer gallon? 199. A gill? 200. For what purpose is the hogshead measure used? 201. How would you find the area of a roof that would fill a cistern of given dimensions, with a fall of $\frac{1}{4}$ inch of rain? 202. What is meant by gross weight? 203. Net weight? 204. Tare? 205. What is meant by a common year? 206. A sidereal or periodic year? 207. A leap year? 208. A civil year? 209. How would you find the number of gallons equivalent to one foot in depth of a cistern? 210. The number of plants in an acre of ground, their distance apart being given? 211. What do you understand by an engine of forty horse power?

III. THE FUNDAMENTAL RULES.

N. B. The pupil should solve all the questions that he can, mentally. Let him never use the slate in obtaining an answer, except when it is absolutely necessary.

5. Examples in Addition and Subtraction.

1–4. Find from the following table, the population and extent of the globe, according to each of the authorities mentioned.[a]

Grand Divisions of the Globe.	According to Balbi.		Weimar Alm. 1840.	
	Population.	English Sq. Miles.	Population.	English Sq. Miles.
Europe - - -	227700000	3700000	233240043	3807195
Asia - - - -	390000000	16045000	608516019	17805146
Africa - - - -	60000000	11254000	101498411	11647428
America - - -	39000000	14730000	48007150	13542400
Oceanica - - -	20300000	4105000	1834194	3347840

5. The population of China in 1743, according to the French missionaries, was 150029855; in 1825, according to Dr. Morrison, 352866002. What is the difference between these two estimates?

6–9. The skull has 8 bones,[b] the face 14,[c] the ear 4,[d] the tongue 1,[e] and there are 32 teeth.[f] How many bones

[a] American Almanac, 1842.

[b] *Os frontis*, two *ossa parietalia*, two *ossa temporum*, *os occipitis*, *os sphenoides*, and *os ethnoides*.

[c] Two *ossa maxillaria superiora*, two *ossa nasi*, two *ossa unguis*, two *ossa malarum*, two *ossa palati*, two *ossa spongiosa inferiora*, the *vomer*, and the *os maxillare inferius*.

[d] *Malleus*, *incus*, *os orbiculare*, and *stapes*.

[e] *Os hyoides*.

[f] Sixteen in each jaw, viz: four *incisores*, two *cuspidati*, four *bicuspides*, and six *molares*.

in the whole head? Which are the most numerous? By how many do they exceed all the others? If you subtract each class of the bones, named above, from the entire number, how many will be left after each subtraction?

10–13. In the trunk there are 24 spinal bones,[a] 24 ribs, the *sternum*, or breast-bone, the *os sacrum*, the *os coccygis*, and two *ossa innominata*. Required the number in the entire trunk? In the trunk and head together? How many more in the trunk than in the skull? How many more in the head than in the trunk?

14–17. Each of the upper extremities contains the following bones, viz: 2 in the shoulders,[b] 3 in the arm,[c] 8 in the *carpus*, or wrist, 5 in the *metacarpus*, or palm, 2 in the thumb, and 3 in each of the fingers.[d] How many in all? How many in both of the upper extremities? In the head, trunk, and upper extremities? How many more in one of the upper extremities than in the face?

18–21. Each of the inferior extremities contains the following bones, viz: 4 in the leg,[e] 7 in the *tarsus*, or ankle, 5 in the *metatarsus*, or foot, and 3 in each of the toes, except the great toe, which has but two. Required the number in each of the lower extremities? In both? In all the extremities? In the whole body?

Ans. to the last, 240.[f]

22–25. How many bones in the whole body, besides those of the head? Besides those of the trunk? Of the upper extremities? Of the lower extremities?

26–29. The skin of the cranium has 3 principal mus-

---

[a] The 24 true vertebræ, 7 *cervical*, 12 *dorsal*, and 5 *lumbar*.

[b] The *clavicle* and *scapula*.

[c] *Humerus*, *radius*, and *ulna*.

[d] Called *phalanges*.

[e] *Os femoris*, *tibia*, *fibula*, and *patella*.

[f] Besides the bones enumerated, are the sesamoid bones, which vary in number. Marjolin counts five for each of the upper, and three for each of the lower extremities, making two hundred and fifty-six in the whole body.—*Wistar's Anatomy*.

cles,[a] each ear has 4, the lids of each eye 2, each eye 6, the nose 2, the mouth 19, the tongue 8, and the lower jaw has 4 pairs. How many in all? How many less than the whole does the mouth contain? How many less than half? What is the difference between the number of bones and the number of muscles in the head?

30–33. The neck and throat contain 50 principal muscles, the trunk 116, each of the superior extremities 46, and each of the inferior extremities 51. How many in all? How many in the entire body? How many less in the head than in the rest of the body? How many more muscles than bones in the whole system?

34–37. Find the difference between the following numbers, commencing the subtraction at the left hand:[b] 84.9108757944362 and 190027.08; 67490083574.00882 and 3375109884726.37095; $16438 and $9728.87758403; 7 and .0019547998027184920625.

38–44. Determine from the following table, the entire population of the United States and Territories, at each of the given dates. In obtaining the results, add two columns of figures at a time.[c]

---

[a] One pair and one single muscle. Most of the muscles are found in pairs, one on each side of the body. There is sometimes a slight difference in the number, in different individuals.

[b] In beginning to subtract at the left hand, if at any point the remaining figures of the subtrahend are greater than those of the minuend, we must add one to the figure we are subtracting. For example, in subtracting 92824 from 164809, say 9 from 16 leaves 7; then as 824 is greater than 809, say 3 from 4 leaves 1; as 24 is greater than 09, say 9 from 18 leaves 9; 2 from 10 leaves 8, 4 from 9 leaves 5. Let the pupil explain the *rationale* of this process.

```
164809
 92824
------
 71985
```

[c] The teacher will find this a valuable exercise, to be performed aloud, at the recitation of the class. In adding two or more columns at once, it will be found most convenient to commence at the left hand, and observe at each step whether there will be one to bring from the right hand figures. Thus in adding 972 and 645, instead of saying 5 and 2 are 7, 4 and 7 are 11, 9 and 6 are 15 and 1 are 16,—say 9 and 6 and 1 are 16, 7 and 4 are 11, 5 and 2 are 7. After a little prac-

TABLE OF POPULATION—FROM THE AMERICAN ALMANAC.

STATES.	1790.	1800.	1810.	1820.	1830.	1840.	1850
Maine	96,540	151,719	228,705	298,335	399,955	501,793	
New Hampshire	141,899	183,762	214,360	244,161	269,328	284,574	
Vermont	85,416	154,465	217,713	235,764	280,652	291,948	
Massachusetts	378,717	423,245	472,040	523,287	610,408	737,699	
Rhode Island	69,110	69,122	77,031	83,059	97,199	108,830	
Connecticut	238,141	251,002	262,042	275,202	297,665	309,978	
New York	340,120	586,756	959,949	1,372,812	1,918,608	2,428,921	
New Jersey	184,139	211,949	249,555	277,575	320,823	373,306	
Pennsylvania	434,373	602,365	810,091	1,049,458	1,348,233	1,724,033	
Delaware	59,098	64,273	72,674	72,749	76,748	78,085	
Maryland	319,728	341,548	380,546	407,350	447,040	469,232	
Dist.of Columbia	......	14,093	24,023	33,039	39,834	43,712	
Virginia	748,308	880,200	974,622	1,065,379	1,211,405	1,239,797	
North Carolina	393,751	478,103	555,500	638,829	737,987	753,419	
South Carolina	249,073	345,591	415,115	502,741	581,185	594,398	
Georgia	82,548	162,101	252,433	340,987	516,823	691,392	
Alabama	......	......	20,845	127,901	309,527	590,756	
Mississippi	......	8,850	40,352	75,448	136,621	375,651	
Louisiana	......	......	76,556	153,407	215,739	352,411	
Florida	......	......	......	....	34,730	54,477	
Texas	......	......	......	....	....	....	
Kentucky	73,077	220,955	406,511	564,317	687,917	779,828	
Tennessee	35,791	105,602	261,727	422,813	681,904	829,210	
Ohio	......	45,365	230,760	581,434	937,903	1,519,467	
Indiana	......	4,875	24,520	147,178	343,031	685,866	
Illinois	......	......	12,282	55,211	157,455	476,183	
Michigan	......	......	4,762	8,896	31,639	212,267	
Iowa	......	......	......	....	....	43,112	
Wisconsin	......	......	......	....	....	30,945	
Missouri	......	......	20,845	66,586	140,445	383,702	
Arkansas	......	......	......	14,273	30,388	97,574	
Minesota Terr.	......	......	......	....	....	....	
Missouri "	......	......	......	....	....	....	
Oregon "	......	......	......	....	....	....	
Indian "	......	......	......	....	....	....	
New Mexico	......	......	......	....	....	....	
California	......	......	......	....	....	....	

45–57. Find the increase in the population from 1790 to 1800; to 1810; 1820; 1830; 1840; 1850; from 1800 to 1810; 1810 to 1820; 1820 to 1830; 1830 to 1840; 1840 to 1850; 1790 to 1820; 1820 to 1850.

58–181. Find the amount of each of the following columns, and also the total amount of receipts and expenditures in each year. Add two columns of figures at a time.

---

tice the pupil will say at once, 972 and 645 are 1616, as naturally as he would say 8 and 6 are 14, without stopping to count his fingers. The numbers to be added should not be mentioned in performing the addition. Thus in adding 8, 5, 9, 3, 7, 9, 6, 4, say 8, 13, 22, 25, 32, 41, 47, 51, instead of saying 8 and 5 are 13, and 9 are 22, and 3 are 25, &c. In adding 65, 48, 27, 92, say 65, 113, 140, 232, and proceed in like manner in all cases.

STATEMENT OF THE RECEIPTS AND EXPENDITURES OF THE UNITED STATES FOR SIXTY YEARS.

FROM THE AMERICAN ALMANAC, AND PUBLIC DOCUMENTS.

	RECEIPTS.			EXPENDITURES.		
Years.	Customs.	Internal and direct taxes.	Lands and Miscellan.	Civil and Miscellan.	Army.	Navy.
1789–91	$4,399,473			$1,083,401	$835,618	$570
1792	3,443,071	$208,943		654,257	1,223,594	53
1793	4,255,306	337,706		472,450	1,237,620	
1794	4,801,065	274,090		705,598	2,733,540	61,409
1795	5,588,461	337,755		1,367,037	2,573,059	410,562
1796	6,567,988	475,290	$4,836	772,485	1,474,661	274,784
1797	7,549,650	575,491	83,541	1,246,904	1,194,055	382,632
1798	7,106,062	644,358	11,963	1,111,038	2,130,837	1,381,348
1799	6,610,449	779,136		1,039,392	2,582,693	2,858,082
1800	9,080,933	1,543,620	444	1,337,613	2,625,041	3,448,716
1801	10,750,779	1,582,377	167,726	1,114,768	1,755,477	2,111,424
1802	12,438,236	828,464	188,628	1,462,929	1,358,589	915,562
1803	10,479,418	287,059	165,676	1,842,636	944,958	1,215,231
1804	11,098,565	101,139	487,527	2,191,009	1,072,017	1,189,833
1805	12,936,487	43,631	540,194	3,768,588	991,136	1,597,500
1806	14,667,698	75,865	765,246	2,891,037	1,540,431	1,649,641
1807	15,845,522	47,784	466,163	1,697,897	1,564,611	1,722,064
1808	16,363,550	27,370	647,939	1,423,286	3,196,985	1,884,068
1809	7,296,021	11,562	442,252	1,215,804	3,771,109	2,427,759
1810	8,583,309	19,879	696,549	1,101,145	2,555,693	1,654,244
1811	13,313,223	9,962	1,040,238	1,367,291	2,259,747	1,965,566
1812	8,958,778	5,762	710,428	1,683,088	12,187,046	3,959,365
1813	13,224,623	8,561	835,655	1,729,435	19,906,362	6,446,600
1814	5,998,772	3,882,482	1,135,971	2,208,029	20,608,366	7,311,291
1815	7,282,942	6,840,733	1,287,959	2,898,871	15,394,700	8,660,000
1816	36,306,875	9,378,344	1,717,985	2,989,742	16,475,412	3,908,278
1817	26,283,348	4,512,288	1,991,226	3,518,937	8,621,075	3,314,598
1818	17,176,385	1,219,613	2,606,565	3,835,839	7,019,140	2,953,695
1819	20,283,609	313,244	3,274,423	3,067,212	9,385,421	3,847,640
1820	15,005,612	137,847	1,635,872	2,592,022	6,154,518	4,387,990
1821	13,004,447	98,377	1,212,966	2,223,122	5,181,114	3,319,243
1822	17,589,762	88,617	1,803,582	1,967,996	5,635,187	2,224,459
1823	19,088,433	44,580	916,523	2,022,094	5,258,295	2,503,766
1824	17,878,326	40,865	984,418	7,155,308	5,270,255	2,904,582
1825	20,098,714	28,102	1,216,090	2,748,544	5,692,831	3,049,084
1826	23,341,332	28,228	1,393,785	2,600,178	6,243,236	4,218,902
1827	19,712,283	22,513	1,495,945	2,314,777	5,675,742	4,263,878
1828	23,205,524	19,671	1,018,309	2,886,052	5,701,203	3,918,786
1829	22,681,966	25,838	1,517,175	3,092,214	6,250,530	3,308,745
1830	21,922,391	29,141	2,329,356	3,228,416	6,752,689	3,239,429
1831	24,224,442	17,440	3,210,815	3,064,346	6,943,239	3,856,183
1832	28,465,237	18,422	2,623,381	4,574,841	7,982,877	3,956,370
1833	29,032,509	3,153	3,967,682	5,051,789	13,096,152	3,901,357
1834	16,214,957	4,216	4,857,601	4,399,779	10,064,428	3,956,260
1835	19,391,311	14,723	4,757,601	3,720,167	9,420,313	3,864,939
1836	23,409,940	1,099	4,877,180	5,388,371	18,466,110	5,800,763
1837	11,169,290		6,863,556	5,524,253	19,417,274	6,852,060
1838	16,158,800		3,214,184	5,666,703	19,936,312	5,975,771
1839	23,137,925		7,261,118	4,994,562	14,268,981	6,225,003
1840	13,499,502		3,494,356	5,581,878	11,621,438	6,124,456
1841	14,487,217		1,470,295	6,490,881	13,704,882	6,001,077
1842	18,187,969		1,456,058	6,775,625	9,188,469	8,397,243
1843 [a]	7,046,844		1,018,482	2,867,289	4,158,384	3,672,718
1844 [b]	26,183,571		2,320,948	5,231,747	8,231,317	6,496,991
1845 [b]	27,528,113		2,241,021	5,618,207	9,533,203	6,228,639
1846 [b]	26,712,668		2,786,579	6,783,000	13,579,428	6,450,862
1847 [b]	23,747,864		2,598,926	6,715,854	41,281,606	7,931,633
1848 [b]	31,757,071		3,679,680	5,585,070	27,820,163	9,406,737
1849 [b]						

[a] 6 months of 1843.

[b] For the year ending June 30.

6. Examples in Addition, Subtraction, Multiplication, and Division.

1. Invoice of 6 bales of Dry Goods:—

Boston, November 27, 1849.

Messrs. Thompson & Allen,

Bought of William Mansfield,[a]

Mark	Items				
T. & A.	A bale containing				
No. 35.	15 pieces Lint Strelitz Osnaburgs, each 130 yds., @ 10c.				
	Wrapper, 20 yds., @ 6c.				
	Bale, cording and packing,	1	25		
36 to 40	Five bales, containing				
	No. 36, 464 yds.				
	No. 37, 481 "				
	No. 38, 437 "				
	No. 39, 475 "				
	No. 40, 470 "				
	2327 " @ 11c.				
	Bales, cording and packing,	5	00		
				$458	42

2. Bill and Receipt:—

Philadelphia, March 25, 1848.

Benjamin Stabler,

To John Farmer, Dr.

To 4 tons of hay,	at $25 50		
115 bushels of oats,	" 30		
95 " corn,	" 94		
Received payment,		$	

John Farmer.

[a] The items should be extended in the left hand column, and the amounts of each package (or all the packages that are included together, as in Nos. 36 to 40), carried out in the right hand column. This is the usual form of an American invoice. The pupil should give the amount for each of the items that is left blank.

3. Bill, unreceipted:—

New York, July 7, 1849.

George Lenox,

Bought of Carlisle & Williams,

1 doz. Long Shawls,	@	$6.75		
3 pieces Sheeting, 30, 31, 33 yds.	@	.09		
1 piece Mousseline de Laine, 28 "	@	.17		
2 pieces Broadcloth, 35 and 40 "	@	4.75		
6 doz. Linen Hdkfs.	@	.25		
12 doz. Cotton Half Hose,	@	1.50		
4 months.			$486	47

4. Statement of Account:—

Albany, Jan. 5, 1849.

Mason, Hamilton & Co.

1848.	To Johnson & Brooks,		Drs.	
Aug. 27.	To Stoves, &c., per bill rendered, 6 mo.		$843	75
Sept. 5.	" " " "		375	69
" 13.	" " " "		118	50
" 30.	" " " "		97	11
Oct. 8.	" " " "		103	88
" 23.	" " " "		491	35
Nov. 19.	" " " "		85	67
	Contra, Cr.			
Oct. 1.	By Cash,	450.00		
Nov. 7.	" acceptance of draft,	200.00		
Dec. 12.	" note at 4 months,	562.50		
	Balance,		$	
	Due by average, June 21, E. & O. E.[a]			

5. The population of the Chinese Empire has been estimated[b] as follows: China proper, 148897000; Corea, 8463000; Thibet and Boutan, 6800000; Mandshuria, Mongolia, &c., 9000000; Colonies, 1000000. Its territory con-

[a] Errors and omissions excepted.

[b] Murray's Encyclopædia of Geography.

tains about 5350000 square miles. Required the entire population, and the number of inhabitants to a square mile.

6. Bill of Lading:—

Shipped, in good order and condition, by Norton & Phillips, in and upon the Brig called the *Margaret*, whereof *Meriam* is master for this present voyage, and now lying in the port of *Charleston*, and bound for *New York*,

Three Boxes and Five Bales of Goods,
Weight 1835 *lbs.*

(Margin: # 1 to 8, B. Van Pelt, Care of L. Haines, 116 Broadway, New York.)

Being marked and numbered as in the margin, to be delivered in the like good order and condition, at the aforesaid port of *New York* (the danger of the seas only excepted), unto *Benjamin Van Pelt*, or to his assigns, he or they paying freight for the said goods, at the rate of 28 cts. per 100 lbs., with Primage and Average accustomed.

In testimony whereof, the Master or Purser of the said Brig hath affirmed to three Bills of Lading, all of this tenor and date, one of which being accomplished, the others to stand void.

Dated at Charleston, the 30th day of March, 1849.

J. M. Meriam.

Required the amount of freight on the above merchandise.

7. Receipt in full:—[a]

Worcester, June 5th, 1849.

James A. Chase,

1848.	To Moses Allen,	Dr.
Oct. 27.	To Mdse. per bill rendered,	27.35
Nov. 3.	" Hire of Horse and Chaise to Leicester,	1.50
1849.		
March 11,	" 1 ton of Coal,	9.75

Received payment in full, $

Moses Allen.

[a] A receipt *in full* is understood to cover all demands.

8. Publisher's Estimate for one thousand copies of a book:

Printing and corrections,	$513.00
Paper and certificate of copyright,	326.00
Binding,	175.00
Advertising,	200.00
	$

1 copy to be deposited in the Clerk's Office,
10 copies to the author,
989 copies for sale at $1.75,
Deduct cost

Balance to cover commissions, interest on capital, and profit to author and publisher, *when all are sold.* $516.75

9. Order for Goods :—

Providence, August 18, 1849.

Messrs. Anthony & Smith,

Please ship to us by first packet:

25 cwt. Sugar (Brown Havana) about $5.75 per cwt.
13 hhd. W. I. Molasses, $23.
27 boxes Louisiana Oranges, $3.80.
Old Java Coffee, say 1300 lbs. at .09.
Honey, about 250 gallons at .67.
18 chests Black Tea, $5.50 per chest.
49 bbls. Genesee Flour, $4.75.

And oblige, Yours, &c.,

L. KENTON & RAY.

What would be the amount of the above order, if all the articles were sent at the prices affixed to them?

10. Determine by the Roman Notation, how many years elapsed from the discovery of America by Columbus, Anno Domini MCCCCXCII, to the adoption of the United States Constitution, A. D. MDCCLXXXIX.[a]

[a] The pupil will probably find no difficulty in adding or subtracting Roman numerals, except when such numbers as IV, IX, XL, XC,

11. The Pilgrims landed at Plymouth in MDCXX, and Benjamin Franklin was born LXXXVI years afterwards. Find by the Roman Notation, in what year he was born.

12. Multiply by the Roman method, and determine the number of weeks that elapsed between the landing of the Pilgrims and the birth of Franklin, allowing LII weeks to each year, and adding XII to the product.

13. Settlement of Account, exhibiting different forms of receipts :—

Account rendered Jan. 1, 1848.

Lowell, Jan. 1, 1848.

Thomas Lawrence

1846.	To Henry Appleton,		Dr.
July 11.	To Mdse. per bill,		$984.32
" 29.	" 5 bbls. Flour, @	$5.25	
Aug. 31.	" 49 yds. Broadcloth, @	4.50	
Oct. 17.	" 56 yds. Carpeting, @	1.15	
	" Balance of interest from Jan. 1, 1847,		16.70
			$
1847.	Cr.		
Jan. 13.	By Cash,	100.00	
Mar. 4.	" Mdse.	27.42	127.42
	Balance,		$1184.75

occur. The difficulty may be removed by changing them to the forms IIII, VIIII, XXXX, LXXXX. Thus, if it were required to add DCCXLIV, CCXCIX, and MDLXVIII, writing them as in the margin, we commence at the right hand, and say, eleven Is are equivalent to two Vs and I; two Vs and two to be added from the Is make four Vs or two Xs; nine Xs and two to be added make eleven Xs, or two Ls and X; two Ls and two to be added make four Ls, or two Cs; four Cs and two Cs make six Cs, or one D and C; two Ds and one D are three Ds, or M and D; M and M are MM.

DCCXXXXIIII
CCLXXXXVIIII
MDLXVIII

MMDCXI

In multiplication, the multiplier may be divided into any convenient number of parts, and after multiplying by each part, the several partial products should be added together.

On settling the account, August 5, 1849, Thomas Lawrence presented the following receipts, viz :—

I. *A Receipt on account.*

"Received of Thomas Lawrence Seventy-five Dollars, on account. Lowell, Jan. 25, 1848.
$75. HENRY APPLETON."

II. *A Receipt through a third person.*

"Received of Thomas Lawrence, per hands of James Brown, One Hundred and Ten Dollars, on account.
$110. HENRY APPLETON,
Lowell, March 4, 1848. per WILLIAM ALLEN."

III. *Receipt for Note on account.*

"Received of Thomas Lawrence his note dated July 5, 1848, @ 8 months date my favor, for Seven Hundred and Fifty Dollars, on account. Lowell, July 7, 1848.
Note $750. HENRY APPLETON."

The receipts being all correct, there is found to be due to Henry Appleton $78.46 interest, in addition to the unsettled balance of account. Required the amount that has been paid, the balance outstanding, and the amount paid at settlement, for which H. A. gives the following

Receipt in full.

"Received of Thomas Lawrence, $\frac{\quad}{100}$
Dollars, in full of all demands to this date.
$ HENRY APPLETON."
Lowell, August 5, 1849.

14. Allowing 365 days and 6 hours to each year, what length of time would be required to count the number of tons of carbonic acid contained in the atmosphere, counting 3 every second, for 12 hours a day?[a]

Ans. 111697y. 124d. 5h. 26m. 40sec.

15. A horse-power is generally estimated as sufficient to raise 33000lbs. 1 foot high in 1 minute; and Desaguliers

[a] See Example 17.

estimates the power of 1 horse as equivalent to that of 5 men. According to the latter estimate, how much would 1000 men be able to raise to the height of 20 feet in 5 minutes?

Ans. 1650000lbs.

16. Account Sales:—

ACCOUNT SALES. 8 Hhds. Molasses, per *Mayflower*, from Trinidad, for account of Williams & Tasistro.

Charges.			Marks	No.	gals.	qt.	
			W T	#29,	113		
Insurance on $150 at			29	30,	112	3	
1 per cent.	$1.50		to	31,	115	3	
Policy,	.50		36	32,	125		
	——	2.00		33,	119	1	
Freight and storage,		19.50		34,	117		
Duties,		65.30		35,	120	3	
Brokerage,		3.88		36,	121	2	
Commission & Guaranty,		16.33					
				Gro.	945	gals.	
Net proceeds, due				Dft.	12	"	
Oct. 5, 1849.		$219.54					
				Net	933	gals.	
Boston, Aug. 28, 1849, E. E.					@ 35 cts.		$

WELLINGTON, CARTER & CO.

17. It has been estimated[a] that the atmosphere contains 3994592925000000 tons of nitrogen, 1233010020000000 tons of oxygen, 5287350000000 tons of carbonic acid, and 54459705000000 tons of aqueous vapor. Required the weight of the whole, in pounds.

Ans. 11843664 trillion pounds, according to the French Notation.[b]

[a] Griffiths.

[b] The old English Notation is still employed in some works, but the French method, which is simpler and more easily learned, is generally adopted, even by the more modern English writers. In the English system, each period consisted of six places. A billion was therefore equivalent to a million million; a trillion, was a million billion, and so on.

18. Invoice of 2 cases merchandise, shipped at Havre on board the ship *Duchesse d' Orleans*, Richardson, Master, bound for New York, purchased for account and risk of Messrs. Hamilton & Co., of Philadelphia, and to them consigned.

Mark	No.	Description	Rate	Amount	Total
◇H	225	One Case.			
		No. 54, 107 C[a] 214 doz. Braces	6.[b]		
		55, 1 " 2 " "	9.50		
		Discount 2 °/o [c]		26.05[d]	
		108 Boxes,	.60		
		Packing,		25.90	F.1367.65
	226	One Case.			
		No. 72, 29 C 58 doz. Braces,	9.50		
		80, 20 " 40 " "	10.		
		87, 15 " 30 " "	14.		
		Discount 8 °/o		109.70	
		64 Boxes,	.60		
		Packing,		22.80	
		Commission 3 °/o			80.70
					F. 2770.85
		E. E. Paris, 25 November, 1849,			
		LEROUX & CIE.			

19. The bulk of carbonic acid produced by a healthy adult in 24 hours, is about 15000 cubic inches, and weighs

[a] Cartons.

[b] In French currency the Franc is considered as the unit, and Centimes are written as hundredths. Calculations are therefore made as in Federal Money. Thus 214 doz. at 6 Fr. = 1284 Frs.; 11 doz. at 1.25, would be 13.75, &c.

[c] This character (°/o) is an abbreviation for *per cent.*

[d] No fraction of a franc is counted, less than 5 centimes.

about 6 ounces.[a] If this were the average rate, how much would be produced daily by the population of a town containing 5000 inhabitants.

Ans. 43402c. ft. 1344c. in.=16cwt. 2qr. 27lb.

7. Table for Additional Exercise.

In every kind of business, *correctness* and *readiness* in the use of figures are of the first importance. These can only be obtained by practice, and to give an opportunity for such practice, a page of figures is here inserted, which may be so divided by the teacher as to make many thousands of examples. Thus, in addition, the pupil may add a single column, or any two, three or more contiguous columns, or parts of columns, or the entire page; and he may be exercised in adding two, three, or more columns at once.[b] In subtraction (in which it is recommended that he should always be required to commence at the left hand),[c] he may take parts or the whole of any two adjoining rows, or he may subtract, (without using his slate,) part of the figures in any row, from the remaining figures. In multiplication and division the examples may be varied in like manner to any desired extent.

By this arrangement a whole class may be engaged in the same operation, without having the question written on the board, and the teacher can save the time that might otherwise be required for setting down examples on the slate. Thus, should any scholar require additional exercise in either of the fundamental rules, he may be told to add columns 19, 20, and 21, or to subtract row 15 from row 14, prefixing a 7 to the upper row; or to multiply row 7 by row 11; or to divide row 3 by the first five figures in row 18; or he may be required to form examples for himself.

---

[a] Griffiths. [b] See note [c], page 35. [c] See note [b], page 35.

1 2 3 4 5 6 7 8 9 0 1 2 3 4 5 6 7 8 9 0 1 2 3 4 5 6 7 8

2 8 7 4 0 5 1 7 4 9 3 9 5 2 1 5 7 8 0 6 9 3 4 2 6 7 9 2

3 4 6 3 7 9 1 0 3 6 7 3 4 5 9 7 1 8 5 0 4 6 8 0 3 5 9 3

4 7 1 2 5 9 5 1 0 7 6 7 8 9 7 7 0 0 6 5 6 9 5 5 4 1 3 4

5 5 3 1 8 9 6 0 5 7 4 0 2 5 6 9 7 0 8 9 6 4 6 3 2 9 1 5

6 7 9 0 4 5 6 1 5 7 8 8 0 0 4 5 6 7 8 9 4 3 1 5 6 3 6 6

7 4 2 9 1 7 0 8 3 6 0 2 0 1 6 2 9 6 8 7 3 0 7 6 9 3 5 7

8 6 3 0 6 9 0 5 7 9 3 8 0 0 8 3 9 6 8 0 3 6 9 7 0 1 6 8

9 9 7 5 0 2 1 8 9 0 6 7 9 2 1 4 6 0 7 3 5 9 2 0 7 4 2 9

10 5 9 6 8 1 0 7 5 2 9 5 7 4 1 0 2 1 5 6 4 8 1 9 6 3 1 0

11 6 6 2 0 7 1 5 6 8 6 1 5 6 2 7 9 5 1 6 2 1 0 8 4 6 2 1

12 1 4 6 5 3 0 9 5 1 9 3 6 2 6 8 0 1 6 9 5 1 6 0 6 7 4 2

13 6 2 0 9 8 6 8 1 4 6 1 0 8 4 0 1 6 8 3 6 8 5 1 6 1 0 3

14 6 0 5 6 3 6 8 5 6 9 6 0 6 8 3 2 8 6 0 9 6 2 4 1 6 3 4

15 2 9 6 2 9 6 0 7 9 6 3 8 0 0 3 6 7 2 9 5 6 1 7 3 3 0 5

16 3 0 8 4 1 0 4 6 6 8 0 6 4 1 7 7 9 4 6 2 6 0 4 7 9 0 6

17 9 6 7 0 8 6 9 3 1 6 7 9 0 6 7 2 4 1 6 0 9 7 0 6 2 4 7

18 1 6 4 3 4 9 0 2 8 1 6 0 7 9 2 1 7 4 2 6 7 7 0 2 6 4 8

19 0 7 5 3 5 9 1 1 6 5 0 2 6 9 1 7 6 9 2 4 6 9 0 2 6 1 9

20 2 3 1 6 7 0 4 6 1 6 0 7 1 5 6 8 5 2 0 7 9 5 1 0 2 6 0

21 2 9 0 5 9 1 7 0 5 7 9 3 7 0 1 3 6 0 7 9 2 5 1 9 5 6 1

22 5 6 0 7 3 5 7 9 0 2 4 5 6 0 0 9 8 2 6 0 3 7 9 1 5 7 2

23 5 9 7 5 9 0 7 0 8 6 9 7 9 0 8 6 3 4 8 4 3 9 7 3 3 5 3

24 5 8 6 8 0 6 9 6 4 8 0 4 6 8 7 9 0 4 7 9 0 8 6 9 0 6 4

25 6 5 0 9 7 6 0 8 2 9 7 6 8 0 6 4 7 0 9 8 7 8 0 8 0 8 5

26 9 7 8 4 9 6 9 9 0 6 7 9 6 8 0 6 0 9 7 6 8 9 0 7 8 9 6

27 7 8 7 9 2 6 9 7 0 6 0 8 7 0 7 8 7 3 0 1 6 3 9 1 7 5 7

28 3 5 0 1 6 9 5 1 7 4 0 1 5 8 3 9 2 5 3 6 5 4 9 5 8 1 8

29 4 6 3 7 0 1 6 7 3 0 4 3 6 8 1 3 6 0 3 2 0 7 2 6 0 3 9

30 0 3 6 5 6 3 2 6 3 5 9 6 4 2 0 5 4 7 3 9 0 2 5 3 6 4 0

31 5 6 1 9 4 1 3 5 7 3 5 9 6 4 0 4 7 3 7 5 9 1 6 3 0 3 1

32 0 3 5 7 0 4 8 5 5 1 4 0 2 6 7 2 0 6 4 8 6 2 5 1 6 5 2

33 0 5 4 2 5 6 0 7 3 4 6 5 3 6 1 5 0 3 6 7 2 5 6 4 0 6 3

34 4 8 4 0 5 1 3 3 0 5 2 7 4 0 3 8 4 2 6 0 3 1 5 3 0 5 4

35 6 3 0 4 8 4 7 3 8 5 4 0 1 6 8 1 0 2 8 1 6 2 1 4 1 0 5

36 6 0 1 6 9 8 3 5 6 1 0 9 2 1 3 0 8 4 5 7 2 9 4 6 1 3 6

8. Fractions and Compound Numbers.

1–10. What part of the entire number of bones in the human body is contained in the skull? [See § 5, Ex. 6–21, and express all the answers in the lowest terms.] What part in the face?—the ear?—the tongue?—the whole head? —the spine?—the ribs?—the whole trunk?—the head and trunk?—the extremities?

11. Set down the answers to the questions in the preceding paragraph, and add all the fractions together.

12–13. If you were to count three every second, for ten hours a day, how long would it take you to count a million? How long would it take to count a billion, at the same rate?

14. The population of Washington in 1840 was 23364. If all the inhabitants were to commence counting at the rate mentioned in the last example, how long would it take them to count a trillion dollars?

15. The salary of the President of the United States is $25000 per annum. How many years' salary would be equivalent to a million dollars?

16. Pure silver is worth at the mint $15.50 a pound. How many tons would it take to be worth a million dollars? Express the answer in tons and fractions of a ton, and also in tons, cwt., qrs., &c.

17. Standard gold being worth $18.60 per ounce, what weight of gold would be required to furnish a half-eagle to every inhabitant of the United States, estimating the population at 21000000? Express the answer as in the last example.

18. What is the amount of the President's salary per day?—per hour?—per minute?

19. The average pulsation of the heart is in childhood about 105 beats in a minute, in youth 90 beats, in middle

age 75, and in old age 60. Estimating the duration of childhood at 14 years, youth at 11 years, middle age at 25 years,[a] how many times would the heart beat in each period? How many times in the whole life?

20. TABLE OF DISCOVERIES AND IMPROVEMENTS.[b]

	B. C.
Astronomical observations first made in Babylon . . .	2234
Lyre invented	1004
Sculpture	1900
Agriculture by Triptolemus	1600
Chariots of War	1500
Alphabetical letters introduced into Europe . . .	1500
The first ship seen in Greece, arrived at Rhodes from Egypt	1485
Iron discovered in Greece, by the burning of Mount Ida .	1406
Seaman's Compass invented in China	1120
Gold and Silver money first coined by Phidon, king of Argos	894
Parchment invented by Attalus, king of Pergamus . .	887
Weights and measures instituted	869
First astronomical observation of an eclipse . . .	721
Ionic order used in building	650
Maps and globes invented by Anaximander . . .	600
Sun-dials invented	558
Signs of the Zodiac, invented by Anaximander . . .	547
Corinthian order of architecture	540
First public library established at Athens	526
Silk brought from Persia to Greece	325
The art of painting brought from Etruria to Rome, by Quintus Pictor	291
Solar quadrants introduced	290
Mirrors in silver invented by Praxiteles	288
Silver money first coined at Rome	269
Water-clocks	245
Hour-glass invented in Alexandria	240
Burning mirrors invented by Archimedes	212
First fabricating of glass	200

---

[a] Encyclopædia Americana, Rees' Encyclopædia.

[b] Memoria Technica, Rees' Encyclopædia, Beckmann, Scientific American.

	B. C.
Brass invented	146
Paper invented in China	105
Rhetoric first taught at Rome	87
Blister-plasters invented	60
Julian year regulated by Cæsar	45
Apple trees brought from Syria and Africa into Italy	9
	A. C.
Vulgate edition of the Bible discovered	218
Porcelain invented in China	274
Water-mills invented by Belisarius	555
Sugar first mentioned, by Paul Eginetta, a physician	625
Quills used for writing	636
Stone buildings first erected in England, by Bennet, a monk	670
The system of couriers, or posts, invented by Charlemagne	808
Figures used by the Arabs, borrowed from the Indians	813
Lanterns invented by king Alfred	890
High towers first erected on churches	1000
Musical notes invented by Guido, of Arezzo	1021
Heraldry originated	1100
Distillation first practised	1150
Glass windows first used in England	1180
Chimneys built in England	1236
Leaden pipes for conveying water, invented	1252
Magic lanterns invented by Roger Bacon	1290
Tallow candles first used	1290
Fulminating powder invented by Roger Bacon	1290
Spectacles invented by Spina	1299
Wind-mills invented	1299
Alum discovered in Syria	1300
Paper made of linen	1302
Coals first used in England	1307
Saw-mills at Augsburg	1322
Woollen cloths first made in England	1331
Gold first coined in England	1344
Painting in oil colors	1410
Muskets used in England	1421
Pumps invented	1425
Printing invented	1440
Glass first made in England	1457
Wood-cuts invented	1460
Almanacs first published in Buda	1460

Printing introduced into England by Caxton . . . 1470
Watches invented at Nuremberg 1477
Stages and post horses established 1483
Tobacco discovered in St. Domingo 1496
Shillings first coined in England 1505
Stops in literature introduced 1520
Spinning-wheel invented at Brunswick 1530
Variation of the Compass first observed . . . 1540
Pins first used in England 1543
Needles first made in England by an East Indian . . 1545
Sextant invented by Tycho Brahe 1550
Lace knit in Germany 1561
Coaches first used in England 1580
Telescopes invented by Jansen 1590
Stocking weaving invented 1590
Decimal arithmetic invented at Bruges 1602
Microscopes used at Naples . . . . , . . 1618
Circulation of the blood discovered by Harvey . . 1619
Coins made with dies in England 1620
Thermometers invented by Drehel 1620
Barometer invented by Torricelli, an Italian . . . 1626
Newspapers first published . . , . . . 1630
Regular posts established in London 1635
Coffee brought to England 1641
Steam engines invented by the Marquis of Worcester . 1649
Pendulums for clocks invented 1649
Air-pumps invented 1650
Air-guns invented by Guter 1656
Spring pocket watches invented by Dr. Hook . . . 1658
Engines invented to extinguish fires 1663
Bayonets invented at Bayonne 1670
Micrometer invented 1677
Telegraphs invented 1687
New style adopted in England 1752
Spinning frame by Arkwright 1761
Cotton first planted in the United States 1769
Steam engine improved by Watts 1769
Georgium Sidus discovered by Herschel 1781
Power looms invented by Cartwright 1783
Steam cotton-mills first erected 1783
Steam grist-mills first erected 1785
Stereotype printing invented by Mr. Ged, Scotland . . 1785

Cotton first spun in America 1787
Mesmerism, or animal magnetism, discovered by Mesmer . 1788
Sunday schools established in Yorkshire, by Robert Raikes 1789
Galvanism, 1767,—its extraordinary effects on animals discovered by Mrs. Galvani 1789
Steam woollen factory first erected, at Leeds . . . 1792
Flax spun by steam 1793
Vaccination introduced by Jenner 1798
Ceres discovered by Piazzi 1801
Pallas discovered by Olbers 1801
Life-boats invented 1802
Juno discovered by Harding 1804
Vesta discovered by Olbers 1807
Steam first used to propel boats by Fulton, in America . 1807
Engraving on steel first invented by Perkins, an American 1818
Gas first used for lighting streets in the U. S., at Baltimore 1821
Egyptian hieroglyphics first decyphered by Champollion 1822
Macadamizing streets commenced in London by McAdam 1824
First locomotive at Liverpool 1829
Electro-magnetic Telegraph invented by Morse, America . 1832
Daguerreotype impressions first taken by Daguerre in France 1839
The existence of the planet Neptune predicted by Adams and Leverrier 1846
Magnetic pendulum used for measuring longitude . . 1848

21–30. How many years elapsed between the invention of the lyre and the invention of musical notes? The earliest known astronomical observations, and the invention of the signs of the Zodiac? The first eclipse on record, and the invention of the telescope? The use of the mariners' compass in China, and the introduction of solar quadrants? The arrival of the first ship in Greece, and the building of the first steamboat? The establishment of the first public library at Athens, and the invention of printing? The invention of the hour-glass, and the invention of watches? The use of paper in China, and the manufacture of paper out of linen? The first manufacture of glass, and the invention of spectacles? The discovery of the Vulgate edition of the Bible, and the establishment of Sunday schools?

31–40. Allow 365¼ days for each year, and reduce to weeks the number of years that elapsed between the manufacture of porcelain in China, and the manufacture of glass in England? The invention of pendulum clocks, and the use of the magnetic pendulum for measuring longitude? The introduction of figures by the Arabs, and the invention of decimal arithmetic? The invention of the steam engine, and the introduction of locomotives? The invention of wind-mills, and the application of steam to grist-mills? The first astronomical observations known to have been made at Babylon, and the prediction of the planet Neptune? The invention of the sun-dial, and the invention of watches? The use of the compass in China, and the discovery of the variation of the needle? The introduction of the art of painting into Rome, and the invention of the daguerreotype? The regulation of the Julian year, and the introduction of the new style into England?

41–43. Give the answer to each of the remaining questions on this page, in years, months, and days,—in years and fractions of a year,—and in years and decimals of a year. What length of time elapsed between the birth of John Calvin, July 10, 1509, and the birth of Oliver Cromwell, April 25, 1599?

44–52. What length of time elapsed between the birth of Fenelon, August 6, 1651, and the birth of William Penn, October 14, 1644? Between the birth of Galileo Galilei, February 19, 1564, and the birth of Sir William Herschel, November 15, 1738? Between the birth of Sir Francis Bacon, January 22, 1561, and the birth of Benjamin Franklin, January 17, 1706?

53–67. What length of time elapsed between the birth of Tycho Brahe, December 19, 1546, and the birth of Sir Isaac Newton, December 25, 1642? The birth of George Washington, February 22, 1732, and Patrick Henry, May 29, 1736? The birth of William Shakspeare, April 23, 1564, and John Milton, December 9, 1608? The birth of John Adams, October 19, 1735, and Thomas Jefferson,

April 2, 1743? The birth of Roger Sherman, April 19, 1721, and Benjamin West, October 10, 1738?[a]

68. How many days have elapsed since the birth of Sir Humphrey Davy, December 17, 1779?

69. The equatorial diameter of the earth is 7970 miles, and the circumference is $3\frac{1}{7}$ times the diameter. If a man 6 feet high were to travel round the earth, how many yards farther would his head travel than his feet?[b]

70. A pound Avoirdupois = 14oz. 11dwt. 16gr. Troy. What part of a Troy ounce is an ounce Avoirdupois?

Ans. $\frac{175}{192}$.

71–75. Find the difference of latitude and longitude[c] between Boston and Philadelphia. New York and St. Louis. Charleston and New Orleans. Cape Horn and the Cape of Good Hope. Paris and St. Petersburg.

76. Assuming 1 ton as the average amount of carbonic acid produced by 6100 persons in 24 hours, and estimating the total population of the globe at 759999000, how many tons would be produced daily by human respiration?

Ans. 124590 tons.

77. The average length of the tropical year is 365d. 5h. 48m. $49\frac{7}{10}$ sec.[d] How many days in 4 centuries, and what fraction of another day?

78. A balance made by Ramsden, for the Royal Society, is capable of weighing 10 pounds avoirdupois, and turns with .01 of a grain.[e] What part of the weight is required to turn the scale?

---

[a] Encyclopædia Americana, Rees' Encyclopædia, Belknap, Biog. American, Allen.

[b] Keith.

[c] See Table, § 9.

[d] Somerville. The tropical or civil year is the time that elapses between two consecutive returns of the sun to the same equinox or solstice.

[e] Ure's Dictionary.

9. TABLE OF LATITUDES AND LONGITUDES,

AND DISTANCES FROM WASHINGTON, OF THE PRINCIPAL CITIES, OBSERVATORIES, NAVAL STATIONS, ETC.

FROM THE UNITED STATES AND AMERICAN ALMANACS.

Place.	State.	Object.	Longitude from Greenwich. ° ′ ″	Latitude. ° ′ ″	Dist. from Wash. Miles
Aberdeen	Scotland	Obs.	W. 2 5 42	57 8 57.8	
Abo	Finland	Obs.	E. 22 17 12	60 26 56.8	
Albany	N. Y.	Capitol	W. 73 42 49	42 39 3	370
Alexandria	D. C.		" 77 6 45	38 49	7
Altona	Denmark	Obs.	E. 9 56 39	53 32 45.3	
Amherst	Mass.	College Ch.	W. 72 31 28	42 22 15.6	383
Annapolis	Md.		" 76 33	38 58 35	40
Apalachicola Bay	Fa.	Light	" 85 5 15	29 37 25	883
Armagh	Ireland	Obs.	" 6 38 53	54 21 12.7	
Auburn	N. Y.		" 76 28	42 55	333
Augusta	Ga.		" 81 54	33 28	575
Augusta	Me.	State House.	" 69 50	44 18 43	595
Baker's Island	Mass.	Light.	" 70 47 37	42 32 11	452
Baltimore	Md.	St. Mary's C.	" 76 37	39 17 55	40
Bangor	Me.	State House.	" 68 47	44 47 50	663
Barnstable	Mass.	New C. H.	" 70 18 34	41 42 6	466
Batavia	N. Y.		" 78 13	42 59	374
Beaufort	S. C.	Arsenal	" 80 41 23	32 25 57	635
Bedford	England	Obs.	" 28 0	52 8 27.6	
Berlin	Prussia	Obs. Old.	E. 13 23 53	52 31 13.5	
Blackheath	England	Obs.	" 41	51 28 2	
Bologna	Italy		" 11 20 53	44 29 54	
Bonn	Germany	Obs.	" 7 6 45	50 44 9.1	
Boston	Mass.	State House.	W. 71 4 20	42 21 23	440
Bremen	Germany	Obs.	E. 8 48 59	53 4 36	
Breslau	Germany	Obs.	" 17 2 30	51 6 30	
Bridgeport	Conn.	Baptist Ch.	W. 73 11 46	41 10 30	284
Bristol	R. I.	Episcopal Ch.	" 71 17 19	41 40 3	409
Brooklyn	N. Y.	Navy Yard.	" 74 0 3	40 42 0	226
Brunswick	Me.	College	" 69 55	43 53	570
Brussels	Belgium	Obs.	E. 4 22 8	50 51 10.8	
Buda	Hungary	Obs.	" 19 3 11	47 29 12.2	
Buffalo	N. Y.		W. 78 55	42 53	381
Burlington	N. J.	Obs.	" 74 52 6	40 4 52	156
Burlington	Vt.		" 73 10	44 27	513
Bushy Heath	England	Obs.	" 20 14	51 37 44.3	
Calais	Me.		" 67 12 15	45 11 18	786
Cambridge	Mass.	Obs. Old.	" 71 7 21	42 22 15	437
Cambridge	England	Obs.	E. 5 51	52 12 51.8	
Camden	S. C.		W. 80 33	34 17	473
Canandaigua	N. Y.		" 77 17	42 54	341
Cape Ann	Mass.	N. Light.	" 70 34 44	42 38 18	470
Cape Elizabeth	Me.	Light	" 70 11 6	43 33 6	
Cape Cod	Me	Light	" 70 4 9	42 2 22	507
Cape Horn	S. America		" 67 16 8	55 58 40 s.	
Cape of Good Hope	Africa	Obs.	E. 18 28 45	33 56 3 s.	
Castine	Me.	Fort	W. 68 59 33	44 22 45	671
Charleston	S. C.	St. M. Ch.	" 79 57 27	32 46 33	540
Charlestown	Mass.	Navy Yard	" 71 3 33	42 22	441
Charlottesville	Va.	Univ. Obs.	" 78 31 30	38 2 3	121
Chicago	Ill.		" 87 30 30	42 0	717
Christiania	Sweden	Obs.	E. 10 44 57	59 54 42.4	
Cincinnati	Ohio	Obs.	W. 84 24	39 5 54	492
Columbia	S. C.		" 81 7	33 57	506
Columbus	Ohio		" 83 3	39 57	393
Concord	N. H.	State House.	" 71 29	43 12 29	481
Constantinople	Turkey	M. St. Sophia	E. 28 59	41 0 16	

Place.	State.	Object.	Longitude from Greenwich.	Latitude.	Dist. from Wash
			° ′ ″	° ′ ″	Miles
Copenhagen	Denmark	Obs.	E. 12 34 57	55 40 53	
Coteau des Prairies	Iowa	Red Quarry	W. 94 19 15	44 0 52	
Coteau du Missouri	Iowa		" 90 15 15	38 37 28	
Cracow	Poland	Obs.	E. 19 57 47	50 3 50	
Dayton	Ohio		W. 84 11	39 44	461
Dedham	Mass	1st Cong. Ch.	" 71 10 59	42 14 57	422
Detroit	Mich		" 82 58	42 24	524
Dorchester	Mass	Obs.	" 71 4 30	42 19 10	438
Dorpat	Russia	Obs.	E. 26 43 45	58 22 47.1	
Dover	N. H.		W. 70 54	43 13	490
Dover	Del.		" 75 30	39 10	120
Dover	Ohio		" 81 29	40 30 52	372
Dublin	Ireland	Obs.	" 6 20 30	53 23 13	
Durham	Scotland	Obs.	" 1 34 30	54 46 14.9	
Easton	Md.	Court House.	" 76 8	38 46 10	80
Eastport	Me.		" 66 56	44 54	769
Edenton	N. C.		" 76 40	36 5	274
Edinburgh	Scotland	Obs.	" 3 10 54	55 57 23.2	
Exeter	N. H.		" 70 55	42 58	474
Florence	Italy		E. 11 15 54	43 46 40.8	
Frankfort	Ky.		W. 84 40	38 14	542
Fredericksburg	Va.		" 77 38	38 34	56
Frederickton	N. B.		" 66 45	46 3	
Frederick	Md.		" 77 18	39 24	43
Galveston	Texas		" 94 47 15	29 15	
Germantown	Pa.	Obs.	" 75 10 9	40 2 40	144
Geneva	Switzerland	Obs.	E. 6 9 22	46 11 59.4	
Georgetown	D. C.	Obs.	W. 77 5 15	38 55	2
Georgetown	S. C.		" 79 17	33 21	488
Gloucester	Mass.	Univ. Church	" 70 40 19	42 36 44	462
Gotha	Germany		E. 10 44 6	50 56 5.2	
Gottingen	Hanover	Obs.	" 9 56 38	51 31 47.9	
Greenfield	Mass.	2d Cong. Ch.	W. 72 36 32	42 35 16	396
Greenwich	England	Obs.	0	51 28 39	
Hagerstown	Md.		" 77 35	39 37	68
Halifax	N. S.		" 63 36 40	44 39 20	936
Hallowell	Me.		" 69 52 30	44 17	593
Hamburg	Germany	Obs.	E. 9 58 32	53 33 5	
Harrisburg	Pa.		W. 76 50	40 16	110
Hartford	Conn.	State House	" 72 40 45	41 45 59	335
Havana	W. I.	Moro Castle	" 82 22 21	23 9 26	
Haverford	Pa.	School	" 75 18 36	40 1 12	130
Helsingfors	Finnland	Obs.	E. 24 57 53	60 9 42.3	
Holmes's Hole	Mass.	Windmill	W. 70 36 38	41 27 15	457
Hudson	N. Y.		" 73 46	42 14	341
Hudson	Ohio	Obs.	" 81 24 54	41 14 42.6	335
Huntsville	Ala.		" 86 57	34 36	708
Indianapolis	Ind.		" 86 5	39 55	571
Ipswich	Mass.	Eastern Light	" 70 46 17	42 41 8	462
Jackson	Miss.		" 90 8	32 23	1010
Jefferson	Mo.		" 92 8	38 36	936
Kasan	Russia	Obs.	E. 49 6 38	55 47 30	
Kensington	England	Obs.	W. 11 43	51 30 12.7	
Key West		Light.	" 81 48 30	24 32 32	
Kingston	U. C.		" 76 40	44 8	456
Knoxville	Tenn.		" 83 54	35 59	498
Konigsberg	East Prussia	Obs.	E. 20 30 8	54 42 50.4	
Kremsmunster	Germany	Obs.	" 14 8 9	48 3 24	
Lancaster	Pa.		W. 76 20 33	40 2 36	111
Lexington	Ky.		" 84 18	38 6	522
Leyden	Holland	Obs.	E. 4 31 53	52 9 28.2	
Little Rock	Ark.		W. 92 12	34 40	1065
Lockport	N. Y.		" 78 46	43 11	403

Place.	State.	Object.	Longitude from Greenwich.	Latitude.	Dist. from Wash
			° ″ ′	° ′ ″	Miles
Louisville	Ky.		W. 85 30	38 3	596
London	England	St. Paul's	" 5 45	50 30 49	
Lowell	Mass.	St. Ann's Ch.	" 71 18 57	42 38 48	444
Lynchburg	Va.		" 79 22	37 36	198
Lynn	Mass.	Church	" 70 57 25	42 27 51	441
Machias	Me.		" 67 22	44 33	
Madras	India	Obs.	E. 80 15 57	13 4 9.2	
Manheim	Germany	Obs.	" 8 27 51	49 29 13.7	
Marblehead	Mass.	Light	W. 70 50 39	42 30 14	458
Marseilles	France	Obs.	" 5 22 15	43 17 49	
Matinicus Rock	Me.	Light.	" 68 57 1	43 46 42	
Mexico	Mexico		" 99 5	19 25 45	
Middletown	Conn.	W. Univ.	" 72 39	41 33 8	326
Milledgeville	Ga.		" 83 20	33 4 30	618
Milan	Italy	Obs.	E. 9 11 48	45 28 0.7	
Mobile	Ala.		W. 88 11	30 40	1013
Mobile Point		Light	" 88 0 36	30 13 38	
Modena		Obs.	E. 10 55 48	44 38 52.8	
Monhegan Island	Me.		W. 69 18 5	43 45 52	
Monomy	Mass.	Pt. Light	" 70 0 6	41 33 31	500
Montpelier	Vt.		" 72 36	44 17	516
Montreal	L. C.		" 73 35	45 31	601
Munich	Bavaria	Obs.	E. 11 36 38	48 8 45	
Nantucket	Mass.	South Tower	W. 70 6 12	41 16 56	490
Naples	Italy	Obs.	E. 14 15 5	40 51 46.6	
Nashville	Tenn.	Univ. Obs.	W. 86 49 3	36 9 33	684
Natchez	Miss.	Castle	" 91 24 42	31 33 48	1110
Newark	N. J.		" 74 10	40 45	215
New Bedford	Mass.		" 70 55 49	41 38 7	434
Newbern	N. C.		" 77 5	35 20	348
Newburg	N. Y.		" 74 1	41 31	286
Newburyport	Mass.	Light	" 70 49 30	42 48 23	478
Newcastle	Del.		" 75 32	39 40	105
New Haven	Conn.	Obs.	" 72 57 30	41 18 30	300
New London	Conn.		" 72 9	41 22	353
New Orleans	La.	N. E. Light	" 89 1 50	29 8 32	1172
Newport	R. I.	Court House	" 71 21 14	41 28 20	408
Newport	Mo.		" 91 7 15	38 33 58	
New York	N. Y.	City Hall	" 74 1 6	40 42 35	225
Nicolaeff	Russia	Obs.	E. 31 58 47	46 58 20.6	
Norfolk	Va.	Farm. Bank	W. 76 18 47	36 50 50	230
Northampton	Mass.	1st Cong. Ch.	" 72 38 21	42 19 8	380
Norwich	Conn.		" 72 7	41 33	357
Ormskirk	England	Obs.	" 2 54	53 34 18	
Oxford	England	Obs.	" 1 15 23	51 45 40	
Padua	Italy	Obs.	E. 11 52 18	45 24 2.5	
Palermo	Italy	Obs. Old.	" 13 21 24	38 6 44	
Paramatta	N. S. Wales	Obs.	" 151 1 35	33 48 50 s.	
Paris	France	Obs.	" 2 20 22	48 50 13	
Pensacola	Fa.	Navy Yard	W. 87 15 21	30 20 30	1050
Petersburg	Va.		" 77 20	37 13 54	140
Philadelphia	Pa.	State House.	" 75 9 45	39 56 58	138
Pittsburg	Pa.		" 79 58	40 26 15	226
Pittsfield	Mass.	1st Cong. Ch.	" 73 15 36	42 26 55	380
Plattsburg	N. Y.		" 73 26	44 42	539
Plymouth	Mass.	Court House.	" 70 40 28	41 57 28	447
Portland	Me.	Tn. House	" 70 20 30	43 39 26	545
Portsmouth	N. H.	Unit. Church.	" 70 45 50	43 4 35	493
Poughkeepsie	N. Y.		" 73 55	41 41	301
Prague	Bohemia	Obs.	E. 14 25 29	50 5 18.5	
Princeton	N. J.	Nassau Hall	W. 74 39 33	40 20 41	177
Providence	R. I.	Univ. Hall	" 71 24 48	41 49 32	400
Quebec	L. C.	Citadel	" 71 16	46 49 12	781

Place.	State.	Object.	Longitude from Greenwich.	Latitude.	Dist. from Wash
			° ′ ″	° ′ ″	Miles
Raleigh	N. C.		W. 78 48	35 47	288
Regent's Park	London	Obs.	" 9 17	51 31 30	
Richmond	Va.	Capitol	" 77 27 28	37 32 17	117
Rochester	N. Y.	Roch. House.	" 77 51	43 8 17	369
Rome	Italy	Obs.	E. 12 28 41	41 53 54	
Roscoe	Ohio		W. 81 52	40 16 40	338
Sackett's Harbour	N. Y.		" 75 57	43 55	415
Saco	Me.		" 70 26	43 31	528
St. Augustine	Fa.		" 81 35	29 48 30	821
St. Croix	W. I.	Obs.	" 64 41 15	17 44 32	
St. Fernando	Spain	Obs.	" 6 12 17	36 27 43	
St. Helena	Africa	Obs.	" 5 42 30	15 55 26 s.	
St. Joseph's Bay	Fa.	Light.	" 85 23 15	29 52	808
St. Louis	Mo.	Cathedral	" 90 15 10	38 37 28	854
St. Mark's	Fa.		" 84 11 0	30 2	662
St. Petersburg	Russia	Pulkova Obs.	E. 30 19 46	59 46 18.7	
Salem	Mass.	E. I. M. Hall	W. 70 53 57	42 31 19	448
Sandwich	Mass.	1st Cong. Ch.	" 70 30 13	41 45 31	456
Savannah	Ga.	Exchange	" 81 7 9	32 4 56	662
Schenectady	N. Y.		" 73 55	42 48	391
Slough	England	Obs.	" 36 0	51 30 20	
South Bend	L. Michigan		" 87 19 39	41 37 6	
Southwick	Mass.	Obs.	" 72 48 57	42 0 47	359
Speyer	Germany	Obs.	E. 8 26 38	49 18 55.2	
Springfield	Mass.	Court House.	W. 72 35 47	42 6 1	363
Springfield	Ill.		" 89 33 12	39 48	780
Stockholm	Sweden	Obs.	E. 18 3 30	59 20 31	
Stratford	Conn.		W. 73 8 45	41 11 7	287
Tallahassee	Fa.		" 84 36	30 28	896
Tampa Bay	Fa.	Egmont Key.	" 82 45 15	27 36 4	
Ti Tanka Tam. Lake	Iowa		" 93 20 54	44 16 41	
Taunton	Mass.		" 71 6 5	41 54 8	420
Toronto	U. C.		" 79 20	43 33	500
Tortugas	Fa.	Light	" 82 52 22	24 37 20	
Trenton	N. J.		" 74 39	40 14	166
Troy	N. Y.		" 73 40	42 44	376
Turin	Italy		E. 7 42 6	45 4 6	
Turtle Island	Lake Erie		W. 83 23 35	41 45 9	
Tuscaloosa	Ala.	Obs.	" 87 42	33 12	858
University of Va.	Va.		" 78 31 29	38 2 3	124
Upsala	Sweden	Obs.	E. 17 38 42	59 51 50	
Utica	N. Y.	Dutch Church	W. 75 13	43 6 49	383
Utrecht	Holland	Obs.	E. 5 7 23	52 5 11	
Vandalia	Ill.		W. 89 2	38 50	781
Vera Cruz	Mexico		" 96 8 30	19 11 52	
Vevay	Ind.		" 84 59	38 46	544
Vienna	Austria	Obs.	E. 16 22 59	48 12 35.5	
Vincennes	Ind.		W. 87 25	38 43	688
WASHINGTON	D. C.	Capitol	" 77 1 30	38 53 23	0
Washington	Miss.		" 91 20	31 36	1104
Warsaw	Poland	Obs.	E. 20 58 23	52 13 1	
Weasel Mountain	N. J.	C. S. St.	W. 74 12 15	40 52 35	
West Hills	N. Y.	C. S. St.	" 73 25 17	40 48 49	260
West Quoddy Head	Me.	Light	" 66 57 19	44 49 4	
Wheeling	Va.		" 80 42	40 7	264
Williamstown	Mass.	C. Church	" 73 13 20	42 42 51	396
Wilmington	Del.		" 75 28	39 41	110
Wilmington	N. C.		" 78 10	34 11	405
Wilna	Poland	Obs.	E. 25 17 59	54 41 0	
Worcester	Mass.	Antiq. Hall	W. 71 48 10	42 16 13	398
York	Me.		" 70 40	43 10	502
York	Pa.		" 76 40	39 58	90
Yorktown	Va.		" 76 34	37 13	

10. Miscellaneous Examples in Integers, Decimals, Fractions, and Compound Numbers.

1–4. Fifteen degrees of difference in longitude are equivalent to an hour's difference of time. What difference of longitude would make a difference of four minutes in the time? A difference of one minute? Of four seconds? Of one second?

5–6. Give a rule in your own words for determining the difference of time when the difference of longitude is known. For the difference of longitude when the difference of time is known.

7–10. What is the difference of longitude between two places, if the difference of time is 2h. 15m. 27sec.? If the difference is 4h. 0m. 16sec.? 39m. 45sec.? 10h. 48m. 49sec.?

11–15. What is the difference of time between Boston and Philadelphia? New York and St. Louis? London and St. Petersburg? Paris and New Orleans? Washington and Stockholm?

16–20. When it is noon at Cincinnati, what time is it at Charleston? At Little Rock? At Florence? At Jefferson? At Raleigh?

21–25. When it is 1h. 20m. 25sec. P. M. at Providence, what time is it at Paramatta? At Vera Cruz? At Albany? At Harrisburg? At Hartford?[a]

26. Estimating the pound sterling at $4.84, in how many years would the agricultural products of the United States pay the British National Debt, the annual amount of the former being about $252,240,779,[b] and the latter amounting in 1847 to £764,608,284?[c]

---

[a] The table in § 9, furnishes materials for an indefinite number of questions, similar to these.

[b] Patent Office Report, 1847.

[c] Annual Register.

27. French Invoice:—

Lyon, 9 Avril, 1848.

Messrs. Merrit & Lamb Doivt.

à Perigord & Lubin.

No. 40, Une caisse par roulage et paquebot Bavaria

P L du 24 Avril.

568[a]	Cravates longues	$6\frac{11}{12}$	42	"	290	50		
569	" ¾ écossais	$11\frac{8}{12}$	36	"				
570	" "	$20\frac{1}{12}$	42	"				
571	" "	$8\frac{9}{12}$						
572	" "	$6\frac{8}{12}$						
573	" "	$5\frac{4}{12}$						
574	" "	8						
		$28\frac{9}{12}$	48	"				
575	" "	19	54	"				
576	" "	5	48	"				
577	" "	$2\frac{6}{12}$	36	"				
578	" "	8	30	"				
					4530			
	Emballage				45			
	Comm. 3%						137	25
						F.	4712	25

28. How many centuries in a trillion seconds, allowing 95 leap years in each 4 centuries, and 24 leap years in each of the remaining centuries, and how many years, days, hours, minutes, and seconds of another century?

29. If a cannon-ball could be fired from the earth to the sun, and move uniformly, in a straight line, at the rate of 1600 feet per second, in what time would it reach the sun, the mean distance being estimated at 95000000 miles?

[a] No. of design.

30. Invoice of Goods shipped by M. Naylor & Sons, of London, in the *Westminster*, Warren, Master, for Boston; by order and for account and risk of Messrs. W. Appleton & Co.

Mark	No.		s. d.	£.	s.	d.	£.	s.	d.
A	86	Two Chests Asafœtida,							
	—	Gr. 5 0 7 tr. 2 24							
	87	" 5 0 24 " 2 24							
		Gr. 10 1 3 tr. 1 1 20							
		1 1 22 dft. 2							
		8 3 9	50						
		Disct. $2\frac{1}{2}$ per cent.			11				
		Cording,			4				
	88	Two boxes Kino,							
	—	Gr. 2 21 tr. 21							
	89	" 2 25 " 23							
		1 1 18 tr. 44							
		1 18 dft. 2							
		1 0 0	45						
		Disct. $2\frac{1}{2}$ per cent.			1	2			
		Cording,			2				
		One Cask Litharge, M. N.							
	90	Gr. 6 1 8							
		Tr. 1 1							
		6 0 7	22 6						
		Cording,			1	6			
		Charges,							
		Customs entries,			6				
		Cartage, wharfage, and shipping,			15				
		B. of Lading 2s. 6d., Postages 2s.			4	6			
							32	3	9
		Brokerage and com., 3 per cent.						19	4
		Insurance £40.	35		14				
		Stamp 2s. Comm. 1s.			3			17	
		E. E. London, 27th Jan., 1850.							
		M. NAYLOR & SONS.							

31. TIME-BOOK.

JUNE, 1849.

NAMES.	25 Monday.	26 Tuesday.	27 Wednesday.	28 Thursday.	29 Friday.	30 Saturday.	Total.	Wages per week.	Amount.	REMARKS.
George Brooks,	1	1	1	1	1	1	6	10.50		Ex. workm'n.
Frank. Jones,	a	½	1	1	1	1	4½	6.00		Rather slow.
John Smith,	1	1	1	1	½	1	5½	5.00		Intemperate.
Wm. Brown,	1	1	a	1	1	a	4	9.00		Good hand.
Thos. Martin,	1	½	½	1	1	1	5	4.00		Careless.

Find the amount of each man's wages, and the total amount of wages for the week.

32. If 150 leaves of paper make a pile an inch high, and each leaf is twice the thickness of a hair, what would be the extent of a quadrillion hair-breadths?

33. In what length of time would light, which moves at the rate of 192000 miles per second, reach us from a star that is one quintillion miles distant?

34–41. Express 1853 by a scale of notation that shall have 9 for its base, instead of 10.[a] Express the same number by a scale of 7; of 5; of 3; of 2; of 8; of 6; of 4.

42–44. In the first three weeks of June, 1849, the number of gallons of water consumed in the city of Philadelphia was as follows: June 1st. 6439650; 2d. 6163665; 3d. 3955785; 4th. 6041005; 5th. 6102335; 6th. 5887015; 7th. 4569085; 8th. 5887680; 9th. 5489035; 10th.

[a] If 9 had been adopted as the base of our system, 9 units of any order would have made one of the next higher order. Therefore if we wish to reduce 14646 to the scale of 9, we divide by 9 and find that there are 1627 units of the 2d order, and 3 of the 1st. The 1627 units of the 2d order are equivalent to 180 units of the 3d order, and 7 of the 2d. The 180 units of the 3d order make 20 units of the 4th order and 0 of the 3d. The 20 units of the 4th order make 2 units of the 5th order and 2 of the 4th. The whole number would therefore be expressed thus: 22073.

```
9)14646
9)1627–3
9)180–7
9)20–0
9)2–2
```

3557140; 11th. 4753075; 12th. 4395096; 13th. 5703690; 14th, 5550365; 15th. 6346655; 16th. 6623640; 17th. 5550365; 18th. 6224995; 19th. 5918345; 20th. 6565310; 21st. 7239945. What was the total consumption of the three weeks? The average daily consumption? If the whole quantity were put into 100-gallon casks, and the casks were arranged side by side, how far would they extend, each cask occupying a space of 3 feet?

45–50. Determine from the following table, the area of the United States and Territories, the amount of exports and imports, in the years 1840 and 1847, and the number of inhabitants to a square mile in each state and territory, by the census of 1840.

STATES.	No. Square Miles.	Exports 1840.	Exports 1847.	Imports 1840.	Imports 1847.
Maine	[a]35,000	[b]$1,018,269	[b]$1,634,203	[b]$628,762	[b]$574,056
New Hampshire	8,030	20,979	1,690	114,647	16,935
Vermont	8,000	305,150	514,298	404,617	239,641
Massachusetts	7,250	10,186,261	11,248,462	16,513,858	34,477,008
Rhode Island	1,200	206,989	192,369	274,534	305,489
Connecticut	4,750	518,210	599,192	277,072	275,823
New York	46,000	34,264,080	49,844,368	60,440,750	84,167,352
New Jersey	6,851	16,076	19,128	19,209	4,837
Pennsylvania	47,000	6,820,145	8,544,391	8,464,882	9,587,516
Delaware	2,120	37,001	235,459	802	12,722
Maryland	11,000	5,768,768	9,762,244	4,910,746	4,432,314
Dist. of Columbia	50	753,923	124,269	119,852	25,049
Virginia	61,352	4,778,220	5,658,374	545,685	386,127
North Carolina	45,500	387,484	284,919	252,532	142,384
South Carolina	28,000	10,036,769	10,431,517	2,058,870	1,580,658
Georgia	58,000	6,862,959	5,712,149	491,428	207,180
Alabama	50,722	12,854,694	9,054,580	574,651	390,161
Mississippi	47,147	....	....	....	336
Louisiana	46,431	34,236,936	42,051,633	10,673,190	9,222,969
Florida	59,268	1,858,850	1,810,538	190,728	143,298
Texas	325,520	....	....	....	29,826
Kentucky	37,680	....	....	2,241	26,956
Tennessee	44,000	....	....	28,938	1,256
Ohio	39,964	991,954	778,944	4,915	90,681
Indiana	33,809	....	....	....	....
Illinois	55,405	....	52,100	....	266
Michigan	56,243	162,229	93,795	138,610	37,603
Iowa	50,914	....	....	....	....
Wisconsin	53,924	....	....	....	....
Missouri	67,380	....	....	10,600	167,195
Arkansas	52,198	....	....	....	....
Minesota Terr.	166,000	....	....	....	....
Missouri "	579,000	....	....	....	....
Oregon "	341,463	....	....	....	....
Indian "	248,851	....	....	....	....
New Mexico	77,387	....	....	....	....
California	448,691	....	....	....	....

[a] U. S. Land Office Documents, 1849; see also, 12th Ann. Rep. Mass. Board of Educ., pp. 34, 35. [b] Am. Almanac.

51. Water is composed of two volumes of hydrogen and one volume of oxygen, a volume of oxygen being of the same weight as 16 volumes of hydrogen. Required the weight of each gas in a cubic foot of water, which weighs, at the temperature of 62° Fahrenheit, 62.5 lb.

52–55. Give abbreviated rules for multiplying, and for dividing[a] by 5; by 25; by 125; by 625.

11. Test Examples.

1. From the sum of 287.53 + 195.7 + 6008 + 7.975 + 3092.06, subtract by a single operation the sum of 948 + 27.008 + 1090.3 + 4726.87 + 95.953.[b]

2–10. Reduce to whole numbers $\frac{6409}{13}$; $\frac{4688}{8}$; $\frac{231}{7}$; $\frac{1001}{11}$; $\frac{1001}{13}$; $\frac{1001}{7}$; $\frac{9875}{25}$; $\frac{9875}{125}$; $\frac{131043}{361}$.

11–15. Reduce 137 to 8ths; to 4ths; to 11ths; to 125ths; to 1016ths.

16–25. Reduce to simple fractions $\frac{2}{3}$ of 7; $\frac{1}{2}$ of $\frac{4}{7}$ of $3\frac{1}{4}$; $\frac{3}{4}$ of $3\frac{1}{2}$ of $2\frac{2}{11}$; $\frac{11}{16}$ of $\frac{4}{33}$ of 90; $\frac{13}{6\frac{2}{3}}$; $\frac{4\frac{7}{9}}{9}$; $\frac{16}{\frac{8}{11}}$; $\frac{\frac{3}{17}}{9}$; $\frac{17\frac{5}{8}}{42\frac{2}{7}}$; $\frac{17}{3}$ of $\frac{11}{17}$ of $\frac{6}{11}$ of $\frac{2\frac{3}{4}}{8\frac{4}{5}}$ of $16\frac{7}{8}$.

---

[a] Since $\frac{1}{25}$=.04 and 25=$\frac{1}{.04}$, we may obtain $\frac{1}{25}$ of a number, (which is equivalent to dividing the number by 25,) by multiplying by .04, and we may obtain 25 times a number, by dividing by .04. By finding the decimal value of $\frac{1}{5}$, &c., similar rules may be formed for each of the other cases.

[b] The *arithmetical complement* of a number is the difference between the number and some power of ten. Such examples as the 1st in this section can be most readily solved by addition, taking the arithmetical complement mentally, of each of the numbers to be subtracted. This may be done by taking each of the figures from 9, except the right hand figure, which should be taken from 10. After adding all the arithmetical complements, we must deduct 1 from the next place at the left. For example, if the sum of 4692.8 — 271.04 — 2637 + 1384.2, were required, writing the numbers as in the margin, we commence by saying 4 from 10 leave 6; 2+9+8=19; 1+4+3+8+2=18; 1+8+6+2+9=26; 2+3+3+7+6=21; 2+1+7—1+4=13; 1—1=0.

    4692.8
    —271.04
    —2637.
    1384.2
    ———————
    3168.96

26–30. Find the value of $15\frac{7}{12}+3\frac{11}{18}+27\frac{5}{6}+\frac{8}{9}+41\frac{17}{20}$; $\frac{3}{4}$ of $3\frac{3}{11}-\frac{2}{7}$ of $\frac{14}{17}$ of $7\frac{2}{7}$; $\frac{3}{7}$ of $\frac{9}{11}$ of $8\frac{5}{9}\times5\frac{1}{2}\times\frac{2}{3}$ of $1\frac{1}{11}$; $\frac{4}{19}\div\frac{3}{7}$; $\frac{2}{5}$ of $\frac{10}{13}\div\frac{7}{13}$ of $\frac{4}{5}$.

31–35. Reduce to mixed numbers $\frac{87}{11}$; $\frac{187}{13}$; $\frac{330}{19}$; $\frac{2002}{31}$; $\frac{40175}{186}$.

36–40. Reduce to a fractional form $49\frac{11}{19}$; $108\frac{27}{56}$; .0932; $231\frac{16}{17}$; 8.087.

41–45. Reduce to the lowest terms $\frac{144}{208}$; $\frac{219}{292}$; $\frac{264}{418}$; $\frac{532}{836}$; $\frac{492}{2964}$.

46–55. Find the numerators and denominators that are indicated by a blank in the following fractions; $\frac{1}{2}=\frac{11}{\quad}$; $\frac{4}{7}=\frac{3}{\quad}$; $\frac{5}{17}=\frac{\quad}{4}$; $\frac{7}{8}=\frac{\quad}{11}$; $\frac{19}{23}=\frac{21}{\quad}$; $8\frac{4}{11}=\frac{\quad}{15}$; $5\frac{3}{5}=\frac{4}{\quad}$; $7\frac{7}{8}=\frac{2}{\quad}$; $\frac{3}{7}$ of $\frac{5}{9}=\frac{\quad}{63}$; $2\frac{1}{2}:6\frac{1}{4}=\frac{\quad}{25}$.

56–60. Reduce to fractions or mixed numbers .18; $2.0\dot{5}$; $.\dot{8}\dot{2}$; $27.8\dot{5}\dot{4}$; $.05\dot{6}\dot{3}$.

61–65. Reduce to decimals $\frac{11}{80}$; $\frac{5}{6}$; $\frac{49}{128}$; $\frac{6}{7}$; $\frac{25}{222}$.

66–70. Reduce to the least common denominator $\frac{27}{35}$ and $\frac{19}{21}$; $\frac{13}{77}$ and $\frac{25}{88}$; $\frac{1}{2}$, $\frac{5}{6}$, $\frac{9}{14}$, and $\frac{17}{21}$; $\frac{9}{10}$, $\frac{3}{14}$, and $\frac{19}{35}$; $\frac{6}{7}$, $\frac{2}{3}$, $\frac{4}{5}$, and $\frac{9}{15}$.

71. A clerk, in balancing his books, found an error of $407.25. What was the probable cause of the error?[a]

72–75. Divide .0072 by 576000; 27.9 by .00124; 46500 by .0002976; 87.002×1.008 by 963×72000.

76–80. Reduce 8s. 7d. 2qr. to the decimal of a £; 2yd. 2ft. 3in. to the fraction of a furlong; $\frac{8}{11}$ miles to fur., r., yd., ft., and in.; 7mo. 21d. to the fraction of a year; $4\frac{5}{9}$d. to seconds.

---

[a] The difference between any number, and the same number transposed in any way, is divisible by 9. Thus, 723—327, 723—372, 723—273, 723—237, are each exactly divisible by 9. Therefore, if we find in comparing the books of a counting-room or banking-house, that they do not agree, and the amount of their disagreement is divisible by 9, we know that it *may* have arisen from a transposition. We shall thus frequently be enabled to discover an error readily, which would otherwise have required a long and tedious examination.

81–85. Multiply in a single line[a] 4793 by 27; by 103; by 251; by 19; by 874.

IV. MEASURES, WEIGHTS, AND CURRENCIES.

THE standard weights of nearly every country, are derived from the linear measures. Coins are made of platina, gold, silver, or copper. As gold and silver are too soft to be used by themselves, some other metal is mixed with them before coining. The metal which is added is called *alloy*.

12. STANDARDS OF THE UNITED STATES.[b]

Congress has never fully exercised the power granted to it by the constitution, of establishing a uniform standard of weights and measures. The standards used at all the Custom-Houses were prepared by Mr. F. R. Hassler, in 1835-6,

---

[a] Multiplication in a single line will be found a very valuable exercise, and in many cases it is much more expeditious than the ordinary method. The accompanying example, of the multiplication of 4872 by 3956, will show the several products that are to be added to obtain each figure of the entire product, and will perhaps render the process as intelligible as it could be made by any formal rule.

$$
\begin{array}{rrrrrrr}
 & & & 4 & 8 & 7 & 2 \\
 & & & 3 & 9 & 5 & 6 \\
\hline
 & & & \multicolumn{4}{r}{6\times4+6\times8+6\times7+6\times2} \\
 & & \multicolumn{4}{r}{5\times4+5\times8+5\times7+5\times2} & \\
 & \multicolumn{4}{r}{9\times4+9\times8+9\times7+9\times2} & & \\
\multicolumn{4}{r}{3\times4+3\times8+3\times7+3\times2} & & & \\
\hline
\underset{\text{m.}}{12}\ + & \underset{\text{h.th.}}{60}\ + & \underset{\text{t.th.}}{113}\ + & \underset{\text{th.}}{133}\ + & \underset{\text{h.}}{101}\ + & \underset{\text{tens}}{52}\ + & \underset{\text{un.}}{12}
\end{array}
$$

To obtain the product in a single line, we say $6\times2=12$. Set down 2 and carry 1. $1+6\times7+5\times2=53$. Set down 3 and carry 5. $5+6\times8+5\times7+9\times2=106$. Set down 6 and carry 10. $10+6\times4+5\times8+9\times7+3\times2=143$. Set down 3 and carry 14. $14+5\times4+9\times8+3\times7=127$. Set down 7 and carry 12. $12+9\times4+3\times8=72$. Set down 2 and carry 7. 7 and $3\times4=19$, which we set down, making the entire product 19273632.

$$
\begin{array}{r}
4872 \\
3956 \\
\hline
19273632
\end{array}
$$

[b] A. D. Bache. Report on Weights, Measures, and Balances.

and are similar to those used in England, anterior to the passage of the "Act of Uniformity," in May, 1834.

Many of the states have attempted to establish uniformity within their own limits, and have passed laws for that purpose. There is, therefore, a slight diversity in the usages of different sections of the Union, but, as nearly all the laws have assumed the English system for their basis, it does not seem desirable to attempt making an abstract of them. The teacher, however, should make his pupils familiar with the laws that have been passed on the subject by the legislatures of their own state.

There is a great discrepancy in the statements of different writers on arithmetic, relative to the government standard. Many of those who have alluded to the subject, seem to have regarded the local customs of their own neighborhood as identical with the practice of the United States' officers, and probably no one has given correct information as to the standards in actual use at the Mint, the Custom-Houses, and in all the departments of the General Government.

In the joint resolution of June 14, 1836, the Secretary of the Treasury is "directed to cause a complete set of all the weights and measures adopted as standards, and now either made, or in the progress of manufacture, for the use of the several Custom-Houses, and for other purposes, to be delivered to the governor of each state in the Union, or such person as he may appoint, for the use of the states respectively, to the end that a uniform standard of weights and measures may be established throughout the United States." In order further to secure this uniformity, Congress directed, in 1838, the preparation and distribution to the states, of balances for adjusting weights and capacity measures. In 1848, twenty-one of the states had received these standards, and a sufficient number had been prepared to meet the demand from the remaining states.

But in these very standards, there is a great want of system. The foot is subdivided decimally, instead of being divided into inches; the decimal multiples of the Troy pound, and the decimal sub-multiples of the Avoirdupois pound are given, although they are never used. The whole matter is, therefore, in great confusion, not only in this country, but in every other country, except France. Some steps have been taken towards the adoption of a uniform international standard, and it is not improbable that some modification of the French system will eventually come into general use throughout the civilized world.[a]

1. LONG MEASURE.

The denominations are Leagues, Miles, Furlongs, Rods, Yards, Feet, and Inches.

Le.		*m.*		*f.*		*r.*		*yd.*		*ft.*		*in.*
1	=	3	=	24	=	960	=	5280	=	15840	=	190080
		1	=	8	=	320	=	1760	=	5280	=	63360
				1	=	40	=	220	=	660	=	7920
						1	=	$5\frac{1}{2}$	=	$16\frac{1}{2}$	=	198
								1	=	3	=	36
										1	=	12

The English standard unit of Long Measure is the yard, which is equivalent to $\frac{360000}{391393}$ of the length of a "pendulum vibrating seconds of mean time in the latitude of London, in a vacuum at the level of the sea."[b] The United States standard, the original, of which the state standards are copies, is a brass scale of 82 inches in length, prepared for the survey of the coast of the United States, by Troughton, of London, and deposited in the Office of Weights and Measures.

The Rod is sometimes called Perch, or Pole.

The Yard, for CLOTH MEASURE, is subdivided into Quarters and Nails.

---

[a] Prof. McCulloh, U. S. Mint. [b] McCulloch.

yd.	*qr.*	*na.*	*in.*
1 =	4 =	16 =	36
	1 =	4 =	9
		1 =	$2\frac{1}{4}$

Surveyors use CHAIN MEASURE, in which the unit is a Chain of 4 Rods. It is subdivided into Poles and Links.

Mile.	*fur.*	*ch.*	*poles.*	*l.*	*in.*
1 =	8 =	80 =	320 =	8000 =	63360
	1 =	10 =	40 =	1000 =	7920
		1 =	4 =	100 =	792
			1 =	25 =	198
				1 =	7.92

A lot of land measuring 10 chains in length and 1 in breadth, contains an acre.

A *palm* = 3 inches; a *hand* = 4 inches; a *span* = 9 inches; a *pace* = 3 feet; a *fathom* = 6 feet; a *knot*, or geographical mile, is $\frac{1}{60}$ of a degree, or $\frac{1}{21600}$ of the earth's circumference, and is equivalent to 1.15257 statute miles, or 6085.56 feet;[a] a *degree* at the equator, is $69\frac{1}{5}$ miles.

The inch is generally subdivided on scales into 10ths or decimal parts, but sometimes into halves, quarters, eighths, and sixteenths. In the work of carpenters and other mechanics, the duodecimal division is sometimes employed. The *inch* = 12 lines, or primes; the *prime* = 12 seconds; the *second* = 12 thirds, &c., &c.

---

[a] This is the value of the geographical or nautical mile, employed in the Topographical Bureau at Washington, in 1849.* The rate of a ship's sailing is determined by the half-minute glass and log-line. The *log* is a piece of board, loaded on one end so that it will stand vertically in the water. The intervals between the knots on the line, are intended to bear the same proportion to a sea mile, as a half minute to an hour. Thus if 8 knots on the line run off of the reel, while the sand is running out of the half-minute glass, the vessel is moving 8 knots an hour.

The length of the knots on the log-line would accordingly be $\frac{1}{120}$ of a nautical mile or 50.713 ft., if perfect accuracy were required. But in order to keep the ship "behind her reckoning," and avoid the danger of running ashore, they are made three or four feet shorter. The length of a knot for a 28 seconds' glass is usually $6\frac{1}{2}$ fathoms.

* **Determined, for the Bureau, by John Downes, from Bessel's Elements of the Terrestrial Spheroid.**

2. SQUARE MEASURE.

The Square Mile is subdivided into Acres, Roods, Rods, Yards, Feet, and Inches.

M.	*A.*	*R.*	*sq. r.*	*sq. yd.*	*sq. ft.*	*sq. in.*
1 =	640 =	2560 =	102400 =	3097600 =	27878400 =	4014489600
	1 =	4 =	160 =	4840 =	43560 =	6272640
		1 =	40 =	1210 =	10890 =	1568160
			1 =	$(5\frac{1}{2})^2$ $30\frac{1}{4}$ =	$(16\frac{1}{2})^2$ $272\frac{1}{4}$ =	$(198)^2$ 39204
				1 =	$(3)^2$ 9 =	$(36)^2$ 1296
					1 =	$(12)^2$ 144

A square piece of land, measuring 209 feet (or nearly 70 paces), on each side, is about equivalent to an acre.

The value of the several denominations in Square and Cubic Measure, is determined by the standards employed in Long Measure.

3. CUBIC MEASURE.

The denominations are Cubic Yards, Cubic Feet, and Cubic Inches.

c. yd.	*c. ft.*	*c. in.*
1 =	$(3)^3$ or 27 =	$(36)^3$ or 46656
	1 =	$(12)^3$ or 1728

A foot of wood is 16 cubic feet. 8 feet of wood, or 128 cubic feet, make a cord. A ton of timber, storage or shipping, is 40 cubic feet. A perch of stone is $24\frac{3}{4}$ cubic feet = 1 perch square and $1\frac{1}{2}$ feet thick. A square of earth is a cube measuring 6 feet on each side, and is equivalent to 216 cubic feet. In measuring round timber, a deduction is sometimes made from the diameter of each stick, to allow for waste in sawing. Therefore, a ton of round timber, although nominally but 40 feet, often contains about 50 feet.[a]

[a] In England 40 c. ft. of round timber, or 50 c. ft. of hewn timber, make 1 *load* or *ton*. In the American lumber yards, the custom is nearly or quite universal of estimating the ton at 40 ft. for both hewn and round timber. But as there are different modes of measurement

Boards are measured by the superficial foot; but if they are more than 1 inch thick, allowance is made for the additional thickness. A board 1½ ft. wide, 20 ft. long, and of any thickness not exceeding 1 inch, contains $20 \times 1\frac{1}{2} = 30$ ft. But if it is 1¼ inches thick, it will contain $20 \times 1\frac{1}{2} \times 1\frac{1}{4} = 37\frac{1}{2}$ ft.

4. LIQUID MEASURE.

The denominations are Hogsheads, Gallons, Quarts, Pints, and Gills.

hhd.	*gall.*	*qt.*	*pt.*	*gi.*
1 =	63 =	252 =	504 =	2016
	1 =	4 =	8 =	32
		1 =	2 =	8
			1 =	4

The United States standard for measuring liquids, is the gallon, which is a vessel containing 58372.2 grains (8.3389 pounds avoirdupois) of the standard pound of distilled water, at the temperature of 39°.83 Fahrenheit, the vessel being weighed in air in which the barometer is 30 inches at 62° Fahrenheit. This corresponds very nearly with the English wine gallon, which contains 231 cubic inches. Milk and malt liquors are sold by Beer Measure in many places, the beer gallon containing 282 cubic inches. The hogshead (measure) is used only in estimating the contents of cisterns, wells, or large bodies of water.

The Gallon (Cong.[a]) is subdivided by apothecaries, into Pints (O[b]), Fluidounces (f℥), Fluidrachms (fʒ), and Minims (♏).

Cong.	O.	f℥	fʒ	♏
1 =	8 =	128 =	1024 =	61440
	1 =	16 =	128 =	7680
		1 =	8 =	480
			1 =	60

---

in use, a ton of round timber will often contain 50 c. ft. Most timber is now sold by Board Measure, the "ton" being nearly obsolete.

[a] From the Latin, *Congiarium*, a gallon.

[b] From *Octans*, an eighth part.

It is sometimes desirable to make an estimate of the weight of fluids. A pint of water weighs a pound;[a] 45 drops make about a fluidrachm; a common teacup holds *about* 4 fluidounces; a common tablespoon *about* half a fluidounce; a teaspoon *about* 1 fluidrachm.[b]

5. DRY MEASURE.

The denominations are Bushels, Pecks, Quarts, and Pints.

bu.	*pk.*	*qt.*	*pt.*
1 =	4 =	32 =	64
	1 =	8 =	16
		1 =	2

The United States standard is the bushel measure, containing 543391.89 standard grains, (77.6274 pounds Avoirdupois,) of distilled water, at the temperature of 39°.83 Fahrenheit, and barometer 30 inches at 62° Fahrenheit. This corresponds very nearly with the Winchester bushel, which is a cylinder 18½ inches in diameter, and 8 inches deep, containing 2150.42 cubic inches.

6. TROY WEIGHT.[c]

The denominations are Pounds, Ounces, Pennyweights, and Grains.

lb.	*oz.*	*dwt.*	*gr.*
1 =	12 =	240 =	5760
	1 =	20 =	480
		1 =	24

The standard Troy pound[d] is equivalent to the weight of

[a] A pint of distilled water at 62° Fahrenheit weighs 1 lb. The difference between distilled water and well-water at any ordinary temperature is so slight, that in making *estimates*, a pint of water may always be considered as weighing a pound.

[b] United States Dispensatory.

[c] "Troy" Weight is said to signify London weight, the name being derived from Troy Novant, the ancient name of London.—*McCulloch.*

[d] The Troy pound was declared to be "the standard and Troy pound of the Mint of the United States, conformably to which the coinage thereof shall be regulated," by an act of Congress of 19th May, 1828.

22.79442 cubic inches of distilled water at its maximum density, the barometer standing at 30 inches; or to 22.8157 c. in. of water at 62° Fahrenheit, barometer at 30 inches.[a] It was copied by Captain Kater, in 1827, from the English Imperial Troy pound.

In APOTHECARIES' WEIGHT, which is used in compounding medicines, the Troy pound is subdivided into Ounces (℥), Drachms (ʒ), Scruples (℈), and Grains.

lb.	℥	ʒ	℈	*gr.*
1 =	12 =	96 =	288 =	5760
	1 =	8 =	24 =	480
		1 =	3 =	60
			1 =	20

In DIAMOND WEIGHT, 4 quarters = 1 grain; 4 grains = 1 carat; 7½ carats = 1 Troy dwt. A diamond weighing 1 carat is worth about $9 if rough, and $36 if cut. The value increases as the square of the weight, unless the weight exceeds 20 carats, in which case the increase is not so rapid. Thus a cut diamond weighing 3 carats, would be worth about $3^2 \times 36 = 324$ dollars.[b] The gold carat grain = $2\frac{1}{5}$ dwt.

At the UNITED STATES MINT, the Troy ounce is adopted as the standard, and all weights are expressed in decimal multiples and submultiples of the ounce. Thus 95lb. 8oz. 15dwt. 15gr. of bullion, would be credited on the Mint books as 1148.65625oz.[c]

7. AVOIRDUPOIS WEIGHT.

The denominations are Tons, Hundred-weights, Quarters, Pounds, Ounces, and Drams.

This is the only *direct* legislation in regard to the adoption of standards, but the joint resolution of June 14th, 1836, indirectly recognises the weights and measures used at the Custom-Houses, as having been "adopted as standards."

[a] McCulloch. [b] Encyclopædia Americana.

[c] Professor McCulloh, of the U. S. Mint.

T.	*cwt.*[a]	*qr.*	*lb.*	*oz.*	*dr.*	*gr. Troy.*
1 =	20 =	80 =	2240 =	35840 =	573440 =	15680000
	1 =	4 =	112 =	1792 =	28672 =	784000
		1 =	28 =	448 =	7168 =	196000
			1 =	16 =	256 =	7000
				1 =	16 =	$437\frac{1}{2}$
					1 =	$27\frac{11}{32}$

The standard Avoirdupois pound is equivalent to 7000 Troy grains, or to the weight of 27.7274 c. in. of distilled water, at 62° Fahrenheit, barometer at 30 inches.[b]

In many of the states, statutes have been enacted, fixing the ton at 2000 lb., the hundred at 100 lb., and the quarter at 25 lb. But in the standard of the general government, the ton of 2240 lb. with its subdivisions, is still retained.[c] Even where the legal ton is 2000 lb., 2240 lb. are often allowed in weighing bulky or cheap materials, such as iron, coal, plaster, &c.

8. FEDERAL MONEY.[d]

The denominations are Eagles, Dollars, Dimes, Cents, and Mills.

E.	*$*	*di.*	*ct.*	*mi.*
1 =	10 =	100 =	1000 =	10000
	1 =	10 =	100 =	1000
		1 =	10 =	100
			1 =	10

The only coins in circulation are the double eagle, the

[a] From c. for *centum* (which signifies one hundred), and wt. *weight*.

[b] The old English pound, which is said to have been the legal standard of weight from the time of William the Conqueror, to that of Henry VII., was derived from the weight of grains of wheat; 32 grains gathered from the middle of the ear, and well dried, made a pennyweight, 20 pennyweights an ounce, and 12 ounces a pound. Henry VII. altered this weight and introduced the present Troy pound, which is $\frac{3}{4}$ of an ounce heavier than the Saxon pound. The Avoirdupois pound was introduced by a statute of 24 Henry VIII.—*Brande.*

[c] See the different Tariff Acts.

[d] Manual of Coins.

eagle, half-eagle, quarter-eagle, and dollar, of gold; the dollar, half-dollar, quarter-dollar, dime, and half-dime, of silver; and the cent and half-cent, of copper. The gold and silver coins contain $\frac{9}{10}$ pure metal, and $\frac{1}{10}$ alloy. The alloy of gold is composed of about $\frac{1}{4}$ silver and $\frac{3}{4}$ copper (not to exceed $\frac{1}{2}$ of silver); the alloy of silver is pure copper. The metal thus alloyed is called *standard.* The eagle contains 258 grains of standard gold, the dollar 412½ grains of standard silver, and the cent 168 grains of copper, and the multiples and subdivisions of all the coins the same proportion. Previous to 1834, the eagle contained 270 grains, of which 22½ grains were alloy. By the act of Congress, June 28th, 1834, it was provided that all gold coins minted anterior to the 31st of July, of that year, should be receivable in all payments at the rate of 94.8 cents per pennyweight. The old eagle is therefore worth $10.665.

By the same act, the following coins were rendered current in the United States:—

The gold coins of Great Britain, Portugal, and Brazil, of not less than 22 carats fine, at 94$\frac{8}{10}$ cents per pennyweight.

The gold coins of France, $\frac{9}{10}$ fine, at 93$\frac{1}{10}$ cents per pennyweight.

The gold coins of Spain, Mexico, and Colombia, of the fineness of 20 carats 3$\frac{7}{16}$ grains, at 88$\frac{9}{10}$ cents per pennyweight.

The silver dollars of Mexico, Peru, Chili, and Central America, of not less weight than 415 grains each, and those re-stamped in Brazil, of the like weight, and of not less fineness than 10 oz. 15 dwt. in the Troy pound, of standard silver, at $1.00 each.

The Five-franc piece of France, when of not less fineness than 10 oz. 16 dwt. in the Troy pound, of standard silver, and weighing not less than 384 grains each, at the rate of 93 cents.

The following table exhibits nearly the value of the principal gold coins of different countries. But, as most of the coins that circulate in the community are more or less worn, the current value is generally a few cents less.

NAMES OF COINS.	Weight.	Contents in pure Gold.	Assay.	New value.
	dw. gr.	grains.	car. gr.	d. c. m.
United States.—Eagle coined before July 31, 1834 .	11 6	247.5	22	10 66 5
Do. coined after July 31, 1834, shares in proportion . . .	10 18	232	21 2¼⅘	10
Austrian Dominions.—Souverain .	3 14	78.6	21 3¾	3 38 7
Double Ducat	4 12	106.4	23 2¾	4 59 3
Hungarian do.	2 5¾	53.3	23 3¼	2 29 7
Bavaria.—Carolin	6 5¼	115	18 2	4 95 7
Max d'or, or Maximilian . .	4 4	77	18 1¾	3 31
Ducat	2 5¾	52.8	23 2¼	2 27 5
Berne.—Ducat, double in proportion	1 23	45.9	23 1¾	1 97 7
Pistole	4 21	105.5	21 2½	4 54 2
**Brazil.*—Johannes, ½ in proportion	18		21 3¾	17 6 4
Dobraon	34 12	759	22	32 70 6
Dobra	18 6	401.5	22	17 30 1
Moidore, ½ in proportion . .	6 22	152.2	22	6 55 7
Crusado	16¼	14.8	21 3⅜	63 8
Brunswick.—Pistole, double in pro'n.	4 21¼	105.7	21 2½	4 55 2
Ducat	2 5¾	51.8	23 0½	2 23 1
Cologne.—Ducat	2 5¾	52.6	23 2	2 26 7
**Colombia.*—Doubloon	17 9	360.5	20 3	15 53 5
Denmark.—Ducat, current . . .	2	42.2	21 0¼	1 81 5
Ducat, specie	2 5¾	52.6	23 2	2 26 7
Christian d'or	4 7	93.3	21 3	4 2 1
East India.—Rupee, Bombay, 1818	7 11	164.7	22 0½	7 9 6
Rupee of Madras, 1818 . . .	7 12	165	22	7 11
Pagoda, Star	2 4¾	41.8	19	1 79 8
**England.*—Guinea, ½ in proportion	5 8½	118.7	22	5 7 5
Sovereign, do.	5 2½	113.1	22	4 83 8
Seven Shilling Piece . . .	1 19	39.6	22	1 69 8
**France.*—Double Louis, coin. b. 1786	10 11	224.9	21 2	9 68 8
Louis, do.	5 5½	112.4	21 2	4 84 3
Double Louis, coined since 1786	9 20	212.6	21 2½	9 16 2
Louis, do.	4 22	106.3	21 2½	4 58 1
Double Napoleon, or 40 francs	8 7	179	21 2¼	7 70 3
Napoleon, or 20 francs . . .	4 3½	89.7	21 2¼	3 86 6
Frankfort on the Main.—Ducat .	2 5¾	52.9	23 2½	2 27 9
Geneva.—Pistole, old	4 7¼	92.5	21 2	3 98 5
Pistole, new	3 15¾	80	21 3½	3 44 6
Genoa.—Sequin	2 5¾	53.4	23 3¼	2 30 2
Hamburg.—Ducat, double in pro'n.	2 5¾	52.9	23 2½	2 27 9
Hanover.—George d'or	4 6½	92.6	21 2¾	3 99
Ducat	2 5¾	53.3	23 3¼	2 29 7
Gold Florin, double in proportion	2 2	39	18 3½	1 69 4
Holland.—Double Ryder . . .	12 21	283.2	22	12 20 5
Ryder	6 9	140.2	22	6 4 3
Ducat	2 5¾	52.8	23 2¼	2 27 5
Ten Guilder Piece, 5 do. in pr'n.	4 7¾	93.2	21 2¼	4 1 6
Malta.—Double Louis	10 16	215.3	20 0¾	9 27 8
Louis	5 8	108	20 1	4 65 3
Demi Louis	2 16	54.5	20 1¾	2 34 8
**Mexico.*—Doubloons, shares in pr'n.	17 9	360.5	20 3	15 53 5
Milan.—Sequin	2 5¾	53.2	23 3	2 29 3

NAMES OF COINS.	Weight.	Contents in pure Gold.	Assay.	New value.
	dw. gr.	grains.	car. gr.	d. c. m.
Doppia or Pistole	4 1½	88.4	21 3	3 80 7
Forty Lire Piece, 1808	8 8	179.7	21 2¼	7 74 2
Naples.—Six Ducat Piece, 1783	5 16	121.9	21 1¾	5 24 9
Two do. or Sequin, 1762	1 20¼	37.4	20 1¼	1 61 3
Three do. or Oncetta, 1818	2 10¼	58.1	23 3½	2 49 6
Netherlands.—Gold Lion or 14 Florin Piece	5 7¾	117.1	22	5 4 6
Ten Florin Piece, 1820	4 7¾	93.2	21 2¼	4 1 6
Parma.—Quadruple Pistole, double in proportion	18 9	386	21	16 62 7
Pistole or Doppia, 1787	4 14	97.4	21 1	4 19 6
Do. do. 1796	4 14	95.9	20 3¾	4 13 5
Maria Theresa, 1818	4 3½	89.7	21 2¼	3 85 1
Piedmont.—Pistole, coined since 1785 half in proportion	5 20	125.6	21 2⅛	5 41 2
Sequin, half in proportion	2 5¾	52.9	23 2½	2 27 9
Carlino, coined since 1785, half in proportion	29 6	634.4	21 2¾	27 33 4
Piece of 20 Francs, ealled Marengo	4 3¼	82.7	20	3 56 4
Poland.—Ducat	2 5¾	52.9	23 2½	2 27 9
**Portugal.*—Dobraon	34 12	759	22	32 70 6
Dobra	18 6	401.5	22	17 30 1
Johannes	18			17 6 4
Moidore, half in proportion	6 22	152.2	21 3¾	6 55 7
Piece of 16 Testoons, or 1600 rees	2 6	49.3	22	2 12 1
Old Crusado of 400 rees	15	13.6	21 3⅝	58 8
New Crusado of 480 rees	16¼	14.8	21 3½	63 7
Milree, coined in 1755	19¾	18.1	21 3⅜	78
New Dobra	17 6		22	16 25 3
Joannese, double in proportion	9 6½		21 3¾	8 76 3
Half in proportion	4 15		21 3¾	4 37 1
Piece of 12 Testoons, or 1200 rees	1 16¼		21 3⅝	1 57 4
Piece of 8 Testoons, or 800 rees	1 4½		21 3⅝	1 12
Prussia.—Ducat, 1748	2 5¾	52.9	23 2½	2 27 9
Ducat, 1787	2 5¾	52.6	23 2	2 26 7
Frederick, double, 1769	8 14	185	21 2¼	7 97 5
Do. do. 1800	8 14	184.5	21 2	7 95 1
Do. single, 1778	4 7	92.8	21 2½	3 99 9
Do. do. 1800	4 7	92.2	21 2	3 97 5
Rome.—Sequin, coined since 1760	2 4½	52.2	23 3½	2 25
Scudo of the Republic	17 0¼	367	21 2¼	15 80 4
Russia.—Ducat, 1796	2 6	53.2	23 2½	2 29
Ducat, 1763	2 5¾	52.6	23 2	2 26 7
Gold Ruble, 1756	1 0½	22.5	22	96 7
Gold Ruble, 1799	18¾	17.1	21 3¾	73 7
Gold Poltin, 1777	9	8.2	22	35 5
Imperial, 1801	7 17¼	181.9	23 2¼	7 83 6
Half do. 1801	3 20½	90.9	23 2¼	3 91 3
Do. do. 1818	4 3½	91.3	22 0⅛	3 94 2
Sardinia.—Carlino, half in proportion	10 7½	219.8	21 1¼	9 47
Saxony.—Ducat, 1784	2 5¾	52.6	23 2	2 26 7
Ducat, 1797	2 5¾	52.9	23 2½	2 27 9

NAMES OF COINS.	Weight.	Contents in pure Gold.	Assay.	New value.
	dw. gr.	grains.	car. gr.	d. c. m.
Augustus, 1754	4 6½	91.2	21 1⅜	3 92 7
Do. 1784	4 6½	92.2	21 2¼	3 97 4
Sicily.—Ounce, 1751	2 20½	58.2	20 1½	2 50 5
Double do. 1758	5 17	117	20 2	5 4 2
**Spain.*—Quadruple Pistole, or Doubloons, 1772, double and single, and shares in proportion . .	17 8½	372	21 2½	16 3 8
Doubloon, 1801	17 9	360.5	20 3	15 53 5
Pistole, 1801	4 8¼	90.1	20 3	3 88 4
Coronilla, Gold Dollar, or Vintem, 1801	1 3	22.8	20 1½	98 3
Sweden.—Ducat	2 5	51.9	23 2	2 23 6
Switzerland.—Pistole of Helvetic Republic, 1800 . .	4 21½	105.9	21 2½	4 56
Treves.—Ducat	2 5¾	52.6	23 2	2 26 7
Turkey.—Sequin Fonducli, of Constantinople, 1773 . .	2 5¾	43.3	19 1½	1 86 8
Do. 1789	2 5¾	42.9	19 0¾	1 84 8
Half Misseir, 1818	18¼	12.2	16 0½	52 1
Sequin Fonducli	2 5	42.5	19 1	1 83 1
Yermeebeshlek	3 1¾	70.3	22 3½	3 2 8
Tuscany.—Zechino, or Sequin .	2 5¾	53.6	23 3¾	2 30 9
Ruspone of the k'm. of Etruria	6 17¼	161	23 3⅞	6 93 9
Venice.—Zechino or Sequin, shares in proportion	2 6	53.6	23 3¼	2 31
Wurtemburg.—Carolin	6 3½	113.7	18 2	4 89 8
Ducat	2 5	51.9	23 2	2 23 7
Zurich.—Ducat, double, and half in proportion	2 5¾	52.6	23 2	2 26 7

The foregoing Table is copied from the American Almanac for 1835. It was originally compiled from the "Manual of Coins," published at the United States Mint, "Kelly's Cambist," and "Moore's Philadelphia Price Current."

The gold coins of the countries to which the star is prefixed, if possessed of the fineness prescribed, are made, by the act already referred to, to "pass current as money, and to be receivable in all payments, by weight, for all debts and demands, from and after the 31st day of July, 1834." The other coins in the Table are not made a legal tender; but they are sold at a certain rate per dwt., according to the purity of the gold.

TABLE OF THE PRINCIPAL SILVER COINS OF DIFFERENT COUNTRIES.

NAMES OF COINS.	Contents in pure Silver.	Value.
	grains.	$ cts.
Austria.—Rix Dollar, or Florin, *Convention* . .	179.6[a]	48
Copftsuck, or 20 Creutzer Piece	59.4	16
Halbe Copftsuck, or 10 Creutzer Piece .	28.8	08
Baden.—Rix Dollar	358.1	96
Bavaria.—Rix Dollar of 1800	345.6	93
Copftsuck	59.4	16
Brunswick.—Rix Dollar, *Convention*	359.2	96
Denmark.—Ryksdaler	388.4	1 05
Mark, specie, or Half Ryksdaler . . .	64.4	17
East Indies.—Sicca Rupee, Calcutta	175.9	47
Company's Rupee (1835)	165.	44
Bombay, new, or Surat (1818)	164.7	44
Fanam, Cananore	32.9	09
" Bombay, old	35.	09
" Pondicherry . , ...	22.8	06
Gulden, Dutch East India Company . .	148.4	40
England.—Crown (*old*)	429.7	1 15
Shilling "	85.9	23
Crown (*new*)	403.6	1 09
Shilling "	80.7	22
France.—Franc	69.4	19
Genoa.—Scudo of 8 Lire	457.4	1 23
Hamburg.—Rix Dollar, specie	397.5	1 07
Double Mark, 32 Schilling	210.3	57
8 Schilling Piece	50.1	13
Hanover.—Rix Dollar, *Constitution*	400.3	1 08
Florin, or piece of ⅔, fine	200.3	54
Holland.—Florin, or Guilder	146.8	40
12 Stiver Piece	92.4	25
Florin of Batavia	141.6	38
Lubeck.—Rix Dollar, specie	391.9	1 05
Mark	105.1	28
Lucca.—Scudo	372.3	1 00
Malta.—Ounce of Emman. Pinto	337.4	91
2 Tari Piece	17.7	05
Milan.—Scudo of 6 Lire	319.6	86
Lira	52.8	14
Modena.—Scudo of 1796	287.4	77
Naples.—Ducat, new	295.4	80
Piece of 10 Carlini	295.1	80
Netherlands.—Florin of 1816	148.4	40
Poland.—Florin, or Gulden	84.	23
Portugal.—New Crusado (1809)	198.2	53
Seis vintems, or Piece of 120 Rees . . .	46.6	13

[a] Brande.

NAMES OF COINS.	Contents in pure Silver.	Value.
	grains.	$ cts.
Testoon	42.5	11
Portuguese Colonies.—Piece of 8 Macutes, of Portuguese Africa	159.8	43
Prussia.—Rix Dollar, *Convention*	359.	96
Florin, or Piece of $\frac{3}{4}$	198.4	51
Rome.—Scudo, or Crown	371.5	1 00
Paolo	37.2	10
Russia.—Rouble	312.1	84
Rouble of Alexander (1805)	278.1	75
20 Copeck Piece	62.6	17
Sardinia.—Scudo, or Crown	324.7	87
Saxony.—Rix Dollar, *Convention*	358.2	96
Sicily.—Scudo	348.2	93
Spain.—Dollar	370.9	1 00
Sweden.—Rix Dollar	388.5	1 05
Switzerland.—Ecu of 4 Franken	407.6	1 10
Turkey.—Piastre of 1818	67.7	18
Tuscany.—Lira	53.4	14
Wurtemburg.—Rix Dollar, specie	359.1	96
Copftsuck	59.8	16

9. ASTRONOMICAL MEASURES.

The denominations of CIRCULAR MEASURE, and of TIME, are the same in all civilized countries.

Every circle is divided into degrees (°), minutes (′), and seconds (″). Seconds are usually subdivided decimally. A *Sign* in Astronomy, is 30 degrees. A *Quadrant* is 90 degrees.

Circ.	°	′	″
1 =	360 =	21600 =	1296000
	1 =	60 =	3600
		1 =	60

The denominations of Time are Years, Months, Weeks, Days, Hours, Minutes, and Seconds.

Yr.	*Dy.*	*h.*	*min.*	*sec.*
1 =	365 =	8760 =	525600 =	31536000
	1 =	24 =	1440 =	86400
		1 =	60 =	3600
			1 =	60

A common year is 365 days. A bissextile, or leap year, is 366 days. A Julian year is 365¼ days. A tropical, solar, or civil year is 365d. 5h. 48m. 49.7sec.[a]

13. Standards of Great Britain.[b]

1. MEASURES.

The linear,[c] superficial, and cubic measures are the same as in the United States. The old wine, beer, and dry measures have been supplanted by the

Imperial Liquid and Dry Measure.

The denominations are Quarters, Cooms, Bushels, Pecks, Gallons, Pottles, Quarts, Pints, and Gills.

Qr.	*C.*	*bu.*	*pk.*	*gal.*	*pot.*	*qt.*	*pt.*	*gi.*
1 =	2 =	8 =	32 =	64 =	128 =	256 =	512 =	2048
	1 =	4 =	16 =	32 =	64 =	128 =	256 =	1024
		1 =	4 =	8 =	16 =	32 =	64 =	256
			1 =	2 =	4 =	8 =	16 =	64
				1 =	2 =	4 =	8 =	32
					1 =	2 =	4 =	16
						1 =	2 =	8
							1 =	4

The imperial standard gallon contains 10 lb. Avoirdupois of distilled water, temperature 62°, barometer 30 inches. Its capacity is therefore 277.274 c. in. The imperial bushel is a cylinder, of which the inner diameter is 18½ inches, and the depth 8¼ in.

The following denominations of Wine Measure have been discarded under the new system, viz.: the Tun = 2 Pipes, the Pipe or Butt = 2 Hhd., the Hogshead = 63 Gal., the Puncheon = 2 Tierces, and the Tierce = 42 Gallons. In the old Beer Measure, 1 Tun = 2 Butts, 1 Butt = 2 Hhd., 1 Hogshead = 1½ Barrels, 1 Puncheon = 2 Barrels, 1

[a] Somerville.

[b] McCulloch, and Man. of Coins.

[c] The Irish mile = 3038 yd.; the Scotch mile = 1984 yd.

Bbl. = 2 Kilderkins, 1 Kilderkin = 2 Firkins, 1 Firkin = 8 Gal. of Ale, or 9 Gal. of Beer.

Coals were formerly sold by the Chaldron, which was subdivided into Vats, Sacks, and Bushels. The coal bushel held 1 qt. more than the Winchester bushel. Twenty-one chaldrons made a Score.

Score.	*Chal.*	*Vats.*	*Sacks.*	*Bu.*
1 =	21 =	84 =	252 =	756
	1 =	4 =	12 =	36
		1 =	3 =	9
			1 =	3

In Dry Measure, a Last is 2 Weys, and a Wey or Load is 5 Quarters.

2. WEIGHTS.

All weights are derived from the Troy and Avoirdupois pounds. The Imperial standards, and the denominations of Troy, Apothecaries', and Avoirdupois weight, are the same as in the United States.

Not only have the English no *natural* standard of weight, but at the present time they have *no* standard, the Imperial Troy pound having been destroyed[a] by the fire which consumed the Houses of Parliament, Oct. 16, 1834.[b] But the bulk of water to which it is equivalent has been so accurately determined (27.7274 c. in.), that it could easily be restored. The length of the seconds' pendulum at Greenwich, may, therefore, be very properly regarded as the present basis of the entire system of weights, measures, and currencies, both of Great Britain and the United States.

Avoirdupois weight may be readily converted into Troy, or Troy into Avoirdupois.

144 lb. Avoirdupois = 175 lb. Troy.
192 oz. Avoirdupois = 175 oz. Troy.

A stone is generally 14 lb. Avoirdupois. But a stone of

---

[a] Brande. [b] Wade.

butcher's meat, or fish, is 8 lb. A stone of glass is 5 lb. A seam of glass is 24 stone. A truss of hay = 56 lb. A truss of new hay, until the 1st of September = 60 lb. A truss of straw = 36 lb. 36 trusses make a load.

In weighing wool, 1 last = 12 sacks, 1 sack = 2 weys, 1 wey = $6\frac{1}{2}$ tods, 1 tod = 2 stone, 1 stone = 2 cloves, 1 clove = 7 lb. A pack of wool contains 240 lb.

8 lb. = 1 clove of cheese or butter, and 56 lb. = 1 firkin of butter. A wey is 32 cloves in Essex, or 42 cloves in Suffolk.

3. ENGLISH OR STERLING MONEY.

The denominations are Pounds, Shillings, Pence, and Quarters, or Farthings.

£		*s.*		*d.*		*qr.*
1	=	20	=	240	=	960
		1	=	12	=	48
				1	=	4

A guinea is 21 shillings. A crown is 5 shillings. The coin which represents the pound is called a sovereign. The coins are the five-guinea piece, the guinea, half-guinea, quarter-guinea, seven-shilling piece, double sovereign, sovereign, and half-sovereign of gold,—the crown, half-crown, shilling, sixpence, fourpence, threepence, twopence, one-and-a-halfpence, and penny of silver,—and the penny, halfpenny, farthing, and half-farthing of copper.[a]

A pound of silver of the English mint standard, contains 11oz. 2dwt. of pure silver, and 18dwt. of alloy. This pound is coined into 66 shillings. A shilling therefore weighs 87.27 grains, and contains 80.727 grains pure silver. The standard for gold is 11 parts fine gold and 1 part alloy. The sovereign weighs 123.274 grains, and contains 113.001

[a] In Scotland, a *bodle* = $\frac{1}{4}$ of an *achison* = $\frac{1}{3}$ of a *bawbee* = $\frac{1}{3}$ of a *plack* = $\frac{1}{6}$ of a penny.—*Gregory.*

grains of pure gold. A sovereign of full weight is therefore worth $4.866.[a] The guinea and its subdivisions have not been coined since 1816.

14.—Standards of France.[b]

1. MEASURES.

The standard unit of linear measure is the Mètre, from which all measures and weights are derived. It is intended to be equivalent to $\frac{1}{10000000}$ of the distance from the pole to the equator, and was determined by measuring the distance from Dunkirk to Rhodes.[c] The names of the multiples and submultiples of the unit, in all the tables, are formed by the following prefixes:

Deca prefixed,	signifies	10	times.	*Deci*	denotes	$\frac{1}{10}$.
Hecto	"	100	"	*Centi*	"	$\frac{1}{100}$.
Kilo	"	1000	"	*Milli*	"	$\frac{1}{1000}$.
Myria	"	10000	"			

Thus the Decamètre = 10 Mètres; the Decistère = $\frac{1}{10}$ Stère; the Hectogramme = 100 Grammes.

The mètre is equivalent to 39.371 inches. The millimètre is sometimes called *trait* (line), the centimètre, *doigt* (finger), the decimètre, *palme*, and the decamètre, *perche*.

The unit of superficial measure is the Are, which is a square decamètre, and is equivalent to 119.6046 square yards. The hectare is often called *arpent* (acre).

---

[a] In 1838 a dispute arose in the settlement of an account, which was submitted to arbitration. The charges were all made in English money, and the point in dispute was the value of the pound sterling, in our own currency. It was decided by the referees, to be $4.858.—*Mer. Mag.*

At the Custom-House, the sovereign is estimated at $4.84; the pound of the British Provinces, Nova Scotia, New Brunswick, Newfoundland and Canada, at $4.00; the pound of Jamaica, Honduras, and Turk's Island, at $3.00; the pound of Nassau, at $2.50.

[b] McCulloch, Enc. Amer., Hunt's Mer. Mag.

[c] About 570 miles.—*Hunt's Mer. Mag.*

The unit of solid measure is the Stère, which is a cubic mètre, and is equivalent to 35.3174 cubic feet, or 1.30805 cubic yards.

The unit of measures of capacity is the Litre, which is a cubic decimètre, and is equivalent to 1.05676 quarts of our standard. The *setier* is equivalent to the hectolitre, and the *muid*, or barrel, to a kilolitre.

In the Systême Usuel,[a] the names of the ancient weights and measures are retained, together with the subdivisions into halves, quarters, eighths, &c. The toise usuelle = 2 mètres; the pied or foot = $\frac{1}{3}$ mètre; the aune = $1\frac{1}{5}$ mètres; the boisseau = 12.5 litres.

2. WEIGHTS.

The unit of weight is the Gramme, which is derived from the cubic centimètre, and is equivalent to 15.434 Troy grains. In the Systême Usuel, the half-kilogramme is called a livre, which is thus subdivided: 4 gros make 1 once; 16 onces make 1 livre usuelle; 2 livres make 1 kilogramme. A *millier* = 1000 kilogrammes = 2205.48 lb. It is used for marine tonnage.

3. MONEY.

Accounts are kept in Francs and Centimes. The old denomination, Sols or Sous, is sometimes used in accounts, 20 sous being rated as a franc. Ten centimes make a decime.

F.		*D.*		*C.*
1	=	10	=	100
		1	=	10

The subdivision of the franc, therefore, resembles our subdivision of the dollar into dimes and cents. Prior to the revolution, the money was divided into louis-d'ors,

[a] Called "usuel," because the common subdivisions are retained but their value is altered to correspond with the new standard.

ecus or crowns, livres tournois, sous, and deniers; the livre, when newly coined, being equivalent to the franc.

L.d'or.	*e.*	*liv.*	*s.*	*d.*
1 =	4 =	24 =	480 =	5760
	1 =	6 =	120 =	1440
		1 =	20 =	240
			1 =	12

The gold coins now in circulation, are the double louis-d'or, the louis-d'or, the double Napoleon (worth 40 francs), the Napoleon, the 40 franc piece, and the 20 franc piece; the silver coins are the crown, half-crown, 30 sous, 15 sous, 6 livres, 5 francs, 2 francs, franc, and quarter-franc; the copper coins are the 5 sous, 2 sous, sol, decime, 5 centimes, 2 centimes, and centime. The mint standard for both gold and silver, is $\frac{9}{10}$ fine metal, and $\frac{1}{10}$ alloy. A kilogramme of standard gold is coined into 155 twenty-franc pieces. A kilogramme of standard silver is coined into 200 francs. The Custom-House valuation of the franc is $0.186.

4. SUBDIVISIONS OF THE CIRCLE AND OF TIME.

With the introduction of the decimal system of measures, weights, and money, an attempt was also made to change the divisions of the circle and of time. Each quadrant of the circle was divided into 100 degrees (making 400° in the entire circle, instead of 360°), each degree into 100 minutes, and each minute into 100 seconds. A series of logarithmic trigonometrical tables, to correspond with this system, was computed by M. Borda, and a compendium of his work, with the logarithms extended to seven places, has been published.

Each of the 12 months was composed of three decades of 10 days each, and at the end of the year, five intercalary days, (in leap years six days,) were added. The day was divided into 10 hours, the hour into 100 minutes, and the minute into 100 seconds. After a short trial, the attempt to introduce these changes entirely failed.

15. Miscellaneous Table[a] of Measures, Weights, and Moneys of Account.

1. Acapulco.—See Mexico.

2. Alexandria.—The yard or *pik* = 26.8 inches. The measures for corn are the *rhebebe* = 4.364 Eng. bushels, and the *quillot* or *kisloz* = 4.729 bushels. The *cantaro* or *quintal* = 100 *rottoli*; but the rottolo has four different values. The *rottolo forforo* = .9347 lb. Av.; 1 *rottolo zaidino* = 1.335 lb. av.; 1 *rottolo zauro* = 2.07 lb. Av.; 1 *rottolo mina* = 1.67 lb. Av.

Money.—Accounts are kept in *current piastres*. 1 piastre = 40 *paras* or *medini*; 1 *medino* = 30 *aspers*, or 8 *borbi*, or 6 *forli*. A *purse* contains 25000 medini. A piastre is worth about 6 cents. Large payments are generally made in Spanish dollars.

3. Alicant.—The yard or *vara* = 4 *palmos* = 29.96 inches. In liquid measure 1 *cantaro* = 8 *medios* = 16 *quartillos* = 3.05 Eng. wine gallons. The *tonnelada* or ton = 2 pipes = 80 arrobas = 100 cantaros. In dry measure 1 *cahiz* = 12 *barchillas* = 96 medios = 192 quartillos = 7 Winch. bush. The *cargo* = 2½ *quintals* = 10 *arrobas*. The arroba = 27 lb. 6oz. Av., and contains 24 large pounds of 18 Castilian ounces, or 36 small pounds of 12oz. each. At the Custom-House, the arroba = 25 lb. of 16oz. each.

Money.—Accounts are kept in *libras* of 20 *sueldos*, each sueldo containing 12 *dineros*. The libra or *peso* = 10 *reals* = 272 *maravedis of plate* or 512 *maravedis vellon* = 78 cents.

4. Amsterdam.—In 1820, the French system of measures and weights was introduced into the Netherlands, the names only being changed. The unit of Long Measure is the *elle*, which equals the French *mètre*. Its decimal divisions are the *palm*, *duim*, and *streep* (corresponding to the decimètre, centimètre, and millimètre), and its decimal multiples, the *roede* and *mijle* (corresponding to the decamètre and kilomètre). The unit of Square Measure is the *vierkante elle* or square elle, which equals the French *centiare* or mètre carré. Its divisions and multiples are the *vierkante palm*, *vierkante duim*, *vierkante streep*, *vierkante roede*, and *vierkante bunder*. The vierkante bunder = 1 Are. In Measures of Capacity 1 *kubicke elle* = 1 *stère*. Its divisions are the *kubicke palm*, *duim*, and *streep*. A *k. elle* of firewood is called *wisse*. In

---

[a] McCulloch, Hunt's Mer. Mag., Am. Al., Boston Custom-House Table, Enc. Amer.

DRY MEASURE, 1 *kop* = 1 *litre*. The *maatje, schepel*, and *mudde*, or *zak*, correspond to the *decilitre, decalitre*, and *hectolitre*. In LIQUID MEASURE, the *kan, maatje, vingerhoede*, and *vat*, are respectively equivalent to the *litre, decilitre, centilitre*, and *hectolitre*. The *last*, or measure for corn, = 27 *mudden*. The *aam*, liquid measure, = 4 *ankers* = 8 *steckans* = 21 *viertels* = 64 *stoopen* = 128 *mingles* = 256 *pintes* = 180 *litres*.

The WEIGHTS are the *wigtje, korrel, lood, ons*, and *pond*, corresponding to the *gramme, decigramme, decagramme, hectogramme*, and *kilogramme*. The *last* for marine tonnage = 2000 *ponds*. The apothecary's new pound = 5787 grains Troy, or 375 grammes, and is subdivided as the English apothecary's pound, into ounces, drams, scruples, and grains. By the old method of calculating, 100 lb. Amsterdam = 108.923 lb. Avoirdupois.

MONEY.—100 cents = 1 florin = $0.40. Accounts are sometimes kept in Flemish money. 1 pound = 6 florins = 20 schillings = 120 stivers = 240 groats = 1920 pennings.

5. ANTWERP.—The same as Amsterdam. Of the old weights, which are still occasionally referred to, the *quintal* of 100 lb. = 103⅓ lb. Av. A schippound is 3 quintals. A stone is 8 pounds. Of the old measures, 1 *last* = 37½ *viertels* = 150 *macken* = about 9⅞ imperial quarters. A barrel = 26½ Eng. gallons.

6. ARABIA.—See BUSSORAH, DJIDDA, MOCHA, MUSCAT.

7. AUSTRIA.—See TRIESTE.

8. BANGKOK.—1 *fathom* = 4 *cubits* = 8 *spans* = 96 *finger-breadths* = about 6½ ft. Eng. 20 fathoms = 1 *sen*, and 100 sen = 1 *yuta*. In weighing, 1 *picul* = 50 *catties* = 133⅓ lb. av.

MONEY.—The currency consists only of silver and cowrie shells. 1 *bat* or *tical* = 4 *salungs* = 8 *fuangs* = 16 *sing-p'hais* = 32 *p'hai-nungs* = 6400 *bia* or *cowries* = 55 cents nearly. 80 ticals make 1 *catty*, and 100 catties make 1 *picul*. Gold and silver are weighed by small weights which have the same denominations as the coins. The p'hai-nung is then divided into 32 *sagas*.

9. BARCELONA.—The yard or *cana* = 8 *palmos* = 32 *quartos* = 21 inches nearly. The *quartera*, or measure for grain = 12 *cortanes* = 48 *picolins* = .235 Winch. quarters. The *carga* or liquid measure = 12 *cortanes* or *arrobas* = 24 *cortarinas* = 72 *mitadellas* = 32.7 wine gallons. 4 cargas = 1 pipe.

MONEY.—1 *libra* = 20 *sueldos* = 240 *dineros* = 480 *mallas* = 53 cents. The libra is likewise divided into $6\frac{2}{3}$ *reales de plata Catalan*, or 10 *reales ardites*.

10. BATAVIA.—The Chinese weights are used, (the picul and catty), but the picul is considered equal to 136 lb. Av. Accounts are kept in florins or guilders, and centimes. The rupee = \$0.44. The rix dollar = 48 stivers = \$0.75. See AMSTERDAM and CANTON.

11. BELGIUM.—See ANTWERP.

12. BENGAL. See CALCUTTA and MADRAS.

13. BILBAO.—See CADIZ.

14. BOMBAY.—1 *guz* = $1\frac{1}{2}$ *haths* = 24 *tussoos* = 27 inches. In salt measure, 1 *rash* = 16 *annas* = 1600 *parahs* = 16800 *adowlies* = 2572176 cub. inches or 40 tons. In grain measure, 1 *candy* = 8 *parahs* = 56 *pailies* = 224 *seers* = 448 *tipprees* = 156lb. 12oz. 12.8dr. av. In liquor measure, 1 *maund* = 50 *seers* = 3000 *rupees* = 76lb. 11oz. 13dr. Av. In weighing all heavy goods except salt, 1 *maund* = 40 *seers* = 2880 *tanks* = 28lb. Av. In pearl weight, 1 *tank* = 24 *ruttees* = 330 *tuckas* = 72 Troy grains. In gold and silver weight, 1 *tola* = 40 *walls* = 179 gr. Troy.

MONEY.—Accounts are kept in rupees. 1 *rupee* = 4 *quarters* = 400 *reas* = 16 *annas* = 50 *pice* = \$0.45. An *urdee* is 2 reas; a *doreea*, 6 reas; a *dooganey*, 4 reas; a *fuddea*, 8 reas; a *paunchea*, 5 rupees; a *gold mohur*, 15 rupees. The annas and reas are imaginary moneys.

15. BRAZIL.—Same as LISBON.

16. BREMEN.—1 ell = 2 feet = 22.76 Eng. inches. 1 *last* = 4 *quarts* = 40 *scheffels* = 160 *viertels* = 640 *spints* = 80.7 Winchester bushels. 1 *oxhoft* = $1\frac{1}{2}$ *tierces* = 6 *ankers* = 30 *viertels* = 264 *quarts* = 58 Eng. wine gal. An *ahm* = 4 ankers. The commercial pound = 2 *marks* = 16 *ounces* = 32 *loths*. 100 lb. Bremen = 109.8 lb. Av. A *shippound* = $2\frac{1}{2}$ *centners* = 290 lb. A *waage* of iron = 120 lb.

MONEY.—1 *thaler* or rix dollar = 72 *grootes* = 360 *swares* = \$0.78$\frac{3}{4}$.

17. BUENOS AYRES.—The same as CADIZ.

18. BURMAH.—See RANGOON.

19. BUSHIRE.—The *league* or *parasang* = 3m. 3fur. 25r. The *royal guz*, or cubit = $37\frac{1}{2}$ inches. The *common guz* = 25 inches.

The *artaba* or principal corn measure = 16 bushels nearly. Pearls are weighed by the *abbas* = 2¼gr. Troy; gold and silver by the *miscal* = 3dwt. very nearly. The *maund shaw* = 2 *maunds tabree* = 13½ lb. Av. at the Custom-House, or 12½ lb. at the bazaar. It is used by dealers in sugar, coffee, copper, and all sorts of drugs. The *maund copra* is 7¾ lb. at the Custom-House, and from 7¼ to 7½ lb. at the bazaar. It is used by dealers in rice and other provisions.

MONEY.—1 *toman* = 50 *abasses* = 100 *mamoodis* = about $2.50. The toman of Gombroon = $5.

20. BUSSORAH.—The Arabian mile = 2148yd. The Aleppo yard, for silks and woollens = 2ft. 2.4in.; the Hadded yard, for cottons and linens, = 2ft. 10.2in.; the Bagdad yard, for all purposes, = 2ft. 7.6in. Gold and silver are weighed by the *cheki* = 100 *miscals* = 7200gr. Troy. 1 *oke of Bagdad* = 2½ *vakias* = 47½oz. Av.; 1 *maund atteree* = 28½ lb. Av.; 1 *maund sofy* or *sesse* = 90¼ lb. Av.; 1 *cutra* of indigo = 138 lb. 15oz. Av. These are the weights used by the European merchants settled at Bussorah; they differ a little from those used by the Arabians.

MONEY.—Accounts are kept in *mamoodis* of 10 *danims*, or 100 *floose*. 1 *toman* = 100 *mamoodis* = $8 nearly.

21. CADIZ.—100 yards or *varas* = 92⅗ English yards. The *common legua* = 800 *varas;* the *legal legua* = 500 *varas.* In corn measure 1 *cahiz* = 12 *fanegas* = 144 *celeminas* = 576 *quartillas* = 1.576 bushels. In liquid measure 1 *cantaro* or *arroba* = 8 *azumbres* = 32 *quartillos.* There are two arrobas, the greater and the less, the former = 4¼, the latter = 3⅜ wine gallons. A *moyo* of wine = 16 arrobas. A *botta* = 30 arrobas of wine, or 38½ of oil. A *pipe* = 27 arrobas of wine, or 34½ of oil. 100 lb. Castile = 101½ lb. Av. The ordinary *quintal* is divided into 4 arrobas, or 100 lbs. of 2 *marcs* each.

MONEY.—Accounts are kept by the *real* of old plate, of which there are 10⅝ in the *peso duro* or hard dollar. The real = 16 *quintos* or 34 *maravedis.* The *ducado de plata,* or ducat of plate, is worth 11 reals. At the U. S. Custom-House the real of plate is estimated at $0.10, and the real vellon at $0.05.

22. CAGLIARI.—The *palm* = 10⅓ inches. The *starello,* or corn measure = 1bu. 1¼pk. Eng. 1 *cantaro* = 4 *rubbi* = 104 *lbs.* = 1248*oz.* = 93 lb. 0oz. 8dr. Av.

MONEY.—1 *lira* = 4 *reali* = 20 *soldi* = \$0.186. 19 reali make 1 *scudo*.

23. CALCUTTA.—1 *coss* = 1000 *fathoms* = 4000 *cubits* = 8000 *spans* = 24000 *hands* = 96000 *fingers* = 288000 *barleycorns* or *jows* = 1m. 1fur. $3\frac{7}{11}$yd. In CLOTH MEASURE 1 *guz* = 2 *hauts* or *cubits* = 16 *gheriahs* = 48 *angullas* = 144 *jorbes* = 1yd. Eng. In SQUARE MEASURE 1 *biggah* = 20 *cottahs* = 320 *chittacks* = 1440 sq. ft. A chittack is 5 cubits or hauts in length, and 4 in breadth. In GRAIN MEASURE 1 *khahoon* = 16 *soallies* = 3200 *pallies* = 12800 *raiks* = 51200 *khaonks* = 30 *bazaar maunds*. In LIQUID MEASURE 1 *bazaar maund* = 8 *pussarees* or *measures* = 40 *seers* = 160 *pouahs* or *pice* = 640 *chittacks* = 3200 *sicca weight*. WEIGHTS.—1 *maund* = 40 *seers* = 640 *chittacks* = 3200 *siccas*. The factory maund = 74 lb. 10oz. $10\frac{2}{3}$dr. Av.; the bazaar maund = 82 lb. 2oz. $1\frac{2}{15}$dr. In GOLD AND SILVER WEIGHT 1 *anna* = $6\frac{1}{4}$ *rutties* = 25 *dhans* or *grains* = 100 *punkhos*; 1 *sicca* = 10 *massas* = 80 *rutties*; 1 *tolah* = 100 *rutties* = 224.588gr. Troy. 1 *mohur* = $166\frac{1}{4}$ *rutties*.

MONEY.—Accounts are kept in sicca, or in current rupees, with their subdivisions, annas and pice. The sicca rupees bear a batta (premium) of 16 per cent. over the current. 1 *gold mohur* = 16 *sicca rupees* = 64 *cahauns* = 256 *annas* = 3072 *pice* or 1024 *punns* = 20480 *gundas*. 4 *cowries* (a species of shell) make 1 *gunda*, and 2560 cowries = 1 current rupee. A current rupee is worth about 44 cents, and a sicca rupee, \$0.50. A *lac* of rupees = 100000 rupees. A *crore* = 100 lacs.

24. CANADA.—See QUEBEC.

25. CANTON.—1 *li* = 180 *fathoms* = 1800 *Chinese feet* = $1897\frac{1}{2}$ Eng. ft. 1 *covid* or *cobre* = 10 *punts* = $14\frac{5}{8}$ inches. There are no liquid or dry measures; all articles that are usually sold by those measures, being sold in Canton by weight. 1 *picul* = 100 *catties* or *gins* = 1600 *taels* or *lyangs* = 16000 *mace* or *tchens* = 160000 *candarines* or *fivans* = 1600000 *cash* or *lis* = $133\frac{1}{3}$ lb. Av. The mace, candarine, and cash are money weights.

MONEY.—Accounts are kept in taels, mace, candarines, and cash. The cash is the only coin made in China. It is composed of 6 parts of copper and 4 of lead, and is cast with a square hole in the middle, so as to be strung on a wire or string. The circulating medium consists principally of cut Spanish dollars. In calculations of prices, and of accounts between foreigners and native merchants, 720 taels = \$1000. In weighing money for payments, 715 taels = \$1000, except to the Company's treasury, when 718 taels = \$1000, or to native merchants, not of the co-hong, or to ship and house compradors, when 715 taels = \$1000. A tael of fine silver should be worth 1000 cash, but on account of their convenience their price is often so much raised that only 750 are given for a tael. In the Custom-House estimate, the tael = \$1.48.

26. CAPE TOWN.—12 *Rhynland inches* = 1 *Rhynland foot;* 27 *Rhynland inches* = 1 *Dutch ell.* In square measure, 1 *morgen* = 600 *roods* = 86400 *square feet* = 12441600 *square inches.* In corn measure, 1 *load* = 10 *muids* = 40 *schepels* = 30bu. 2pk. 5qt. very nearly. In liquid measure 1 *leaguer* = 4 *aams* = 16 *ankers* = 256 *flasks* = 152 wine gallons. The weights are derived from the standard pound of Amsterdam, the *loot,* = $\frac{1}{32}$ of a Dutch pound, being regarded as the unit.

MONEY.—Accounts are kept either in pounds, shillings, pence, and farthings, (as in Great Britain,) or in rix dollars, schillings, and stivers. 1 *rix dollar* = 8 *schillings* = 48 *stivers* = 1s. 6d.

27. CEYLON.—See COLUMBO.

28. CHILI.—See VALPARAISO.

29. CHINA.—See CANTON.

30. CHRISTIANIA.—Measures and weights, same as at Copenhagen.

MONEY.—1 *species dollar* = 120 *skillings* = \$1.06. There are no gold coins made in Norway.

31. CIVITA VECCHIA.—The Roman foot = 11.72 inches English. The *canna* = 78.34in. The builders' canna = 87.96in. The *barrel* = 12.841 imp. gallons of wine, or 12.64 imp. gals. of oil. The *soma* of oil = 36.13 imp. gals. The *rubbio* of corn = 8.143 imp. gals. The *libra* or pound = 12 *onci* = 6912 *grani* = 5234gr. Troy. There are three *cantaros* or quintals,—of 100, 160, and 250 lb. The *migliajo* = 1000 *libre.*

MONEY.—1 *scudo* = 10 *paoli* = 100 *bajocchi* = \$1.00.

32. COLUMBO.—The principal dry measures are the *seer,* which is a perfect cylinder, 4.35in. deep, and 4.35in. diameter,—and the *parrah,* which is a perfect cube, its internal dimensions being 11.57in. on every side. The liquid measure, the weights, and the money, are the same as in Great Britain. A *leaguer* or *legger* = 150 gallons. A *candy* or *bahar* = 500 lb. Av. A *rix dollar* = 1s. 6d.

33. CONSTANTINOPLE.—The *pik* or *pike* is generally estimated at

$\frac{3}{4}$ of a yard Eng. The *berri* = 1826yd.; the Turkish mile = 1409yd. In corn measure 1 *fortin* = 4 *kisloz* = 3.764 bushels. Oil and other liquids are sold by the *alma* or *meter* = 1gal. 3pt.

WEIGHTS.—1 *rottolo* = 176 *drams;* 1 oke = 2.272 *rottoli;* 1 *quintal* or *cantaro* = $7\frac{1}{3}$ *batmans* = 44 *okes* = 124 lb. $7\frac{2}{5}$oz. Av. The quintal of cotton is 45 okes = 127.2 lb. Av.

MONEY.—1 *piastre* = 40 *paras* = 120 *aspers* = about 4 cents. The piastre is exceedingly variable in its value, those coined in 1764 being worth $0.60, and those coined in 1832 being worth only $0.03. A bag of silver = 500 piastres. A bag of gold = 30000 piastres.

34. COPENHAGEN.—The *Danish ell* = 2 *Rhineland feet* = about 25 inches. The Danish mile = 8244yd. In dry measure 1 *last* = 12 *toendes* or *tons* = 96 *scheffels* = 384 *viertels* = $47\frac{1}{2}$ bushels. In liquid measure 1 *quarter* = 2 *pipes* = 4 *hogsheads* = 6 *ahms* or *ohms* = 24 *ankers* = 240 gallons, very nearly. A *fuder* of wine = 930 *pots* = $237\frac{1}{6}$ gallons. 1 *shippound* = 20 *lispounds* = 320 *pounds* = 352.8 lb. Av.

MONEY.—1 *rix dollar* = 6 *marcs* = 96 *skillings*. The *rigsbank dollar* = $0.52. But the money generally used in commercial transactions is bank money, which is at a heavy discount. The old rix dollar, or *species daler*, is worth $1.05.

35. CUBA.—See HAVANA.

36. DANTZIC.—1 *ell* = 2 *Dantzic feet* = 22.6 Eng. inches. The *Rhineland* or *Prussian* foot = 12.356 Eng. in. The Prussian or Berlin ell = $25\frac{1}{2}$ Prussian inches. The Prussian mile = 4.8 English miles. The *last* of grain = $3\frac{3}{4}$ *malters* = 60 *scheffels* = 240 *viertels* = 960 *metzen* = 91 bushels. The *last* of beer = 2 *fuder* = 4 *both* = 8 *hogsheads* = 12 *ahms* = 48 *ankers* = 240 *quarts* = 620gals. 1 *pipe* = 2 *ahms*. The ahm of wine = $39\frac{5}{8}$ gallons. 1 *lispound* = $16\frac{1}{2}$ *pounds* = 264 *ounces* = 8448 *loths* = 17 lb. Av. 100 lb. Dantzic = 103.3 lb. Av. 1 *shippound* = 3 *centners* = 330 *pounds*.

MONEY.—1 *thaler* or dollar = 30 *silver groschen* = 360 *pfennings* = $0.69. Accounts are still sometimes kept in *guldens*, *guilders*, or *florins*. 1 *rix dollar* = 3 *florins* = 90 *groschen* = 270 *schillings* = 1620 *pfennings* = $0.49.

37. DENMARK.—See COPENHAGEN, ELSINEUR.

38. DJIDDA.—See ALEXANDRIA.

39. EAST INDIES.—See BOMBAY, CALCUTTA, MADRAS, TATTA.

40. EGYPT.—See ALEXANDRIA.

41. ELSINEUR.—The same as COPENHAGEN, except that the rix dollar is divided into 4 *orts* instead of 6 *marcs*.

42. GALACZ.—See CONSTANTINOPLE.

43. GENOA.—The *palmo* = 9.725 inches. The *canna piccola*, used by tradesmen and manufacturers, = 9 *palmi;* the *canna grossa*, used by merchants, = 12 *palmi;* the Custom-House *canna* = 10 *palmi*. The *braccio* = $2\frac{1}{3}$ *palmi*. In dry measure 1 *mina* = 8 *quarte* = 96 *gombette* = $3\frac{1}{2}$ bushels nearly. Salt is sold by the *mondino* of 8 *mine*. In liquid measure 1 *mezzarola* = 2 *barilla* = 200 *pinte* = $39\frac{1}{4}$ gallons. The *barilla* of oil = 17 gallons. The pound is of two sorts; the *peso sottile* = $4891\frac{1}{2}$gr. Troy, for weighing gold and silver, and commodities of small bulk,—and the *peso grosso*, for weighing bulky articles. The cantaro of 100 lb. peso grosso = 76 lb. 14oz. Avoir.

MONEY.—1 *lira Italiana* = 100 *centesimi* = 1 French franc. The lira was formerly divided into 20 *soldi*, and the soldo into 12 *denari*. Sales of merchandise continue to be made, for the most part, in the old currency. 6 old *lire di banco* = 5 *new lire* very nearly.

44. GERMANY.—For weights and measures, see BREMEN, DANTZIC, and TRIESTE. The German short mile = 6859 yards; the German long mile = 10126yd.; the Hanover mile = 11559yd.; the Hessian mile = 10547yd.; the mile of Saxony = 9905yd.

MONEY.—The gold ducat = \$2.24. The *florin* or *guilder* = 60 *kreutzers* = 240 *pfennings*. By the Custom-House valuation, the florin of Nuremburg, Frankfort, and the Southern States of Germany = \$0.40; the florin of Augsburg, Austria, and Bohemia, = \$0.485; the florin of St. Gall = \$0.4036. The *thaler* or *rix dollar* = 30 *groschen* = 360 *pfennings*. The thaler of Saxony, Prussia, and the Northern States of Germany = \$0.69. The *guilder* of Surinam, Curaçoa, Essequibo, and Demarara, is divided into 20 *stivers* of 12 *pfennings* each. Its value is fluctuating, but does not differ materially from that of the German florin.

45. GIBRALTAR.—Weights and measures same as in England, except the *arroba* = 25lb. Av., and the *fanega* for grain, = $1\frac{3}{5}$bu. Wine is sold by the gallon, 100 of which = 109.4 U. S. gallons.

MONEY.—Accounts are kept in current dollars, (*pesos*,) divided into 8 *reals* of 16 *quartos* each. 12 reals currency make a *cob*, or hard dollar, by which goods are bought and sold; and 3 of these reals = 5 Spanish *reals vellon.*

46. Greece.—See Patras.

47. Hamburgh.—The Hamburgh foot = 11.289 inches. The Rhineland foot, used by engineers and land surveyors, = 12.36in. The Brabant ell, used in the measurement of piece goods, = 27.585 in. A ton of shipping = 40c. ft. Dry Measures.—1 *stock* = 1½ *last* = 3 *wisps* = 30 *scheffels* = 90 *fass* = 180 *himtems* = 720 *spints* = = 134.4bu. Liquid Measures.—1 *fuder* = 6 *ahms* = 24 *ankers* or 30 *eimers* = 120 *viertels* = 240 *stubgens* = 480 *kanens* = 960 *quartiers* = 1920 *oessels* = 229½gal. U. S. A *fass* of wine = 4 *oxhofts* = 6 *tierces.* An oxhoft or hogshead of French wine = 62 to 64 stubgens; an oxhoft of brandy = 60 stubgens. A pipe of Spanish wine = 96 to 100 stubgens. A tun of beer = 48 stubgens. A pipe of oil = 820 lb. Whale oil is sold by the barrel of 6 *steckan* = 32 gallons U. S. Weights.—1 *shippound* = 2½ *centners* = 20 *lispounds* = 280 *pounds* = 4480 *ounces* = 8960 *loths;* 100 Hamburgh pounds = 106.8 lb. Av. In estimating the carriage of goods, the shippound is reckoned at 380lbs. In things sold by number, a *gross thousand* = 1200; a *ring* = 2 *gross hundred* = 240; a *small thousand* = 1000; a *shock* = 3 *steigs* = 60; a *gross* = 12 dozen.

Money.—1 *marc* = 16 *sols* or *schillings lubs*[a] = 192 *pfennings lubs.* Accounts are also kept, particularly in exchanges, in *pounds, schillings,* and *pence,* or *grotes* Flemish. The pound consists of 2½ *crowns,* 3¾ *thalers,* 7½ *marcs,* 20 *schillings Flem.* or 240 *grotes Flem.* The moneys in circulation are divided into *banco* and *current money.* The former consists of sums credited by the bank to those who have deposited bullion or specie, and is worth an *agio* or premium over current money. This *agio* is usually about 23 per cent., but is constantly varying. Of the coins in circulation, the *rix dollar banco* = about $1, and the *rix dollar current* = $0.80, are the most common. The Hamburgh gold ducat = $2.07. The *marc banco* = $0.35, according to the Custom-House valuation.

48. Havana.—108 *varas* = 100 yards. 1 *fanega* = 3 bushels nearly, or 100 lb. Spanish. 1 *quintal* = 4 *arrobas* = 101¾ lb. Av. An arroba of wine or spirits = 4.1 U. S. gal. nearly.

Money.—1 *dollar* = 8 *reals plate* = 20 *reals vellon* = $1.00. A *doubloon* = $17.

49. Hayti.—See Port au Prince.

---

[a] *Lubs* is a contraction for money of Lubeck.

50. HOLLAND.—See AMSTERDAM.

51. JAPAN.—See NANGASACKI.

52. JAVA.—See BATAVIA.

53. KÖNIGSBERG.—See DANTZIC.

54. LAGUAYRA.—Weights and measures the same as in SPAIN, with the exception of the British Imperial gallon.

MONEY.—The currency consists of silver money, called *macuquena*. 1 *dollar* = 8 *reals* = \$0.75. The money is very unequal in weight and purity.

55. LEGHORN.—1 *braccio* = 20 *soldi* = 60 *quattrini* = 240 *denari* = 22.98 inches. The *canna* = 4 *bracci*. The Tuscan mile = 1808 yards. 1 *barile* = 20 *fiaschi* = 40 *boccali* = 80 *mezzette* = 12 U. S. gal. The *barile* of oil = 16 *fiaschi* = 8.83 U. S. gal.; it weighs about 16 lb. Av. A large jar of oil contains 30 gallons; a small one 15; and a box with 30 bottles contains 4gal. Corn is sold by the sack or *sacco* = 2.0739bu. U. S. The pound is divided into 12 ounces, 96 drachms, 288 denari, and 6912 grani, and is equal to 5240gr. Troy. The quintal or *centinajo* = 100 *pounds* = 74.884 lb. Av.; but in mercantile transactions, on account of tares and other allowances, it is usual to estimate 100 lb. of Leghorn = 77 lb. Av. The *cantaro* is generally 150 lb., but a cantaro of sugar = 151 lb.; of oil = 88 lb.; of brandy = 120 lb.; of stock fish, and some other articles = 160 lb. The *rottolo* = 3 lb.

MONEY.—Accounts are principally kept in *pezze di otto reali*, (or dollars of 8 reals,) the *pezza* being divided into 20 *soldi* or 240 *denari*. The *lira* is another money of account, chiefly used in inferior transactions, and subdivided like the *pezza*. 1 *pezza* = 5¾ *lire*. The moneys of Leghorn have two values, *moneta buona*, or the effective money of the place, and *moneta lunga*, which is worth $\frac{1}{23}$ more than *moneta buona*. The *pezza* of account = \$0.87; the *pezza lunga* = \$0.9076. The Tuscan *scudo* or crown = \$1.05. It was subdivided like the *pezza*.

56. LIMA.—See CADIZ.

57. LISBON.—1 *branca* = 2 *varas* = 3⅓ *covados* or cubits = 10 *palmes* = 86.4 inches. The *palme* = 2 *pes* or feet. The *legoa* = 6760yd. In liquid measure, 1 *tonnelada* = 2 *pipes* = 52 *almudes*; 1 *baril* = 18 *almudes*; 1 *almude* = 2 *potes* = 12 *canadas* = 48 *quartellos* = 4.37gal. U. S. The value of the almude in different parts of Portugal varies from 4⅓ to 6⅝ gallons. The principal dry meas-

ure is the *moyo* = 15 *fanegas* = 60 *alquiéres* = 240 *quartos* = 480 *selemis* = 23.03bu. U. S. The alquiére varies in different parts of Portugal from 3.07 to 3⅞ dry gallons. 1 *quintal* = 4 *arrobas* = 88 *arratels* or pounds = 176 *marcs* = 1408 *ounces* = 89.047 lb. Av.

MONEY.—1 *milree* = 1000 *rees* = $1.12 in *silver*. The milree of the Azores = $0.83⅓; the milree of Madeira = $1.00; the milree of Brazil is fluctuating in value. In the notation of accounts, the milrees are separated from the rees by a crossed cypher (⊕), and the milrees from the millions by a colon (:), thus, Rs. 2:700⊕500 = 2700 milrees and 500 rees. The *crusado of exchange* or old crusado = 400 rees; the *new crusado* = 480 rees; the *testoon* = 100 rees; the *vintem* = 20 rees.

58. MADRAS.—The *garce*, corn measure, = 80 *parahs* = 400 *marcals* = 137bu. U. S. nearly. The *marcal* = 8 *puddies* = 64 *ollucks*. When grain is sold by weight, the garce = 9256½ lbs. Goods are weighed by the *candy* of 20 *maunds* = 500 lb. Av. 1 *maund* = 8 *vis* = 40 *seers* = 320 *pollams* = 3200 *pagodas*. These are the weights adopted by the English, but those used in the Jaghire, (the territory round Madras belonging to the Company,) and in most other parts of the Coromandel coast, are called the Malabar weights. They are the *gursay* or *garce* = 20 *baruays* or *candies* = 400 *manunghs* or *maunds* = 9645½ lb. Av. The *maund* = 8 *visay* or *vis* = 320 *pollams* = 3200 *varahuns*.

MONEY.—The East India Company and European merchants keep their accounts at 12 fanams the rupee. 1 *pagoda* = 3½ *rupees* = 42 *fanams* = 3360 *cash*. The *star* (or *current*) *pagoda* = $1.84.

59. MALABAR.—See MADRAS.

60. MALACCA.—See SINGAPORE.

61. MALAGA.—The *arroba* or *cantara* = 4.19gal. U. S. The regular pipe of Malaga wine contains 35 arrobas, but is reckoned only at 34; a *bota* of Pedro Ximenes wines = 53½ arrobas; a bota of oil is 43, and a pipe 35 arrobas; a *carga* of raisins is 2 baskets, or 7 arrobas; a cask contains as much, though only called 4 arrobas. For other measures, weights, and coins, see CADIZ. Accounts are kept in reals of 34 maravedis vellon.

62. MALTA.—1 *canna* = 8 *palmi* = 2$\frac{2}{7}$yd.[a] The Maltese foot = 11⅙ inches. The *caffiso* or measure for oil = 5½gal. U. S. The *salma* of corn, stricken measure, = 8.22bu. U. S.; heaped measure is

[a] This is the allowance usually made by merchants, in converting Malta into English measure. In reality 1 *canna* = 8.19 inches; 1 *cantaro* = 174⅓ lb. Av.

reckoned 16 per cent. more. The *cantaro* = 100 *rottoli* or pounds = 3000 *oncie* = 175 lb. Av.[a]

MONEY.—In 1825 British silver money was introduced into Malta; the Spanish dollar being made legal tender at 4s. 4d; the Sicilian dollar at 4s. 2d.—and the Maltese scudo at 1s. 8d. The *scudo* = 12 *tari* = 240 *grani* = $0.40.

63. MANILLA.—The same as CADIZ, except that weights are estimated by *piastres*. 16 piastres are estimated = 1 Spanish pound, though they are not quite so much. 1 *tale* of silk = 11 piastres or ounces; 1 *catty* = 22 *piastres;* 1 *marc* of silver = 8 piastres; 1 tale of gold = 10 piastres; 1 *picul* = 100 *catties* = 133⅓ lb. Av.

64. MAURITIUS.—See PORT LOUIS.

65. MECKLENBURG.—See ROSTOCK.

66. MEXICO.—See CADIZ.

67. MOCHA.—The *guz* = 25 inches; the *land covid* = 18in.; the long iron covid = 27in. 1 *cuda*, liquid measure, = 8 *nusseahs* = 128 vakias = about 2gal. U. S. Grain is measured by the *kellah*, 40 of which = 1 *tomand* = about 170 lb. Av. 1 *bahar* = 15 *frazels* = 150 *maunds* = 400 *rottoli* = 6000 *vakias* = 450 lb. Av. There is also a small maund of only 30 vakias; 1 Mocha bahar = 16½ Bombay maunds = 13 Surat maunds = 15.123 seers.

MONEY.—The current coins of the country are *carats* and *commassees;* 1 Spanish dollar = 8 Mocha dollars = 60 commassees = 420 carats.

68. MOGADORE.—The *canna* or cubit = 21 inches. The corn measures are, for the most part, similar to those of Spain. The commercial pound is generally regulated by the weight of 20 Spanish dollars, therefore the quintal of 100 lb. = 119 lb. Av. The market pound for provisions is 50 per cent. heavier.

MONEY.—1 *nutkeel* or ducat = 10 *ounces* = 40 *blankeels* = 960 *fluce* = $0.75.

69. MOLDAVIA.—See GALACZ.

70. MONTEVIDEO.—For weights and measures, see CADIZ. The current coins are the Brazilian patacon and Spanish dollar. 1 *hard* dollar = 1¼ *current* dollars = 960 *centesimos* or cents = $1.00. 1 *real* = 100 *centesimos.*

71. MOROCCO.—See MOGADORE.

---

[a] See Note on Page 97.

72. MUSCAT.—1 *maund* = 24 *cuchas* = 8¾ lb. Av.

MONEY.—1 *mamoody* = 20 *goz* = about $0.05. The coins in circulation are generally sold by weight.

73. NANGASACKI.—The *inc* is about 4 Chinese cubits, or 6½ft. 2½ Japanese leagues are computed to be about 1 Dutch league. The revenues are estimated by two measures of rice, the *man* and *kolf;* the former contains 10000 kolfs, each 3000 bales or bags of rice. The *picul* = 100 *catties* = 1600 *taels* = 16000 *mace* = 160000 *candarines* = about 130 lb. Av. It is, however, generally estimated at 133⅓ lb.

MONEY.—Accounts are kept in taels, mace, and candarines. The Dutch reckon the tael at 3½ florins, or $1.40. The coins in circulation, are the old and new *itjib*, and *cobangs* or *copangs*, of gold,—the *nandiogin*, *itaganne*, and *kodama*, of silver,—and the *seni*, of copper, brass, and iron. Most of them are without any determined value, and are therefore always weighed by the merchants. The *schuit* is a silver piece, 11oz. fine, weighing 4oz. 18dwt. 16gr. Troy.

74. NAPLES.—1 *canna* = 8 *palmi* = 96 *onzie* = 6ft. 11in. 1 *salma* of oil = 16 *staje* = 256 *quarti* = 1536 *mismette*. At Naples, the salma = 42¾gal. U. S.; at Gallipoli it is from 3 to 4 per cent. less; at Bari it is a little larger. The *carro* of wine = 2 *botti* or pipes = 24 *barili* = 1440 *caraffe* = 264gal. U. S. The *carro* of corn = 36 *tomoli* = 52.2bu. U. S. The *cantaro grosso* = 100 *rottoli* = 196½ lb. Av. The *cantaro piccolo* = 106 lb. Av.

MONEY.—1 *ducato* di regno = 10 *carlini* = 100 *grani* = $0.80. The *scudo* of 12 carlini = $0.95.

75. NORWAY.—See CHRISTIANIA.

76. PALERMO.—The yard or *canna* = 8 *palmi* = 3⅕yd. U. S. The *tonna* of liquids = 2 *caffisi* = 4 *barili* = 8 *quartare* = 160 *quartucci* = 9⅝gal. U. S. The *salma grossa* = 9.48bu.; the *salma generale* = 7.62bu. The *cantaro grosso* = 100 *rottoli grossi* of 33 *onzie*, or 110 *rottoli sottili* of 30 *onzie*. The *cantaro sottile* = 100 *rottoli sottili*, or 250 lb. of 12 *onzie*. The rottolo grosso = 1.93 lb. Av.; the rottolo sottile = 1.75 lb. Av.; 100 Sicilian pounds of 12oz. = 70lb. Av.

MONEY.—1 *ducato* = 10 *piccioli* = 100 *bajocchi* = $0.80. Accounts are generally kept in oncie, tari, and grani. 1 *oncia* = 30 *tari* = 600 *grani* = 3 *ducati*.

77. PAPAL STATES.—See CIVITA VECCHIA.

78. PATRAS.—The long *pic*, for measuring linens and woollens, =27in. The short pic, for measuring silks, =25in. The *staro* of corn=2⅓bu. U. S. The quintal is divided into 44 okes, or 132 lb. 100 lb. of Patras =88 lb. Av. Silk weight is $\frac{1}{5}$ heavier.

MONEY.—1 *phœnix* or *drachmè* =100 *lepta*. The phœnix is a silver coin, which should contain $\frac{9}{10}$ of pure metal, and be worth about 16 cents. The lepton is a copper coin. The coinage has been greatly debased.

79. PERSIA.—See BUSHIRE.

80. PERU.—See LIMA.

81. PETERSBURG.—1 *sashen* or fathom =3 *arsheens* =48 *wershok* =7ft. 100 Russian feet=114½ Eng. ft. The *verst*, or Russian mile =500 *sashen* =5fur. 12r. The Polish short mile = 6075yd.; the Polish long mile = 8101yd. The English inch and foot are used throughout Russia, chiefly, however, in measuring timber. In liquid measure, 1 *sorokovy* = 40 *wedros* = 320 *krashkas* = 3520 *tsharkys* = 130gal. U. S. 1 *pipe* = 2 *oxhofts* = 12 *ankers* = 36 *wedros* = 480 *bottles*. 1 *chetwert* of corn = 2 *osmins* = 4 *pajocks* = 8 *chetwericks* = 64 *garnitz* = 5.952bu. U. S. 1 *berkovitz* = 10 *poods* = 400 *pounds* = 12800 *loths* = 38400 *zolotnicks* = 360 lb. Av.[a]

MONEY.—Accounts are kept in bank roubles of 100 copecks. The silver rouble = $0.75, and was declared, by a ukase issued in 1829, to be worth 360 copecks, but the value of the paper rouble fluctuates with the exchange. At the Custom-House, it is estimated at $0.214.

82. PHILIPPINE ISLANDS.—See MANILLA.

83. PORT AU PRINCE.—The measures are the same as in FRANCE. They are divided as in Avoirdupois and Apothecaries' weights, but are about 8 per cent. heavier. The value of the dollar is about $0.33, but is constantly fluctuating.

84. PORT LOUIS.—The measures and weights are those of FRANCE, previous to the Revolution. 100 lb. Fr. = 108 lb. Eng.; 16 Fr. ft. = 15 Eng. ft. The commercial velte = 2 gallons.

MONEY.—Government accounts are kept in sterling money, the franc being received for 10d., and the Spanish dollar for 4s. 4d.

---

[a] The pood is reckoned by merchants at 36 lb. 100 lb. Russian = 90.26 lb. Av., according to Dr. Kelly, or 90.19 lb. according to Nelkenbrecher.

Merchants keep their accounts in dollars and cents, or in dollars livres, and sous.

85. PORTO RICO.—See HAVANA.

86. PORTUGAL.—See LISBON.

87. PRUSSIA.—See DANTZIC.

88. QUEBEC.—The Paris foot is used for all measures of lands granted previous to the conquest, and all measures of length, unless a contrary agreement is made. The English foot, for measuring lands granted since the conquest, and whenever specially agreed upon. The English yard for cloth measure. The English ell of 5qr. when specially agreed upon. The Canada *minot* = $1\frac{1}{8}$bu. for dry measure, except when it is specially agreed that the Winchester bushel shall be used. The Eng. imp. gallon is used for liquids. Weights and currency, as in England. The pound = $4.00.

89. RANGOON.—1 *ten* or basket = 4 *saits* = 8 *sarots* = 16 *pyis* = 64 *salés* = 128 *lamés* = 256 *lamyets*. A *ten* of clean rice ought to weigh 16 *vis* or 58.4 lb. Av. 1 *paiktha* or *vis* = 100 *kyats* or *ticals* = 400 *mat'hs* = 800 *mus* = 1600 *bais* = 6400 *large rwés* = 12800 *small rwés* = 3.65 lb. Av.

MONEY.—Lead is used for small payments; gold and silver, (principally the latter,) for larger ones. There are no coins, but the metal must be weighed, and, very generally, assayed at every payment. Every new assay of silver costs the owner $2\frac{1}{2}$ per cent.

90. RIGA.—1 *clafter* = 3 *ells* = 6 *feet* = 64.74in. 1 *fuder*, for liquids, = 6 *ahms* = 24 *ankers* = 120 *quarts* = 720 *stoofs* = 248gal. U. S. The *loof* for grain = 1.9375bu. A *last* = 48 loofs of wheat, barley, or linseed,—45 loofs of rye,—or 60 loofs of oats, malt, or beans. 1 *shippound* = 20 *lispounds* = 400 *pounds* = 800 *marcs* = 12800 *loths* = 368.68 lb. Av.

MONEY.—See PETERSBURG. The current rix dollar of Riga = 90 groschen = $0.69.

91. ROSTOCK.—1 *ell* = 2 *feet* = 22.76 Eng. in. The *last* = 96 *scheffels* = 116 imp. bu. of oats, or 104 imp. bu. of other grain. The commercial weights are the same as those of Hamburg. There are other weights, 5 per cent. heavier than these, which are principally used in the trade with Russia.

MONEY.—The *rix dollar*, new, = 32 *schillings*, and contains 399.1gr. pure silver.

92. RUSSIA.—See PETERSBURG and RIGA.

93. SALONICA.—Weights and measures same as at Constantinople, except the *kisloz* or *killow*, which = 3.78 kisloz of Smyrna.

MONEY.—Accounts are kept in *piastres* of 40 *paras*, or 120 *aspers*. The coins are those of Constantinople.

94. SARDINIA.—See CAGLIARI and GENOA.

95. SIAM.—See BANGKOK.

96. SICILY.—See PALERMO.

97. SINDE.—See TATTA.

98. SINGAPORE.—English weights and measures are frequently used in reference to European commodities. Piece goods and many other articles, are sold by the *corge* or *score*. A *coyan* of rice or salt = 40 *piculs*. Nearly everything is sold by weight, as in China. The *picul* = 100 *catties* = 133⅓ lb. Av. Gold dust is sold by a Malay weight, called the *bungkal*, = 832gr. Troy. Bengal rice, wheat, and pulses, are sold by the bag, containing 2 Bengal maunds, or 164¼ lb. Av.

MONEY.—Merchants' accounts are kept in Spanish dollars, divided into 100 parts, represented either by Dutch doits or by English copper coins of the same value.

99. SMYRNA.—See CONSTANTINOPLE. The *kisloz* = 1.456 bushels.

100. SOUTH AFRICA.—See CAPE TOWN.

101. SPAIN.—See ALICANT, BARCELONA, CADIZ, GIBRALTAR, and MALAGA. *Vellon* is the old copper coin of Castile, and is of but half the value of the *plate* or silver currency.

102. STOCKHOLM.—The *rod* = 2⅔ *fathoms* = 8 *ells* or *alnas* = 16 *feet* = 15.58 Eng. ft. 1 *pipe* = 2 *oxhofts* = 3 *ahms* = 6 *eimers* = 12 *ankers* = 180 *kannor* = 360 *stup* = 124¼gal. U. S. In corn measure 1 *tun* or barrel = 2 *spann* = 8 *quarts* = 32 *kappor* = 4⅛bu. U. S. The victuali or commercial weights are 1 *skippund* = 20 *lispunds* = 400 *punds* = 375 lb. Av. The iron weights are four-fifths of the victuali. 1 iron *skippund* = 20 *mark punds* = 400 *marks* = 300 lb. Av.

MONEY.—For many years, there were no coins except copper in circulation, but both silver and gold are now coined. 1 *rix dollar* = 48 *skillings* = 576 *rundstycks* or *'ore*. The *banco* currency is worth 30 per cent. of the silver currency. A rix dollar banco is worth about $0.40. A silver rix dollar = $1.06 = 2 dollars 32 skillings banco.

103. SWEDEN.—See STOCKHOLM.

104. TATTA.—1 *guz* = 16 *garces* = 32 inches. But 1 guz of cloth, at Tatta = 34 inches. 1 *carval* of wheat = 60 *cossas* = 240 *twiers* = 960 *puttoes* = 22 Pucca maunds, or 21 Bombay parahs. The gross weights are 1 *maund* = 40 *seers* = 640 *annas* = 2560 *pice* = 74 lb. 5oz. 7dr. Av. The small weights, 1 *tolah* = 12 *massas* = 72 *ruttees* = 576 *hubbahs* = 1728 *moons* = about 6 dwt. Troy.

MONEY.—1 *rupee* = 50 *carivals* = 600 *pice* = 28800 *cowries* = \$0.55.

105. TRIESTE.—The Bohemian mile = 10137 Eng. yd.; the Hungarian mile = 9113 Eng. yd. The ell for woollens = 26.6in.; the ell for silk = 25.2in. The *orna* or *eimer*, liquid measure, = 40 *boccali* = about 15gal. U. S. The *barile* = 173½gal. U. S. The orna of oil = 5½ *caffisi* = 17gal. U. S. The principal dry measure is the *stajo* or *staro* = 2.34bu. U. S. Sometimes the Vienna *metzen* is used, = 2 *polonicks* = 1.723bu. The commercial *pound* = 4 *quarters* = 16 *ounces* = 32 *loths* = 8639gr. Troy.

MONEY.—Contracts are usually in silver; gold coins, not being legal tender, pass only as merchandise. Mercantile accounts are usually kept in *convention* money, so called from an agreement made by some of the German princes in 1763. The current coins are *dollars*, *half-dollars* or *florins*, and *zwanzigers*, or pieces of 20 kreutzers. Ten dollars are coined out of the Cologne marc = 3608gr. Troy, of pure silver. The *florin* = 60 *kreutzers* = 240 *pfennings* = \$0.48½.

106. TUNIS.—The *pic* for woollens = 26.5in.; the silk pic = 24.8in.; the linen pic = 18.6in. The principal oil measure is the *metal* or *mettar* = about 5⅛gal., but it is of different dimensions in different parts of the country. The wine measure is the *millerolle* of Marseilles = 6½ *mitres* = 14.1 imp. gal. The principal corn measure is the *cafiz* = 16 *whibas* = 192 *sahas* = 14½ imp. bu. The *cantaro* = 100 *rottoli* or pounds = 111.05 lb. Av. Gold, silver, and pearls are weighed by the *ounce* of 8 *meticols;* 16 of these ounces make the Tunis pound = 7773.5gr. Troy.

MONEY.—1 *piastre* = 16 *carobas* = 52 *aspers* = 624 *burbine* = \$0.24. The piastres coined since 1828 are worth but \$0.125.

107. TURKEY.—See CONSTANTINOPLE and SALONICA.

108. TUSCANY.—See LEGHORN.

109 URUGUAY.—See MONTEVIDEO.

110. VALPARAISO.—The *vara* = 33.384in. The *fanega*, or principal corn measure, contains 3439 c. in. The *quintal* = 4 *arrobas* = 100 *pounds* = 101.44 lb. Av. See CADIZ.

111. VENEZUELA.—See LAGUAYRA.

112. VENICE.—The woollen *braccio* = 26.6in.; the silk braccio = 24.8in.; the Venice foot = 13.68in. The *botta* = 2 *migliaje* = 5 *bigonzi* = 80 *miri*. The *miro* of oil = 4.028gal. U. S., or 25 lb. *peso grosso*. The *anfora* of wine = 4 *bigonzi* = 8 *mastelli* = 48 *secchii* = 192 *bozze* = 768 *quartuzzi* = 137gal. U. S. The *moggio*, for corn, = 4 *staje* = 16 *quarte* = 64 *quartaroli* = 9.08bu. U. S. 100 lb. *peso grosso* = 105.186 lb. Av.; 100 lb. *peso sottile* = 66.428 lb. Av. The *libra Italiana* is sometimes used, and is equivalent to the French kilogramme.

MONEY.—1 *lira Italiana* = 100 *centesimi* = 1000 *millesimi*. The lira is supposed to be of the same value as the franc, but the lire actually in circulation, are worth only about $0.09. At the Custom-House, the lira is estimated at $0.16.

16. ANCIENT MEASURES, WEIGHTS, AND COINS.[a]

1. SCRIPTURE LONG MEASURE.—1 *schœnus* = 10 *Arabian poles* = 13⅓ *Ezekiel's reeds* = 20 *fathoms* = 80 *cubits* = 160 *spans* = 480 *palms* = 1920 *digits* = 46.2 yards. A *day's journey* = 8 *parasangs* = 24 *eastern miles* = 48 *Sabbath-day's journeys* = 240 *stadia* = 96000 *cubits* = 33.264 miles. LIQUID MEASURE.—1 *coron* or *chomer* = 10 *bath* or *epha* = 30 *seah* = 60 *hin* = 180 *cab* = 720 *log* = 960 *caph* = 103.35 gallons. DRY MEASURE.—1 *coron* or *chomer* = 2 *latech* = 10 *epha* = 30 *seah* = 100 *gomor* = 180 *cab* = 3600 *gachal* = 10.9616 bushels. WEIGHTS.—1 *talent* = 50 *maneh* = 3000 *shekel* = 114 lb. Troy, nearly. MONEY.—1 *talent* = 60 *maneh* = 3000 *shekels* = 6000 *bekah* = 60000 *gerah* = £342 3s. 9d. A *talent of gold* = £5475. The *solidus aureus* or *sextula* = 12s. 0d. 2qr. The *siclus aureus* = £1 16s. 6d.

2. GRECIAN LONG MEASURE.—1 *milion* or mile = 8 *stadios, dulos*, or furlongs = 800 *orgya* or paces = 3200 *pechys* or larger cubits = 3840 *pygon* = 4266⅔ *pygme* or cubits = 4800 *pous* or feet = 6400 *spithame* = $6981\frac{9}{11}$ *orthodoron* = 7680 *dichas* = 19200 *doron* or *dochme* = 76800 *dactylos* or digits = 4835 feet. LIQUID MEASURE.—1 *metretes* = 12 *chous* = 72 *xestes* = 144 *cotyle* = 576 *oxybaphon* = 864

[a] McCulloch, Gregory, Brande, Lavoisne, Enc. Amer.

cyathos = 1728 *conche* = 3456 *mystron* = 4320 *cheme* = 8640 *cochliarion* = 10.335 gallons. Dry Measure.—1 *medimnos* = 48 *chœnix* = 72 *xestes* = 144 *cotyle* = 576 *oxybaphon* = 864 *cyathos* = 8640 *cochliarion* = 1.0906 bushels. Weights.—1 *talent* = 60 *minæ* = 6000 *drachma* = 36000 *obolos* = 56 lb. Av. Money.—1 *tetradrachma* or *stater* = 2 *didrachma* = 4 *drachma* = 6 *tetrobolon* = 12 *diobolon* = 24 *obolos* = 48 *hemiobolos* = 96 *dichalcos* = 192 *chalcos* = 1344 *lepton* = 2s. 7d. The *pentadrachma* = 5 *drachma*. The *drachma* and its multiples, were of silver,—the other coins, mostly of brass. The *stater aureus* = 25 Attic drachmas of silver. The *stater Cyzicenus*, *stater Philippicus*, and *stater Alexandrinus* = 28 silver drachmas. The *stater Daricus* and *stater Crœsius* = 50 silver drachmas.

3. Roman Long Measure.—1 *milliare* = 8 *stadia* = 1000 *passus* = 2000 *gradus* = 3333⅓ *cubits* = 4000 *palmipedes* = 5000 *pedes* or feet = 20000 *palmi minores* = 60000 *unciæ* = 80000 *digiti transversi* = 4835 feet. Square Measure.—1 *as* = $1\frac{1}{11}$ *deunx* = $1\frac{1}{5}$ *dextans* = $1\frac{1}{3}$ *dodrans* = $1\frac{1}{2}$ *bes* = $1\frac{5}{7}$ *septunx* = 2 *semis* or *actus major* = $2\frac{2}{5}$ *quincunx* = 3 *triens* = 4 *quadrans* = 6 *sextans* or *actus minimus* = 8 *clima* or *sescuncia* = 12 *uncia* = 28800sq. feet. Liquid Measure.—1 *culeus* = 20 *amphoræ* = 40 *urnæ* = 160 *congii* = 960 *sextarii* = 1920 *heminæ* = 3840 *quartarii* = 7680 *acetabula* = 11520 *cyathi* = 46080 *ligulæ* = 143.4258 gallons. Dry Measure.—1 *modius* = 2 *semimodii* = 16 *sextarii* = 32 *heminæ* = 128 *acetabula* = 192 *cyathi* = 768 *ligulæ* = 1.0141 pecks. Weights.—1 *libra* = 12 *unciæ* = 36 *duellæ* = 48 *sicilici* = 72 *sextulæ* = 5246gr. Troy. Money.—1 *denarius* = 2 *quinarii* or *victoriati* = 4 *sestertii* = 10 *libellæ* or *asses* = 20 *sembellæ* = 40 *teruncii* = 7d. 3qr. All above the as, (and sometimes the as also,) were of silver, the rest of brass. The gold coin was the *aureus*, which generally weighed double the *denarius*, and was about equivalent to 18 denarii in value.

4. Various Ancient Measures.—The *Olympic stadium* = 202 yards; the *Alexandrian stadium* = 110yd.; the *Egyptian stadium* = 245yd.; the *stadium of Aristotle* = 115yd.; the *Persian parasang* = 6440yd.; the *Egyptian schœne* = 11120yd.; the ancient *Spanish mile* = 1374yd.; the ancient *British mile* = 148yd.; the *league* of Gaul = 2446yd.; the *rasta* of Germany = 4705yd.

[These measures are given principally on the authority of Lavoisne.]

§ 17. Examples for the Pupil

1. The great bell of Moscow was cast in 1653, during the reign of the Empress Anne. Its weight is estimated at 198tons 2cwt. 1qr.[a] To what would this be equivalent in Troy weight?

2. In the first year after the discovery of gold in California, the amount collected is supposed to have been about $5000000. Estimating its average value at $16 per oz. Troy, what would be the weight of the whole in Avoirdupois?

3. Find the weight in Troy grains of the French twenty-franc piece of gold, and also of the silver five-franc piece.

4. The distance from Boston to Liverpool[b] is about 2883 statute miles. At the average rate of 8 knots an hour, in what time would a vessel sail from one of these places to the other?

5–10. On the Cathedral of Paris a bell was placed in 1680, which weighed 340cwt.; a bell was cast in Vienna, in 1711, weighing 354cwt.; in Olmütz is one of 358cwt.; the *Susanne,* a fine-toned bell at Erfurt, which has a large proportion of silver in its composition, weighs 275cwt., and has a clapper weighing 11cwt.[a] Reduce each of these weights to kilogrammes, and find the amount of the whole.

Ans. for the amount, 67966.308kil.

11–15. Great Tom, of Christ Church, Oxford, weighs 17000 lb.; Great Tom of Lincoln, 9894 lb.; the bell of St. Paul's, London, 8400 lb.; a bell at Nankin, China, is said to weigh 50000 lb.; and seven at Pekin, 120000 lb. each.[a] Required the weight of each in Amsterdam ponds.

16–25. Reduce the span and height of each of the following bridges,[c] to the measure required in the right-hand column.

---

[a] Enc. Amer. [b] Cunard Steamers. [c] Brande's Enc.

Name of Bridge.	Material	River.	Place.	Widest Arch.		Date.	Measure.
				Span.	Height.		
Colossus	Wood	Schuylkill	Philadelphia	340 ft.[a]	20 ft.	1813	Tur'h piks.
Piscataqua...	"	Piscataqua	Portsmouth	250 "	27⅓"	1794	Swedish ft.
Southwark...	Iron	Thames	London	240 "	24 "	1818	Ams.palms.
Sunderland...	"	Wear	Sunderland	240 "	30 "	1796	Prussian ft.
Bamberg	Wood	Regnitz	Germany	208 "	17⅓"	1809	Bremen ft.
Trenton	"	Delaware	New Jersey	200 "	32 "	1804	Venice ft.
Vielle Brioude	Stone	Allier	Brioude	183¼"	70¼"	1454	Fr. mètres.
Ulm	"	Danube	Ulm	181⅙"	22¼"	1806	Dantzic ft.
Waterloo	"	Thames	London	120 "	32 "	1816	Script. cub.
Blackfriars...	"	Thames	London	100 "	41½"	1771	Russian ft.

26. Reduce 2 tons 13cwt. 3qr. 17lb. to Canton weight.

27. Reduce 159 taels 3 mace 5 candarines 7 cash to Troy ounces. *Ans.* 193.663oz.

28. In some places milk is sold both by beer measure and by wine measure. To how many wine gallons would the difference amount in a year, in a family that takes 2qt. per day?

29–38. At $4.84 per pound sterling, what is the cost per c. ft. in English money, of each of the following public buildings?

U S. Public Buildings.	Contents.[b]	Cost.[b]
Capitol	4147400 c. ft.	$2250000
Treasury Building	1944740 "	648743
Patent Office	1466660 "	417550
General Post Office	1071252 "	452765
Girard College	2545485 "	1427800
New York Custom-House . .	906000 "	960000
Boston Custom-House	730000 "	776000
Philadelphia Custom-House .	530613 "	257452
Trinity Church, New York .	821070 "	338000
Smithsonian Institution . . .	1545000 "	215000

39. If you had a pair of scales, but no weights, how would you weigh 3lb. 7oz. by using water?

40. How many cents weigh a pound Avoirdupois? How many eagles? How many half-dollars?

[a] This was the largest single span in the world. The bridge was destroyed by an incendiary, September 1, 1838, and a wire suspension bridge has since been erected in its place.

[b] R. Dale Owen.

41. At $0.186 per franc, what would be the value in francs, of a diamond weighing 15 carats, at $36 for the first carat? *Ans.* 43548fr. 40c.[a]

42. What is the value in Federal Money of the French twenty-franc piece, provided the gold is of the present mint standard, and of full weight?

43–44. The largest known diamond was found in Golconda, in 1550, and is in the possession of the Great Mogul. It is half the size of a hen's egg, and is said to weigh 900 carats.[b] In the usual mode of estimating diamonds, what would be its value, at $9 for the first carat? What is its weight in ounces Avoirdupois?

45–50. The Greeks reckoned by the era of the Olympiad, which began at the summer solstice, 776 B. C. The Roman era commenced with the building of the city, April 24, B. C. 753. The Julian era dates from the reformation of the calendar by Julius Cæsar, B. C. 45. The Mohammedan era dates from the Hegira, July 16, A. D. 622. The era of Sulwanah, used in a great part of India, corresponds with A. D. 78. The era of Yezdegird, used in Persia, began June 16th, A. D. 632.[b] Find the time that elapsed between each of these dates and the commencement of the Jewish era, 3760 B. C.

51–67. Find the value of 100 lb. Av. in the commercial weights of each of the following places: Alexandria, Amsterdam, Bombay, Bremen, Cadiz, Calcutta, Canton, Civita Vecchia, Constantinople, Hamburg, Madras, Naples, Patras, Petersburg, Stockholm, Trieste, Venice.

68. Newly burned charcoal will absorb 90 times its bulk of ammonia, or 35 times its bulk of carbonic acid.[c] How many cubic inches of each of these gases would be absorbed by 17 cubic feet of new charcoal?

---

[a] No fraction of a franc is counted, less than 5 centimes. For the method of estimating the value of diamonds, see § 12, p. 73.

[b] Brande. [c] Carpenter.

69. A spider's thread is in some instances no more than $\frac{1}{3000}$ of an inch in diameter. Each thread is formed of from 4 to 6 filaments, and each filament of not less than 1000 fibrils.[a] How many fibrils would occupy a foot in breadth, allowing 1100 to a filament, 5 filaments to a thread, and 3000 threads to an inch?

70. A large sunflower has been observed to lose 1 lb. 4oz. during 24 hours by evaporation, and a cabbage lost 1 lb. 3oz. during the same time.[b] At this rate, in what length of time would the evaporation from each plant amount to 1cwt.?

Ans. Sunflower, 89d. $14\frac{2}{5}$h.
Cabbage, 94d. $7\frac{1}{19}$h.

71. Find the value of each of the French weights and measures, in the weights and measures of the United States.

72–81. Only 4 of the 55 chemical elements[a] seem to be essential to the constitution of living matter, viz: Carbon, Hydrogen, Oxygen, and Nitrogen. What weight of each element is contained in 1cwt. of each of the following substances, the weight of 1 part of Carbon being represented by 6, 1 part of Hydrogen by 1, 1 part of Oxygen by 8, and 1 part of Nitrogen by 14?

	Gum.		*Sugar.*	*Starch.*	*Lignin.*	*Wax.*
Carbon,	414	parts.	421	428	500	806
Hydrogen,	65	"	64	63	56	114
Oxygen,	521	"	515	509	444	80
	1000	"	1000	1000	1000	1000

	Gluten.		*Gelatin.*	*Albumen.*	*Fibrin.*	*Fat.*
Carbon,	557	parts.	483	516	520	790
Hydrogen,	78	"	80	75	72	118
Oxygen,	220	"	276	259	250	92
Nitrogen,	145	"	161	150	158	00
	1000	"	1000	1000	1000	1000

[a] Carpenter. [b] Griffiths.

82. Human hair is from $\frac{1}{250}$ to $\frac{1}{600}$ of an inch in diameter. How many hair's breadths in 49 yards, if there are 370 in an inch?

83. Estimating the entire population of the world at 1000000000, what would be the size of a field that would hold the whole, if each person occupied 1 sq. yd.?

Ans. 322 sq. m. 531A. 2760 sq. yd.

84. The average weight of the lignin, or woody fibre, of an oak, has been estimated at 60 tons, 30 tons of which are carbon.[a] According to this estimate, how many square miles of oak forest would yield an amount of carbon equivalent to that contained in the atmosphere, supposing the oaks to stand 1 rod apart,—carbonic acid consisting of $\frac{3}{11}$ carbon, and $\frac{8}{11}$ oxygen?[b]

85. What is the value of the labor expended in manufacturing 1 lb. Avoirdupois of watch-springs, each spring being worth $2, and weighing $\frac{1}{10}$ of a grain, and the value of the iron employed, being 1 cent?

V. THE FARM.

18. RULES FOR DETERMINING THE WEIGHT OF LIVE CATTLE.

1. Measure in inches, the girth round the breast, just behind the shoulder-blade, and the length of the back from the tail to the fore-part of the shoulder-blade. Multiply the girth by the length and divide by 144. If the girth is less than 3ft., multiply the quotient by 11; if between 3ft. and 5ft., multiply by 16; if between 5ft. and 7ft., multiply by 23; if between 7ft. and 9ft., multiply by 31.[c] If the animal is lean, deduct $\frac{1}{20}$ from the result. Or,

---

[a] Carpenter. [b] See Ex. 17, Sect. 6.

[c] Chambers's Information for the People,—Brit. Husbandry.

2. Take the girth and length in feet. Multiply the square of the girth by the length, and multiply the product by 3.36. The result will be the answer in pounds. Either of these rules will be found useful in making approximate estimates of the weight of cattle, but experienced judges depend more upon observation than upon arbitrary rules. The live weight, multiplied by .605, gives a near approximation to the net weight.

19. Rules for Measuring Grain.

1. If the grain is heaped on a floor, in the form of a cone, measure the depth and slant height in inches. Multiply the difference of the squares of these two measurements by .0005 of the depth, for the contents in bushels.

2. If the grain is heaped against the side of a building, take ½ of the result obtained by the first rule. If heaped against an inner corner of a building, take ¼,—and if against an outer corner, take ¾ of the same result.

3. If the grain is in barrels or common casks, the quantity may be found with tolerable accuracy by the following rule: "To .4 of the square of the bung diameter, add ¼ of the square of the head diameter, and ¼ of the product of the two diameters; multiply the amount of these three quantities by the length of the cask, and divide the product by 265 for wine gallons, by 324 for ale gallons, or by 309 for half-pecks."[a]

4. If the grain is in a bin or crib, .8 of the number of cubic feet, will give the number of bushels. Two bushels of corn on the cob will give about $1\frac{1}{24}$bu. when shelled. A "barrel" of shelled corn, (in the Southern States,) is 5 bushels.[b]

---

[a] Simplified from Dr. Hutton's rule. For more accurate rules, see the article on gauging.

[b] Scientific American.

5. If the grain is not of uniform depth or breadth, take the dimensions in several different places, and take the average length, breadth, and height, with which proceed as in Rule 4.

6. To estimate the yield per acre, measure the quantity produced on 1 sq. rod, and multiply by 160, or measure the produce of 11 yards square, and multiply by 40.

20. WEIGHT OF GRAIN AND HAY.

1. The standard weights of grain[a] in Great Britain, and in many parts of the United States, are as follows: Wheat, 60 lb. per bushel; Indian corn and rye, 56 lb.; barley, 48 lb.; oats, 32 lb. In Indiana, a bushel of rye is 46 lb., and a bushel of oats, 33 lb.[b] Beans and clover seed weigh nearly the same as wheat; castor oil beans, and timothy seed, the same as corn; potatoes, 40 lb. per bushel; buckwheat, 52 lb.; salt, 50 lb.; bran, 20 lb.; dried peaches, 24 lb.; dried apples, 22 lb.; blue grass seed, 14 lb.[c]

2. Ten solid yards of hay which has settled in the stack during winter, weigh about 1 ton. In stacks more than a year old, 9yd., and sometimes 8yd. make a ton. Clover takes 11 or 12yd., and sometimes when it has been stacked very dry, 13yd. to a ton. In the barn, 400 c. ft. in the bay, or 500 c. ft. on the scaffold, make about 1 ton.[d] If a stack of hay has a circular base, its contents may be found very nearly by the following rule: Measure the circumference at the bottom of the stack, and the height of the stack. Multiply the square of the circumference by the height, and divide the product by 40. The contents may also be measured by finding the average length, breadth, and height, and forming the continued product of the three dimensions.[e]

---

[a] Brit. Husbandry. [b] Mer. Mag. [c] Scientific Amer.

[d] Colman, and private information.

[e] A product that is formed by multiplying together three or more factors, is called the "continued product" of those factors. Thus, the continued product of 3, 17, and 5, is $3\times17\times5=255$.

21. Measurement of Land.

1. If the field is square, or oblong, multiply the length by the breadth, and the contents will be the area. If the dimensions are taken in paces, the area will be in square yards. Multiply the number of square yards by 4, and divide the product by 121, and you will obtain the area in square rods; or divide the number of square yards by 484, and point off one figure from the right of the quotient, and the result will be the area in acres.

2. If the field is irregular, divide it into triangles. In each triangle, multiply the length of the longest side by the shortest distance from that side to the opposite angle. Add all the results, divide their sum by 2, and the quotient will be the area. If any portion of the boundary is curved, straight lines should be drawn in such manner as to enclose the same area as the curve.

22. Examples for the Pupil.[a]

1. An ox measured 6ft. 3in. in girth, and the length of his back was 5ft. 4in. Determine his weight by each of the rules.

2–5. There are four heaps of grain; one in an open field, one against the side of a granary, one against an inner corner, and one against an outer corner of a barn. The depth of each heap is 4ft. 8in., and the slant height 6ft. 7in. Required the contents in bushels.

6. A bin 8ft. 6in. long, 3ft. 4in. wide, and 4ft. 9in. deep, is filled with ears of corn. How many bushels are there, and how many will there be when shelled?

7. How much hay in the bay of a barn, the bay being 18ft. square, and the hay 10ft. deep?

---

[a] Colman, Loudon, G. B. Emerson, Johnson, Brit. Husb., and private information.

8

8. How many furrows 9 inches in width, will make an acre in a field that is 6 rods wide? In a field 10 rods wide? 8 rods wide? 40 rods wide?

9. The amount of wheat annually sown in France, on 10863959 acres, is estimated at 32491978 bushels. If, under improved cultivation, $\frac{1}{3}$ of this amount could be saved, and the crops could be increased 5 bushels per acre, what would be the total amount gained?

10. If, by expending $50 per acre in the cultivation of potatoes, one farmer can secure a crop of 300 bushels, worth 25 cents a bushel, and another by expending $43.60 on an acre of corn, obtains a crop worth $67, how much per cent. does the former gain, by cultivating potatoes instead of corn?[a]

11. How much net weight must a pig gain per day, to pay the expense of keeping, when corn sells at $1.00 per bushel, and pork at 10 cents per lb., supposing him to consume 3qt. of corn and 2cts. worth of other food per day?

12. If 3 lb. live weight are equivalent to 2 lb. net weight, what would be the weekly profit of a pig that gains 2 lb. live weight per day, the expense of keeping and the price of pork being the same as in the last example?

13. Produce of 3¼ acres in 1841: 21½bu. wheat, at 8s.; 44bu. oats, at 2s. 9d.; 80bu. potatoes, at 1s.; fodder for the support of 2 calves, at £2 15s.; 423¼ lb. butter, at 1s.; milk sold or given to the pigs, £10. Required the value of the produce per acre, at $4.84 per £.

14. It is estimated that in Great Britain and Ireland, 7085370 acres are annually sown with wheat, at the rate of 2½bu. per acre. How many quarters of 8 bushels are required for seed? If 3½pk. per acre would be sufficient,

[a] Although the crop of potatoes may be more profitable for a single year than corn, yet a rotation of crops is necessary to avoid exhausting the land.

how much is annually wasted, and what is its value, at $1.12½ per bushel?

15. Find, from the following estimate of the expense of an acre of corn, the balance in favor of the crop.

Expenses.—Ploughing, $2.50; harrowing, $2.50; holeing, 50cts; 6bu. leeched ashes, 10cts; 1bu. plaster, 65cts.; 10qt. seed, @ 10cts.; putting on ashes and plaster, and planting, $1.20; harrowing, 30cts.; weeding, $1.50; cultivating, 45cts.; second and third hoeings, $2.30; gathering and husking, $5.00; gathering stalks, $1.50.

Proceeds.—Crop.—50bu. corn @ $1.00; corn fodder, equal to 1 ton of hay, $10.

16. How many hop-vines, and how many poles, will be required to an acre, allowing two poles to each hill, and two vines to each pole, the hills being eight feet apart;—and what would be the expense of the poles at $3¼ per hundred?[a]

Partial Ans. 2720 vines.

17. Required the net returns on an acre for a six years' rotation of crops, raising corn for two years, wheat the third year, and hay for three years; the balance in favor of corn each year being $17.75, the balance in favor of wheat, $21, hay, fourth and fifth years, each $17.50, sixth year, $11.50;—expenses to be deducted,—grass-seed, $1.87½, and six years' compound interest on land, valued at $100 per acre.

18. A yoke of oxen weighed December 15, 4220 lb.; January 15, 4410 lb.; March 7, 4730 lb. Required the total gain from December 15, to March 7, and the average gain per day from December 15 to January 15, from January 15 to March 7, and from December 15 to March 7.

19. If each ox in the preceding example consumed daily 14 lb. of hay, ½bu. of potatoes, and 8qt. of Indian meal,

---

[a] To determine the number of hills or plants in a field, divide the area of the field by the area occupied by each hill or plant.

what was the cost of the gain per lb., supposing hay to be worth $10 per ton, potatoes, 20cts. per bushel, and Indian meal 60cts. per bushel?

20. If an acre of land yields 1½ tons of hay, or 15 tons of carrots or Swedish turnips, or 60 bushels of Indian corn; and if a working horse would consume 3cwt. of hay, or 3cwt. of carrots, or 1bu. of Indian corn, per week, how many acres of each crop would be required to support a horse for the year?

21. Bought December 1, a pair of oxen for $65, and sold the same February 26 at $5 per 100 lb., net weight, their net weight being 1846 lb. Required the amount lost on the sale, the following being the expenses of keeping:—73bu. turnips at 10cts.; 36¼bu. Indian meal at 60cts.; 65¼bu. potatoes at 25cts., and 25 lb. hay per day at 25cts. per 100 lb.; interest on cost, at 6 per cent., 94¼cts. *Ans.* $24.44¼.

22. What is the value of pasture land per acre, 50 acres of which will pasture 8 cows at 25cts. per week, 4 oxen at 50cts., and 75 sheep at 3cts. per week for twenty weeks in the year, the pasturage being estimated at 6 per cent. of the value of the land?

23. Required the yearly expense of keeping a sheep, allowing 155 days for the time of foddering in the barn, and 30 weeks pasturage at 3cts. a week, the food consumed while in the barn being 2 lb. of hay per day at $8 per ton of 2000 lb., and 1½bu. of rutabaga at 10cts. a bushel.

24. Required the amount of profit in the following experiment in stall-feeding sheep. Bought 118 at $2.50, 2 at $3, and 60 at 3\frac{1}{16}$; commissions for purchase and driving, 25cts. each; interest and risk estimated at $9.27; produce consumed, 519bu. turnips at 8cts., 151bu. corn at 75cts., 2 lb. of hay each per day, at $8 per ton of 2000 lb. The sheep were all put up on December 1st, and 125 were sold February 11th at $5 each, and the others, February 18th, at $5.25 each.

25. Estimating the expense of cultivating Indian corn at \$25 per acre, wheat at \$10, oats at \$5, rye at \$8, hay at \$2; and the harvest per acre of corn, 40bu. at 65cts., and fodder worth \$10; wheat, 14bu. at \$1.50, and straw worth \$3; oats, 56bu. at 42cts., and straw worth \$2.40; rye, 12bu. at \$1, and straw worth \$2.40; and hay, 2 tons per acre, at \$10; what would be the profit on a farm containing 7½ acres of corn, 5¼ acres of wheat, 5 acres of oats, 5 acres of rye, and 21 acres of meadow?

26. When hay sells at \$16 per ton of 2240 lb., and corn-stalks at 1ct. per bundle, how much can be saved per week with 6 cows, by substituting corn-stalks for hay, if the average consumption of each cow is 5 bundles of stalks, or 25 lb. of hay per day?

27. Two hogs were fed from April 30th to May 20th, exclusively upon Indian hasty-pudding. The pudding used during the interval, took 4½ bu. of meal, at 78cts. a bushel, and the weight gained by the hogs was 105 lb. Allowing ⅘ of the gain to be net weight, what should be the price of the pork per lb. to gain 20 per cent. on the cost?

28. From the following account of sales, on a farm of 25 acres, for ten years, find the average amount received annually for each item.

Date.	Vegeta's.	Fruit.	Vinegar.	Meat.	Hay.	Stock.	Milk.	Corn and Barley.	Total.
1811	\$132.06	\$126.76	\$183.94	\$14.38	....	....	....	....	
1812	181.91	145.48	12.83	81.92	\$130.18	\$32.00	\$17.13	\$32.05	
1813	112.93	68.07	93.47	69.67	202.68	30.00	44.97	.75	
1814	180.38	63.71	254.92	41.21	253.08	112.00	15.00	103.50	
1815	162.38	151.56	206.60	61.37	506.69	....	16.99	....	
1816	165.36	169.73	187.17	60.69	399.69	37.00	12.71	128.50	
1817	132.49	240.37	295.31	48.24	329.93	....	24.75	....	
1818	84.34	116.71	246.30	77.99	162.21	30.02	98.47	....	
1819	103.85	280.68	131.95	25.84	185.00	....	94.04	....	
1820	111.62	248.88	191.83	87.21	207.43	....	128.27	....	
Tot.									

29. What was the average net annual income of the above farm, estimating the amount of farm produce consumed by the family, and not included in the above account, at \$454, and the expenses at \$638.30 per annum?

30. What is the annual gain on a five-acre wood lot, which cost $50, the yearly growth being 1 cord of wood per acre, the market price of wood averaging $3.12½ per cord, and the expense of cutting and hauling 62½cts. per cord?

31. Required the profit on an acre of hops, the yield being 700 lb., worth 15½cts. a lb., and the expenses of cultivation as follows: Renewal and setting of poles, $12; planting, $1; tying up, $1; hoeing 3 times, at $1.50; 4 loads of manure, at $1.12½; picking, 1ct. a lb.; a man to tend the pickers, $7; board of the laborers, $1.50; kiln-drying and packing, $1 per 100 lb.; bale, 45cts.

32. If a horse consumes the produce of 6 acres of land before he is fit for work, and can afterwards be kept constantly employed for 12 years, by annually consuming the produce of 4 acres,—and a pair of oxen consume 10 acres' produce before they are fit to work, and can be employed only 3 years by consuming the produce of 2½ acres per annum each, the produce of how many acres will represent the difference between the expense of keeping a horse and a yoke of oxen for 12 years? *Ans.* 46 acres.

33. Estimating the value of a horse after 12 years' farm service, at $15, and supposing that every ox can be fattened after 3 years' service by the produce of 1½ acres, so as to be worth $65, what should be the value of 1 acre's produce, to make the expense of keeping, the same for a horse as for a yoke of oxen? *Ans.* $8.71 nearly.

N. B. Add to 46 acres, the number of acres' produce necessary to fatten the oxen, and you will obtain the number of acres' produce equivalent to the difference of value between the horse and the oxen.

34. Find the expense of keeping 8 oxen one year, estimating their total consumption at 1 ton of hay per week from January 1 to May 15, and from October 30th to January 1, the hay being worth $10.75 per ton, pasturage

from May 15 to October 30, at 62½cts. per week each, repairs of yokes and bows $3.75, wear of ploughs, chains, &c. $15.

35. What should be the price of charcoal per bushel, to make 25 per cent. on the following estimate of the cost of burning a kiln of 30 cords, the wages of labor being $1.25 per day? Cost of standing wood, $1.75 per cord; 2½ cords make 100 bushels of coal; 1 man can chop 2 cords for the kiln in a day; collecting, drawing together, covering, and burning, require 20 days' work of 1 man; expense of hauling the whole to market, and selling, $8.75.

Ans. $10\frac{15}{16}$ cents.

36. The whole amount of hay purchased and used at a stage stable, from April 1 to October 1, 1816, was 32T. 4cwt. 10 lb. at an average cost of $25 per ton. At the same stable, from October 1, 1816, to April 1, 1817, there was consumed by the same number of horses, 16T. 13cwt. 3qr. 10 lb. of straw, at $9.75 per ton, and 9T. 14cwt. 1qr. of hay at $25 per ton. A straw-cutter was employed during four months of the latter period. Required the amount of money saved by its use.

37. In an experiment on transplanting the layers of a single stalk of wheat, made by Mr. Miller, of Cambridge, Eng., one grain of wheat produced 3¾ pecks, weighing 47 lb. 7oz. Supposing a cubic inch to contain 200 grains, how many grains were there in the whole? How many in a pound?

38. If 6 men, with cradles, can cut as much grain in a day as 12 men with common sickles, and if the 6 cradlers require 3 men to bind, and 3 boys to assist, while the 12 reapers require only 2 men to bind, how much can be saved per day by cradling, allowing $1.25 for each man, and 50 cents for each boy?

39. If 105 gallons of milk yield 36 lb. of butter, worth

18¾cts. a pound, and 60 lb. skim-milk cheese, worth 6cts. a pound, and if 140 gallons of milk of the same quality, with the same amount of labor, yield 141 lb. of cheese, worth 9cts. a lb., and whey worth $1⅜, which is the more profitable, the manufacture of butter or cheese?

40. It is supposed that a pair of tame pigeons consume a pint of grain per day, for 280 days in the year. As it has been estimated that there are 1250000 pairs in Great Britain, how much would they consume per annum at this rate, and what would be its value, at an average of 50 cents a bushel?

41. The cost of improving a 12-acre field was as follows: Blasting large stones, $44.25; trenching, $20; drains, $3.87½; lime, 64bu. per acre, at 40cts. per bu.; ditch for intercepting hill water, $37.62½. Required the total cost, and the average per acre.

42. What should be the market price of milk per quart, when butter is worth 25cts. a pound, and skim-milk cheese is worth 5cts. a pound,—supposing that 100 gallons of milk will make 34 lb. of butter, and 74 lb. of skim-milk cheese?

43. The late Duke of Athol planted 6500 Scotch acres[a] of mountain ground with the larch, which, in 72 years from the time of planting, will be a forest of timber fit for building ships of the largest class. Supposing the plantation by that time to be thinned out to 400 trees per acre, and each tree to contain one load of 50 c. ft., what will be the value of the wood per acre, and the value of the whole plantation at 25cts. per c. ft., which is about ½ its present value?

44. A hectare in the Isle of Bourbon, produces 76000 kilogrammes of cane, which give 9200 kilogrammes of sugar, at an expense of 2500 francs for labor. A hectare of beet-root produces 40000 kilogrammes of roots, which

---

[a] A Scotch acre = about 1¼ English acres.

will yield 2400 kilogrammes of sugar, at an expense of 354 francs. What per centage of sugar is yielded by the cane, and by the beet-root? What is the cost of each kind of sugar per lb., estimating merely the labor bestowed on each crop?

45. Two men mow a square meadow, but one being a faster mower than the other, agrees to take the outside swath, and cut off all the corners. What part of the whole will each mow, there being twelve swaths in each side of the field? *Ans.* $\frac{31}{72}$, $\frac{41}{72}$.

The following rule will furnish the answer to all questions of this kind. Let the pupil endeavor to prove its accuracy.

Square the number of swaths in the side of the field for a denominator. Then if the number of swaths is odd, multiply it by 2, and diminish the product by 1; or, if it is even, multiply it by 2, and diminish the product by 4, for a numerator. The fraction thus obtained, will show what part of the field the outer man will mow more than the inner one.

46. The amount of hay necessary to sustain oxen, is about .02 of their weight, daily. When fattening, they require about .04 of their weight per day. At $10.50 per ton of 2000 lb., what would be the cost per week of the extra hay consumed in fattening two oxen, one of which weighs 834 lb., and the other 917 lb.?

47. In England and Wales, it has been estimated that there are 3252000 acres of wheat, 1250000A. of barley and rye, 3200000A. of oats, beans and peas, 1200000A. of clover and artificial grasses, 1200000A. of field-roots, 2100000A. fallow, 48000A. of hops, 17300000A. of meadow and pasture, 1200000A. of hedgerows, copses, woods, and wastes, and that the annual agricultural income is £216817624 At $4.84 per pound, what is the average income per acre?

48. What would be the cost of seed in the last example, supposing that there are 500000 acres of barley, and that the following is the average quantity used, per acre? Wheat,

5pk. @ $1.10 per bu.; barley, 3bu. @ $0.50; rye, 1bu. @ $0.80; oats, beans, and peas, 3bu. @ $0.95; grasses, 3pk. @ $2.00 per bu.; field-roots, 13bu. @ $0.45.

49. A farmer wishing to determine the area of his field, finds that it may be divided into four triangles, the dimensions of which are as follows: 1st. Base 240 paces, altitude 87 paces. 2d. Base 213 paces, altitude 95 paces. 3d. Base 107 paces, altitude 28 paces. 4th. Base 92 paces, altitude 79 paces. Required the entire area.

Ans. 5A. 1R. 9¼r. nearly.

50. I have a barrel 20 inches in diameter at the middle point, 16 inches at the head, and 27 inches long. Required its contents in ale, wine, and dry measure.

Ans. 31gal. *nearly*, wine meas.
25gal. 1⅜qt. beer meas.
3bu. 1pk. 2qt. dry meas.

51.[a] A heap of grain piled against the outer corner of a barn, contains 63 bushels, and its depth is 5ft. Required the slant height. *Ans.* 6ft. 8in.

52. A granary holds 1310.4 bushels. The dimensions of the floor are 15ft. 9in. by 12ft. Required the average depth of the grain. *Ans.* 8ft. 8in.

53. A stack of hay with a circular base, has settled so much that 250 c. ft. are estimated as equivalent to 1 ton, and according to this estimate, it is found to contain 1.44 tons. What is the height of the stack, the circumference of the base being 30ft.? *Ans.* 16ft.

54. The area of an irregular field is 2A. 3R., and the average length is 121 paces. What is the average breadth?

Ans. 110 paces.

---

[a] The remaining examples in this section are introduced to test the skill of the pupil in reversing the rules.

VI. THE GARDEN.

23. Examples for the Pupil.[a]

1–5. How many plants would be required for an acre, if they were placed $1\frac{1}{2}$ft. apart? If placed 2ft. apart? 3ft. apart? 6ft. apart? $16\frac{1}{2}$ft. apart?

Ans. to the first, 19360.

6–10. How long a strip of land will be required to contain an acre, if the width is 100ft.? If the width is $3\frac{3}{4}$ rods? 97.6ft.? $18\frac{3}{11}$ft.? 2r. $5\frac{1}{2}$ft.?

11. Allowing 1pt. of early peas to a row 20yd. long, 1pt. of marrowfat peas to a row 32yd. long, 1pt. of string beans to a row 27yd. long, 1pt. of runners to a row 36yd. long, and 1pt. of dwarf kidney beans to a row 26yd. long, how much seed would be required to plant 4 rows of early peas, 6 rows of marrowfats, 3 rows of string beans, 5 rows of runners, and 8 rows of dwarf kidney beans, each row measuring 28 yards?

12. A quarter acre of land cost $25. Expended for labor in improving it, $157; for seed potatoes, $15; rye and grass seed, $1.17; 6 cords of manure, at $5 per cord; 2 casks of lime at $1; cost of other seeds and gathering in the crop, $17.50. The products of the first year were 327 bushels of potatoes, at 60cts.; 5bu. rye, at $1.25; $8\frac{1}{2}$bu. corn, at $1; 100bu. rutabaga, at 30cts.; hay, $12; 600 cabbages, at 50cts. a dozen; 2000 lb. squashes, at 1ct.; fuel taken off, $25.[b] Estimating the value of the land as doubled, what was the first year's balance in favor of the improvement? *Ans.* $125.28.

13–15. How many times must a spade be thrust into the ground in digging a square rod, supposing the surface removed

---

[a] Buist, Colman, Loudon, Johnson. [b] Colman.

at each time to measure 7 by 8 inches? The spade weighing 8 lb., and a spadeful of earth 17 lb., how many pounds must the gardener lift in a day, 10 sq. rods being a day's work? How many tons must he lift in spading an acre?

16. What amount is annually gained by expending $16 per acre to drain a garden of $1\frac{3}{4}$ acres, money being worth 6 per cent. per annum, and the land renting before it was drained for $3 per acre, but after it was drained, for $8 per acre?

17. What distance would a man walk in ploughing a garden containing $\frac{3}{4}$ of an acre, if each furrow was 9 inches wide, adding 1 rod to every 18 for the ground travelled over in turning?

18. What must be the length of a strawberry bed, to contain $\frac{3}{16}$ of an acre, the width being 33 feet?

19. Required the cost of digging a trench 200ft. long, 3ft. broad, and 3ft. deep, a laborer being able to remove 1 cubic yard per hour,—allowing for wages $1.25 per day of 10 hours.

20. Money being worth 6 per cent. simple interest, what is the present value of an orchard containing 150 trees, which will be all in full bearing in 5 years, and will then be worth $20 apiece, the land which they occupy being now worth $1450? *Ans.* $3757.69.

21. In a garden of $1\frac{1}{2}$ acres the following articles were raised in the year 1836:—3500bu. onions, at 5cts.; 45bbl. beets, at $1.50; 14bu. parsnips, at 75cts.; 2bu. beans, at $2; 20bu. potatoes, at $\$\frac{1}{2}$; $100 worth of cabbages; and vegetables for the family, estimated at $100. Allowing one-half of the proceeds for the expense of cultivation, and estimating the value of the land at $450 an acre, what profit was made on the whole?

22. When ordinary pears are sold at 50 cents a bushel, and choice varieties bring 3 cents apiece, how much more profitable is a tree grafted with one of the best kinds, than a common pear-tree, the yield of each being eight bushels, and the choice pears averaging 45 to a peck?

23. In a garden 225ft. long, and 144ft. broad, is a gravel walk 6ft. wide, extending around the whole, at a uniform distance of 8ft. from the boundary line, intersected by three walks each 4ft. wide, parallel to the shortest sides. What amount of land is taken up by the walks?

Ans. $650 \times 6 + 348 \times 4 = 5292$ sq. ft.

24. Throughout the whole extent of the above walks is a trench 10in. wide and 8in. deep, filled with broken stone, and covered with a layer of broken bricks and other rubbish, 3in. thick over the whole width of the walks, the top-dressing of gravel being of the average depth of 4 inches. How many loads of 27c. ft. each, were required of the broken stone? How many of rubbish? How many of gravel? What was the cost of the whole at 50cts. for each load?

Ans. $20\frac{130}{243}$ loads of stone, 49 of rubbish, $65\frac{1}{3}$ of gravel; cost, $

25. What would be the expense of making a hedge of roses 468ft. long, setting the bushes with a distance of $6\frac{1}{2}$ inches between the centres, the cost of the bushes being 28cts. apiece, and the expense of setting $2.25 per hundred?

26. I wish to make a hot-bed, composed of 3 parts manure and 1 part oak leaves. The frame is 12ft. long and 6ft. wide. The bed is to be $3\frac{1}{2}$ ft. deep, and is to project 8 inches beyond the frame, on every side. How many cubic yards of manure will be required, and what will it cost at $1.75 per yard?

27. Two banks, each 104ft. long, and $5\frac{1}{4}$ft. high, and a grass plat 162ft. long and 157ft. wide, are to be covered with turf. Estimating the average thickness of the turf at 3in.,

what would be the cost of the whole at $1.25 per c. yd? At 1¼cts. per sq. ft.?

28. An English market-gardener received for the produce of a single acre in one year, for radishes, £10; cauliflowers, £60; cabbages, £30; celery, £90; endive, £30. Allowing £3 10s. for rent, and £90 for expenses of cultivation and marketing, what was the profit per sq. rod, in Federal Money, estimating the value of the sovereign at $4.84?

29. Frederick Tudor, Esq., of Nahant, Mass., in the year 1841, raised 42284 lb. of sugar-beets on 93 rods of land. Allowing 56 lb. per bushel, how many bushels could be raised per acre at this rate, and what would be the value of the whole at 50cts. per bushel?

30. If 2 men digging can keep 1 man employed in wheeling to the distance of 20 yards, and if each digger removes 15 c. yd. per day, what will it cost to remove 3340 c. yd. to the distance of 160 yards, the wages of each laborer being 75cts. per day? *Ans.* $835.

N. B. First find the number of men required to wheel the dirt away, and add the two diggers, and you will obtain the whole number employed.

31. What number of bulbs will be required for a crocus bed, the bed being 23in. wide, and the rows 6 inches apart, allowing 150 bulbs to each row; and what will they cost, ⅓ of the whole number being bought at 75cts. per 100, ¼ at $1.37½, and the rest at $1.62½?

32. When the double hyacinths were first brought into notice, some of the roots were sold at 2000 guilders apiece. Equally fine varieties can now be bought for $4 a dozen. At 40cts. per guilder, how large a bed could now be stocked with the choicest varieties, at the original cost of a single root, the roots being placed 8 inches apart, and allowing 6in. for border?

33. What will be the cost, at $12.75 per 100, of stocking a tulip bed 20ft. long and 4ft. wide, the bulbs being planted

in rows, allowing 6in. between the plants, and 7in. between the rows? *Ans.* $35.70.

34. In order to introduce enough fresh air to fill a hot-house twice in 24 hours, how many feet of air heated to the proper temperature, must be introduced hourly into a house 40ft. long, and 16ft. wide, the height in front being 6ft., and at the back 18ft.? *Ans.* 320 c. ft.

35. What would be the daily expense of raising the temperature of the hot-house in the last example 15°, if $\frac{1}{200000}$ lb. of coal will raise the heat of a c. ft. of air 1°, coal being worth $6.50 per ton of 2000 lb.?

36. A mechanic has a vacant space in his garden, 18ft. long and 4ft. wide, which he wishes to occupy as an asparagus bed. He trenches the whole to the depth of 2ft., and fills the trench half full of manure before returning the earth. He afterwards plants the whole with roots, setting the roots 9 inches apart. The land is worth 12½cts. a square foot, and he will be obliged to wait 3 years before the bed will be in a condition to be cut. Estimating the labor bestowed on the bed at 4 hours, at 10cts. per hour, the cost of manure at $1.25 per solid yard, the cost of plants at $1.00 per 100, and interest on the whole at 6 per cent. per annum, what will be the entire cost of the bed when it comes into bearing?

37. Allowing 1oz. of onion, carrot, or parsnip seed for sowing 15 sq. yd., ½oz. of cabbage or cauliflower seed for 4 sq. yd., ½oz. of turnip seed for 11 sq. yd., and 160 asparagus plants for a bed 5ft. by 30, what quantity of each will be required to plant a garden of 40 sq. rods with onions, carrots, parsnips, cabbages, cauliflowers, turnips, and asparagus, allowing the same quantity of ground to each?

38. In 1637 a collection of 120 tulips was sold in Holland for 9000 guilders; in England, at the present day, £50 is frequently given for a single bulb no finer than some

of the varieties which can be purchased for less than a dollar. Estimating the florin at 40 cents, and the sovereign at $4.84, how many of the Dutch bulbs would amount to the same as 15 of the English bulbs, at the above rates?

39. Apples should stand 35 feet apart, pears 20ft. apart, and plums 18ft. apart. How large an orchard would be required for 180 apple-trees, 315 pear-trees, and 350 plum-trees, there being 5 rows of each sort, allowing 28ft. between the apples and pears, and 20ft. between the pears and plums? *Ans.* 11A. 3R. 31r. 12¼yd.

40–42. How many hills of corn in a rectangular piece of land containing a quarter of an acre, if the hills are 3ft. apart one way, and 2ft. 9in. the other? if the hills are 2ft. 6in. × 3ft?—2ft. × 2ft. 3in.?

VII. THE HOUSEHOLD.

24. GENERAL INFORMATION.

1. THE grains yield nearly the following quantities of meal and bread per bushel.[a] Wheat weighing 60 lb.—flour, 48 lb.; bread, 64 lb. Rye weighing 54 lb.—meal, 42 lb.; bread, 56 lb. Barley weighing 48 lb.—meal, 37½ lb.; bread, 50 lb. Oats weighing 40 lb.—meal, 22½ lb.; bread, 30 lb.

2. The following weights and measures nearly correspond.[b] Wheat flour, 14oz.=1pt. Indian meal, 18oz.=1pt. Butter, 15oz.=1pt. Loaf sugar, broken, 1 lb.=1pt. White sugar, crushed, 17oz.=1pt. Brown sugar, 18oz.=1pt. Eggs, 10=1 lb. A chaldron of soft coal=58⅔ c. ft. Stone coal, 1bu.=70 lb., or 42 c. ft.=1 ton. A common tumbler holds ½pt. or ½ lb. of water. A common wine-glass holds

---

[a] Brit. Husbandry.

[b] Sci. American, Brit. Husb., Enc. Brit., and private information.

$\frac{1}{2}$ gill or 2oz. of water. Currants, 14oz. = 1pt. Cherries, 12oz. = 1pt. Honey, 23oz. = 1pt. Lard, tallow, or spermaceti, 15oz. = 1pt. Milk, 1 lb. = 1pt. Oil, 15oz. = 1pt. Salæratus, dry, 23oz. = 1pt.

25. Household Mensuration.

1. *To find the contents of any cylindrical vessel, in pints.*—Measure the diameter and height in inches. Then, for *wine measure*, multiply the square of the diameter by the height, and divide the product by 37; for *beer measure*, multiply the square of twice the diameter by twice the height, and divide the product by 359; for *dry measure*, multiply the square of twice the diameter by the height, and divide the product by 171.

Let D represent the diameter, H the height, and C the contents in pints. Then, if the contents are taken in wine measure, $H = 37C \div D^2$; $D = \sqrt{37C \div H}$. For beer measure substitute $44\frac{7}{8}$, and for dry measure $42\frac{3}{4}$, in the place of 37, in each formula.

2. *To find the contents of any vessel with a circular base, and tapering sides.*[a]—Take one half of the sum of the greater and less diameters, for a mean diameter, and proceed with this mean diameter and the height, as in the preceding problem. If great accuracy is required, proceed as in Problem 1, substituting for "the square of the diameter," "the product of the greater and less diameters, added to one third of the square of their difference." In beer and dry measure, use four times this product, in the place of "the square of twice the diameter."

The height and *mean* diameter, may be found as in Prob. 1. Let G be the greater diameter, L the less diameter, and M the mean diameter. Then

$$G = \sqrt{3M^2 - \tfrac{3}{4} \text{ of } L^2} - \tfrac{1}{2}L;$$
$$L = \sqrt{3M^2 - \tfrac{3}{4} \text{ of } G^2} - \tfrac{1}{2}G.$$

[a] Such a vessel forms a *frustum of a cone.*

3. *To find the contents of a bowl, in pints.*—Measure the diameter of the top and the depth, in inches. To three times the square of half the top diameter, add the square of the depth; multiply this sum by the depth, and divide the product by 55 for wine measure, by 67 for beer measure, or by 64 for dry measure.

4. *To find the contents of a well or cylindrical cistern in hogsheads.*—Measure the dimensions in feet, multiply the square of the diameter by the depth, and divide the product by 11. If great accuracy is required, measure the dimensions in inches; multiply the square of the diameter by the depth; multiply the product by 54, and cut off six figures from the right hand.

Let D be the diameter in inches, d the depth, and C the contents in hogsheads. Then $D=\sqrt{1000000\,C\div 54\,d}$; $d=1000000\,C\div 54\,D^2$.

5. *To determine the height to which any cylindrical vessel will be filled by a gallon.*—For wine measure, divide 294; for beer measure, divide 359; for dry measure, divide 342, by the square of the diameter, measured in inches.

One quart would evidently fill the vessel $\frac{1}{4}$ as high; one pint would fill it $\frac{1}{8}$ as high; and one gill would fill it $\frac{1}{32}$ as high. Any cylindrical vessel can therefore be marked by this rule, so as to serve the purpose of a set of measures.

Let N be the number of the required measure which will be equivalent to a gallon, D the diameter, and H the height. Then, for wine measure $D=\sqrt{294\div(N\times H)}$; for beer measure $D=\sqrt{359\div(N\times H)}$; for dry measure $D=\sqrt{342\div(N\times H.)}$

6. *To test the accuracy of cylindrical dry measures.*—Measure the diameter in inches, and divide 2738 by the square of the diameter. The quotient will be the depth for 1 bushel. The depth for a half bushel, will be $\frac{1}{2}$ the

quotient; for a peck, $\frac{1}{4}$ of the quotient; for a half peck, $\frac{1}{8}$ of the quotient, &c.

26. Examples for the Pupil.

1–3. I have a cylindrical tin dish, $6\frac{1}{2}$in. in diameter, and 6in. deep. Required its contents in wine, in beer, and in dry measure. *Ans.* 3qt. 3.4gi. wine meas.; 2qt. 1.65pt. beer meas.; 3qt. *nearly*, dry meas.

4–6. At what height should marks be placed on the inside of the above dish, to indicate a quart of each measure?

Ans. $1\frac{3}{4}$ in. *nearly* for wine meas.; $2\frac{1}{8}$in. *nearly* for beer meas.; 2in. for dry meas.

7–9. Find the contents in each measure, of a pan, the height being 5 inches, the top diameter 17in., and the bottom diameter 9in.

Ans. by the accurate rule, 23.56pt. wine meas.; 19.42pt. beer meas.; 20.39pt. dry meas.

10–12. Find the contents in each measure, of a bowl, the diameter of which is 6in., and the depth 3in.

Ans. 1.96pt. wine meas.; 1.61pt. beer meas.; 1.69pt. dry meas.

13. A cylindrical cistern is 5 feet in diameter, and 6ft. deep. How many hogsheads does it hold?

14–15. A cylindrical cistern is 7ft. 6in. in diameter, and 6ft. 9in. deep. Find its contents by each rule.

Ans. by Rule 1, $34\frac{91}{176}$hhd.; by Rule 2, 35.43hhd.

16–17. The house to which the above cistern belongs, is 40ft. long, and 34ft. wide, and is supplied with eave-troughs which convey all the water that falls on the roof to the cistern. To what depth would the cistern be filled by a single shower, in which there is a fall of $\frac{3}{4}$in. of rain, allowing 6in. on each side for the projection of the eaves? How many inches of rain would fill the cistern?

18. A man has the following cylindrical measures: One

designed for a bushel, the diameter of which is 18½ inches; one designed for a half-bushel, diameter 14½in.; one designed for a peck, diameter 12in.; one designed for a half-peck, diameter 9¼in.; and one designed for a quarter-peck, diameter 7½in. What should be the depth of each?

Ans. for the half-bushel, 6½in.

19. I send a barrel of flour to a baker, and he agrees to furnish me an equal weight of bread in return. The flour weighs 1cwt. 3qr., and cost me $7.50; the barrel, (which the baker keeps,) is worth 25cts. How much do I pay the baker for his trouble? *Ans.* $2.12½.

N. B. If 48 lb. of flour make 64 lb. of bread, how many lb. of flour will make 1cwt. 3qr. of bread? And what will be the value of the flour that is left, at $7.50 per bbl.?

20. A cow gave 350 gallons of milk in a year, and consumed in the same time two tons of hay, at $9.75; 12bu. of Indian meal, at 60cts.; 50bu. of beets, at 40cts.; and her pasturage cost $8.75. Estimating the value of her milk at 5cts. a quart, and allowing 10 per cent. on $25 for the interest of her worth, and risk of keeping, and $2.50 for labor and attendance, what amount of profit did she yield?

21. Find the cost of 100 lb. of bread, made from each of the following grains, the grain being of the full weight given in § 24, and the cost of manufacture in each instance, being 56cts.; wheat at $1.10 per bushel; rye, at 90cts. per bu.; barley, at 75cts. per bu.; oats, at 40cts. per bu.

22. How much crushed sugar, by measure, should be used in preserving 6qt. of currants, in order that there may be equal weights of sugar and fruit?[a]

23. How would you measure the following ingredients?—2 lb. flour, 2 lb. sugar, 1 lb. butter, and 1 lb. eggs.[a]

24. A common watch vibrates 5 times in a second; if from any cause each vibration is $\frac{1}{3600}$ less than its proper time, how much will the watch gain per day?

---

[a] See Section 24.

25. A cloak containing $6\frac{7}{8}$yd. of broadcloth that is $1\frac{1}{2}$yd. wide, is to be lined with silk that is $\frac{7}{8}$yd. wide. How much silk will be required?

26. A parlor 35ft. 6in. long, and 19ft. 6in. wide, is to be carpeted. How much carpeting, that is a yard wide, will be required, provided there is no waste in matching the figures? How much will be required, if $\frac{1}{4}$yd. is lost in matching each breadth?

27. A man wishes to build in his cellar a potato bin that will hold 40 bushels. It is to be 6ft. long, and 3ft. 6 in. high. What must be the width of the bin?

28. If a family of 3 persons consume a barrel of flour in 11 weeks, what is the average amount of bread eaten daily by each person, 3 pounds of flour being sufficient to make 4 pounds of bread? *Ans.* $1\frac{13}{99}$ lb.

29. Required the entire quantity, and the value at 22cts. a gallon, of the milk taken by D. N. Breed, of Lynn, Mass., from one cow, in 11 months of 1839–40. The daily average was as follows:—from April 15 to April 30, 1839, 6qt.; in May, 14qt.; in June, 16qt.; in July, 13qt.; in August, 12qt.; in Sept., 11qt.; in Oct., 10qt.; in Nov., 10qt.; in Dec., 9qt.; in Jan., 1840, 9qt.; in Feb., 7qt., and from March 1 to March 15, 2qt.

30. Alfred the Great had large candles made with marks upon them, so that he might judge of the time by the quantity that had burned.[a] If a candle were lighted at noon, of such size that only 9in. would be consumed in 24 hours, what would be the time when $2\frac{1}{2}$in. had burned?

31. How much coal, at 2240 lb. to the ton, would fill a bin that holds 160 bushels of potatoes?

32. How many wine gallons in a pail 12in. in diameter, and $14\frac{1}{2}$in. deep?

---

[a] Carpenter. The Clepsydra had not been introduced into England.

33. A cylindrical vessel is 4 inches in diameter. At what heights must marks be placed for 1qt. wine measure; for 1pt. beer measure; and for 1 half-peck dry measure?

34. Required the contents in wine measure, of a water-pail, the height being 8⅛in., the top diameter 11¼in., and the bottom diameter 9in.

35. How many rolls of paper hangings, each 9yd. long, and ½yd. wide, would be required to paper one side of a room that is 19ft. long and 9ft. 6in. high?

36. A saving of $1 per annum, invested at 6 per cent. compound interest, will amount in 40 years to $154.761966. If a young man at 20 years of age, commences laying up 6¼cts. per day, how much will he be worth when he is 60 years old?

37. A man expends all his income, but in looking over his accounts at the end of the year, he finds that $125 has been laid out foolishly. If he resolves to retrench, and saves that amount annually, what will he be worth in 40 years?

38. Find the ratio of illumination between two lights, which throw shadows of equal intensity, if placed at the distances of 9ft. and 5½ft. respectively.[a]

---

[a] "The following method of measuring the comparative illuminating power of different lights, is founded on the law that the amount of rays thrown on a given surface, is inversely as the square of the distance of the illuminating body. Place two lights, which are to be compared with each other, at the distance of a few feet, or yards, from a screen of white paper, or a white wall. On holding a small card near the wall, two shadows will be projected on it. Bring the fainter light nearer the card, or remove the brighter light farther from it, till both shadows acquire the same intensity. Measure now the distances of the two lights from the wall or screen, and the squares of these distances will give the ratio of illumination. In this experiment the spectator should be equidistant from each shadow." — *Bigelow's Technology.*

39. If a cylindrical cistern is 7ft. 6in. in diameter, what must be its depth, to hold 45 hogsheads? To hold 30 hogsheads?

40. If wood is sawed 2ft. long, how many cords will there be in a pile 50ft. long, and 8ft. 3in. high?

41. How many potatoes are there in a cellar, there being 5 barrels full, and one half-full, each barrel holding 2bu. 3pk., and a bin that is 10ft. long and 5ft. wide, being filled to the height of 3ft. 6in.

42. An imperial gallon of sperm oil, burned in an Argand lamp, which yields a light equivalent to 5 candles, (6 to a lb.) will burn about 100 hours. The solar lamp, with an imperial gallon of whale oil, yielding a light equivalent to $4\frac{3}{4}$ candles, (6 to a lb.) will burn about 90 hours.[a] When sperm oil is $1.25 per gallon, and whale oil $62\frac{1}{2}$cts. per gallon, how much can be saved on every gallon by the use of whale oil? *Ans.* $44\frac{3}{8}$ cents.

43. A cylindrical cup contains 3qt. 3.4gi. wine measure. Required the height, the diameter being $6\frac{1}{2}$ inches. *Ans.* 6 inches.

44. What is the greater diameter of a pan, the less diameter being 9in., the height 5in., and the contents 19.42pt. beer measure? *Ans.* 17 inches.

45. A cylindrical cistern, 6ft. 9in. deep, contains 35.43 hogsheads. What is its diameter?

46. A cylindrical cistern, $7\frac{1}{2}$ft. in diameter, holds 47.24 hogsheads. What is its depth? *Ans.* 9ft.

47. What should be the diameter of a cylinder that is $6\frac{1}{2}$in. deep, to hold a half-bushel? *Ans.* $14\frac{1}{2}$ inches.

48. What should be the diameter of a cylinder that is $2\frac{1}{2}$in. deep, to hold a half-pint, beer measure? *Ans.* 3in. *nearly.*

---

[a] Parnell.

VIII. ARTIFICERS' WORK.[a]

27. THE CARPENTER AND JOINER.

1. *The contents of a board* are found by multiplying the length by the *mean* breadth, provided the thickness does not exceed 1 inch. If the board tapers regularly, the mean breadth is half the sum of the two end breadths. If it is of irregular shape, the breadth should be taken at a number of different points at equal intervals, and their average should be regarded as the mean breadth. If the thickness exceeds 1 inch, multiply the number of feet in the area by the number of inches in the thickness.

If the contents are given, and either of the dimensions are required, divide the contents by the product of the given dimensions.

Examples.—A board 12ft. 6in. long, 1ft. 3in. broad, and $\frac{3}{4}$in. thick, contains $12\frac{1}{2} \times 1\frac{1}{4} = 15\frac{5}{8}$ft. board measure.

A board 13ft. 4in. long, 1ft. 8in. broad, and $1\frac{1}{2}$ in. thick, contains $13\frac{1}{3} \times 1\frac{2}{3} \times 1\frac{1}{2} = 33\frac{1}{3}$ft. board measure.

2. *The contents of square or hewn timber* are found by multiplying the mean breadth by the mean thickness, and their product by the length. The mean breadth and thickness are found in the same manner as in measuring boards. Sometimes the contents are found by squaring $\frac{1}{4}$ of the girt, and multiplying by the length. This method is erroneous, and always gives the contents too great. Timber is often sold by board measure. Cubic feet can be reduced to board feet by multiplying by 12, (provided the thickness exceeds 1 inch,) or the contents can be found at once in board measure, by taking one of the dimensions in inches, and the other two in feet. If the contents are given, and any two of the dimensions, the other dimension may be found,

---

[a] Ingram, Gillespie, Crossley and Martin, Nicholson, Pratt, and private information.

by dividing the contents by the product of both the given dimensions.

Examples.—A hewn log 19ft. 6in. long, 2ft. 2in. broad, and 2ft. 1in. thick, contains $19\frac{1}{2}\times 2\frac{1}{6}\times 2\frac{1}{12}=88\frac{1}{48}$ c. ft., or $1056\frac{1}{4}$ft. board measure.

A hewn log 30ft. 8in. long, 1ft. 10in. wide, and 1ft. 3in. thick, contains $30\frac{2}{3} \times 22 \times 1\frac{1}{4} = 168\frac{2}{3}$ board feet.

3. *The contents of round timber* are usually found by squaring $\frac{1}{4}$ of the mean girt, and multiplying it by the length. If the tree is covered with bark, $\frac{1}{4}$in. should be deducted from the quarter-girt before squaring. If the bark is very thick, more than $\frac{1}{4}$in. is sometimes allowed. No rough timber is considered measurable, if the diameter is less than 6 inches.[a]

This rule gives the contents too small, nearly in the proportion of 11 to 14. A *ton* of timber is considered as equivalent to 40 c. ft., but 40 c. ft. of round timber as generally measured, really contains 50 c. ft. The statement usually given without any explanation in Arithmetical tables, that "40ft. of round timber, or 50ft. of hewn timber make 1 ton," is therefore erroneous. The allowance was originally introduced as a partial compensation to the purchaser of round timber, for the waste occasioned in squaring it.

If the *true* content is required, it can be found very nearly, by squaring $\frac{1}{5}$ of the girt, and multiplying by twice the length.[b]

Example.—A log 47ft. 8in. long, and girting at the ends 18ft. and 6ft., has for its mean quarter girt $\frac{1}{4}$ of $(18+6)\div 2=3$. Its contents are therefore $3^2\times 47\frac{2}{3}=429$ c. ft.

4. *Flooring, partitioning, and roofing,* and all large and plain work, in which a uniform quantity of materials

---

[a] Let G be the mean girt in feet, L the length, and C the contents in c. ft. Then $L=166\div G^2$; $G=4\times\sqrt{C\div L}$.

[b] In employing this rule, as well as in the usual method, 1in. should be deducted from the girt if the tree is covered with bark.

and labor is expended, are generally measured by the "square"=100 sq. ft. Some work is measured by the linear foot or yard, some by the square foot or yard, and some by the cubic foot. For some of the more difficult kinds of work, it is usual to allow "measure and half," or "double measure," but the custom varies so much in different places, that it is impossible to give any general rules of measurement. Shingles are generally 18 inches long, and of the average width of 4 inches. When nailed to the roof $\frac{1}{3}$ is usually left out to the weather, and 6 shingles are therefore required to a square foot. But on account of waste and defects, 1000 should be allowed to a "square." The weight of a square of partitioning may be estimated at from 1500 to 2000 lb.; a square of single-joisted flooring, at from 1200 to 2000 lb.; a square of framed flooring, at from 2700 to 4500 lb.; a square of deafening, at about 1500 lb.[a] When a floor is covered with people, 120 lb. per sq. ft. should be added to the weight.[b]

5. *The sliding rule* is often used by carpenters and other artificers, in the measurement of timber and work. The foot is divided in the usual way, into inches and eighths of an inch, and it is also subdivided decimally and logarithmically, so as to facilitate the labor of computing. The use of the rule can only be learned by practice.

The carpenter's square is used in determining whether the corners of boards or buildings are square. In framing buildings, the corners are sometimes "squared," by measuring 8ft. on one timber, and 6ft. on the other, and placing the extremities of the measured lines 10ft. apart. This mode of operation is founded on the property of right-angled triangles, that the square of the hypothenuse is equal to the sum of the squares of the other two sides. A roof is said to have a *true pitch*, when the length of each rafter is $\frac{3}{4}$ of

---

[a] Hatfield. [b] Tredgold.

the breadth of the building. The two sides of the roof then form nearly a right angle.

6. *The area of posts.*—All rules for determining the resistance of timber, should be based on the supposition that the timber is of "merchantable" quality, straight-grained, seasoned, and free from large knots, splits, decay, or other defects. When the height of a piece of timber exceeds about ten times its thickness, it will bend before crushing. To find the area of a post that will safely bear a given weight, when the height of the post is less than ten times its least thickness;—Divide the given weight in pounds by 1000 for pine, or by 1400 for oak, and the quotient will be the least area of the post in inches.

EXAMPLES.

1. Required the contents of a board 13ft. 7in. long, ⅞in. thick, and 1ft. 6in. wide, and its value at 9½cts. per foot.
Partial Ans. Value $1.94.

2. The length of a board is 11ft. 8in., the thickness 1½in., and the breadths measured at five different points are as follows; 1ft. 6in., 2ft. 3in., 1ft. 9in., 2ft. 6in., and 2ft. What are the contents? *Ans.* 35ft.

3. How many cubic feet, and how many feet board measure, in a log 21ft. 6in. long, the mean breadth being 1ft. 4in., and the mean thickness 1ft. 3in.?

4. A log is 42ft. 9in. long, and the girts at four points, outside of the bark, are 56, 45½, 58, and 64½ inches. What are its contents by the ordinary rule, and by the correct rule?

5. How many squares of flooring in a four story house, 42ft. 6in. by 28ft 4in. within the walls, deducting from each floor the vacancy for the stairway, 13ft. by 7ft. 6in.; and what is the cost of the whole, at $3.87½ per square?

6. A board measures at five different points, 1ft. 6in.,

1ft. 9in., 1ft. 8in., 1ft. 4in., and 1ft. 3in., in breadth. What is its length, the area being $18\frac{3}{4}$ft.? *Ans.* 12ft. 6'.

7. A stick of timber 9ft. 6in. long, measures 1 ton $14\frac{5}{8}$ c. ft. What is its mean area? *Ans.* 5ft. 9'.

28. The Mason.

1. *Rubble*[a] *walls* are generally measured by the perch, which is $16\frac{1}{2}$ft. long, 1ft. deep, and $1\frac{1}{2}$ft. thick, and is therefore equivalent to $24\frac{3}{4}$ c. ft. In some places, 25 c. ft. is allowed to the perch, in measuring stone before it is laid, and 22 c. ft. after it is laid in the wall. When the wall is not of uniform height, the height should be measured at several places, from the bottom of the foundation to the top of the wall, and the mean height employed in computing the solid contents. Nine pecks of good lime and 3 one horse loads of sand, will make mortar for 3 perches of wall. Rough stone and marble are often measured by the cubic foot. In measuring workmanship, linear feet and yards, and square feet and yards are employed.

2. The *rood* of 36 sq. yd. is sometimes employed in measuring walls that are more than 18in. thick. The wall should first be reduced to 2ft. thick. Thus a wall 90ft. long, 8ft. high, and 21in. thick, is equivalent to $\frac{7}{8}$ of a wall 90ft. long, 8ft. high, and 2ft. thick; $\frac{7}{8}\times90\times8=630$ sq. ft; $630\div9=70$ sq. yd.; $70\div36=1$ rood 34yd.

3. *Cisterns* can be measured accurately by finding the solid contents in cubic inches, and dividing by the number of cubic inches in a hogshead, (63×231.) But, in measuring *circular* cisterns, the rules given in Sect. 25, are much more convenient than this method, and are sufficiently correct for ordinary purposes. If great accuracy is required, measure the diameter and depth in inches, multiply the square of the

[a] Rough stone work is called *rubble work*.

diameter by the depth, and multiply the product by either of the following numbers, to obtain the contents :—

In cubic feet	In ale gallons	In wine gallons	In cubic inches
.00045451	.002785	.0034	.785398
In hogsheads	In bushels	In lbs. of water	In Imperial gal.
.000054	$\frac{1}{2738}$	.028326	.0028326

4. *Arches* are measured by applying a line close to the surface in taking the dimensions. If the arch is not of uniform length, breadth, and thickness, the dimensions may be measured at several points, and the mean of all the measurements taken. A pointed arch will sustain almost any weight on its crown, provided the lowest stones do not give way. Therefore the Gothic arch is stronger for lofty buildings than the circular, but the circular arch is far better adapted than the Gothic, for bridges or other works, where every part of the arch may be exposed to equal, or nearly equal pressures.

EXAMPLES.

1. How many cubic feet in a block of marble, 4ft. 6in. long, 3ft. 8in. wide, and 2ft. 4in. thick?

2. How many perches of 24¾ c. ft. in a wall 97ft. long, and 2ft. thick, the heights at five different points being 4ft. 8in., 3ft. 8in., 3ft. 9in., 4ft., and 5ft. 2in. and how much lime will be required to make mortar for the whole?

3. Required the cost at 50cts. per perch of 25 c. ft., of making a cellar wall 6ft. 3in. high, and 1ft. 6in. thick, the outside measurement being 47ft. long, and 22ft. 6in. wide?

4. How many hogsheads in a cylindrical cistern 13ft. 6in. in diameter, and 11ft. 7in. deep?

5. How much hewn work in 18 lintels and sills, each 4ft. by 1ft. 6in., and in chimney coping, 58ft. 6in. by 19in., and what is the cost of the whole at 10½cts. per foot?

Partial Ans. Cost, $21.06.

6. Find the contents in cubic feet, in hogsheads, and in Imperial gallons, of a cylindrical cistern 10ft. 6in. in diameter, and 9ft. 2in. deep.

1st *Ans.* 793.738 c. ft.

29. The Bricklayer.

1. *Brickwork* is measured either by the square yard, or by the square rod, and is usually estimated at 1½ bricks thick, =12 inches, the actual thickness being reduced to the standard of 1½ bricks. 272 sq. ft. is generally counted as a rod, the fraction of ¼ sq. ft. being rejected.[a] The rood of 18ft. sq., or 324 sq. ft., and the rod of 16½ sq. ft., are also used. But the common practice is now to reckon bricks by the 1000, the number required depending on the size of the bricks. In estimating the number, an allowance of $\frac{1}{10}$ of the solid contents should be made for the space occupied by the mortar.

2. The *usual dimensions* of bricks, are; length 8in., breadth 4in., thickness 2in. Whatever the length may be, the breadth is generally ½, and the thickness ¼ as great; thus a brick 9 inches long, would be 4½ in. wide, and 2¼ in. thick. In some places, all walls are charged as solid, no allowance being made for doors, windows, or other openings; in others, the openings are deducted in charging for materials, but the workmanship is estimated as if the walls were solid; in others an allowance of one half is made for all openings; and in others the actual materials employed, and workmanship expended, are charged.

3. The *number of rods* in any wall may be found by multiplying the area of the surface by ⅓ of the number of half-bricks[b] in the thickness, and dividing the product by the number of square feet allowed to a rod. A load of sand= 30 struck bushels; a ton = 24 c. ft.

---

[a] The weight of a rod of brickwork may be estimated at 16 tons.

[b] A half-brick = 4 inches.

EXAMPLES.

1. How many bricks of the usual size, will be required to make a wall 40ft. long, 16ft. high, and 2 bricks thick, making allowance for mortar?

Ans. $\frac{9}{10}$ of $40\times16\times1\frac{1}{3}\times27=20736$.

2. How many roods of 324ft. in a wall 96ft. long, 10ft. high, and 3 bricks thick, and what is the cost of the bricks at $7.75 per M? *Ans.* Cost, $361.58.

3. A garden contains 1¼ acres, and is 150ft. wide. Required the cost of enclosing it with a brick wall 9ft. 4in. high and 3 bricks thick, at $7.50 per M., deducting 2 doors, each 6ft. 3in. by 4ft., and a gateway 12ft. wide?

Ans. $3404.19.

30. The Plasterer.

Plain plastering is measured either by the square foot, the square yard, or the "square" of 100 sq. ft. The number of coats, and the quality of the finishings, should be stated in the bill. Cornices and mouldings, if 12 inches or more in girt, are sometimes estimated by the square foot; if less than 12 inches, they are usually measured by the linear foot. Plastering on walls is called *rendering; ceiling*, is plastering on laths. The custom varies as to the proper allowance for doors and windows.

EXAMPLES.

1. What will be the cost of plastering a ceiling 21ft. 6in. long, and 19ft. 6in. wide, at 11cts. per sq. yd.?

2. How much plastering on a partition 22ft. 3in. long, and 7ft. 9in. high, deducting two doors, each 6ft. by 3ft. 2in., and what will it cost at 12½cts. per sq. yd.?

Ans. Cost, $1.87.

3. How many squares of ceiling, and of rendering, and

how many feet of cornice, in a hall 60ft. long, 28ft. 6in. wide, and 10ft. high, deducting 17yd. 5ft. 6′ for doors and windows? *Ans.* 16 sq. 1yd. 2ft. 6′ of rendering.
17 sq. 1yd. 1ft. of ceiling.
177ft. of cornice.

31. THE PAINTER AND GLAZIER.

Painting is measured by the square yard. In taking the dimensions, the measuring line is laid into all the mouldings, so as to reach every point which the brush touches.

Glazing is sometimes measured by the square foot, sometimes by the piece, or by the light. In estimating by the square foot, it is customary to include the whole sash. Circular or oval windows are measured as if they were square.

EXAMPLES.

1. A room is 24ft. long, 18ft. 6in. wide, and 9ft. 6in. high. How many yards of painting are in it, deducting a fire-place 4ft. 6in. by 4ft., and 3 windows, each 6ft. by 3ft. 3in.? *Ans.* 81yd. 2ft.

2. At 12½cts. per yard, what will it cost to paint a wall 15ft. 8′ by 8ft. 3′? *Ans.* $1.78.

3. How many feet of glazing in an oval window 4ft. 6′ by 2ft. 8′?

4. At 18¾cts. per foot, what will it cost to glaze 3 stories of a house, with 8 windows in each story, the breadth of each window being 3ft., and the height 5ft. 4in.?
Ans. $72.

32. THE PAVER, SLATER AND TILER.

Paving is measured by the square foot, the square yard, or the rood of 36 sq. yds. If the pavement is grooved, the grooves are added to the surface measure.

Slating and *Tiling* are measured by the square yard, by

the rood, or by the "square" of 100 sq. ft. It is not usual to make any deductions for chimneys, skylights, or other apertures. In measuring the girt of slate roofs, allowance must be made for the double row at the eaves.[a]

EXAMPLES.

1. A yard is 80ft. long, and 28ft. 6in. wide. What will it cost to pave it at 70cts. per square yard?

2. The side of a square court measures 120 feet. What will it cost to pave it, leaving a flagged walk 6ft. wide around the outside, at 75cts. per yd., and paving the rest with bricks at $8.00 per M., allowing 4 bricks to a square foot? *Ans.* $601.25.

3. How many roods of tiling in a roof 52ft. 8′ long, and 45ft. 6′ in girt?

4. At $1.30 per yard, what will be the expense of slating a roof 49ft. 6′ long, and girting $46\frac{2}{3}$ft.? *Ans.* $$333\frac{2}{3}$.

33. THE PLUMBER.

Plumbers' work is generally done by the pound or hundred-weight. A square foot of sheet lead, $\frac{1}{10}$in. thick, weighs 5.899 lb. From this value the weight of a square foot of any other thickness can readily be determined. Lead pipes of $\frac{3}{4}$in. bore, weigh about 10 lb. per yd.; 1in. bore, 12 lb.; $1\frac{1}{4}$in. bore, 16 lb.; $1\frac{1}{2}$in. bore, 18 lb.; $1\frac{3}{4}$in. bore, 21 lb.; 2in. bore, 24 lb.

The contents of lead pipe of any given dimensions, may be found by the table in Sect. 28, for measuring cisterns.

EXAMPLES.

1. What is the weight of 1 sq. ft. of lead, the thickness being $\frac{1}{8}$ of an inch?

[a] The weight of a square foot of slating may be estimated at $11\frac{1}{4}$ lb. The greatest force of the wind on a roof, is about 40 lb. per square foot.—*Tredgold.*

2. Find the weight of lead necessary to cover one side of a roof 36ft. 9in. long, and 18ft. 3in. wide, at $8\frac{1}{2}$ lb. per square foot.

3. What is the cost of covering and guttering a roof, at 18s. per cwt., the length of the roof being 43ft., and the girt 32ft.; 57ft. of guttering 2ft. wide; weight of roofing, 9.831 lb. per sq. ft., and of guttering 7.373 lb. per sq. ft.?

Ans. £115 9s. $1\frac{1}{2}$d.

4. Required the expense of a leaden pipe, $1\frac{3}{4}$in. bore, and 185ft. long, at 11cts. per lb.

5. A roof 40ft. long, and 57ft. girt, is covered with lead $\frac{1}{7}$in. thick; the water pipe is $1\frac{1}{4}$in. bore and 52ft. long, and the waste pipe is 2in. bore and 40ft. long; the water cistern is 4ft. 3in. long, 3ft. 6in. wide, and 3ft. deep, and lined with lead $\frac{1}{5}$in. thick. What is the amount of the plumber's bill, rating the sheet lead at $7.50 per cwt., and the pipe at 10cts. per pound?

34. SPECIFICATION AND ESTIMATES.

1. *Specification* of materials to be provided, and labor to be performed, in the construction and finishing of a Schoolhouse for the City of Worcester, to be erected on Summit Street in said City, according to plans drawn by E. Boyden, Architect, and herewith presented.

Size of House.—58 feet long by 50 feet wide, not including the projection of the pilasters.

Height of stories as figured on Section of Front Elevation.

The location of the cellar, and depth of excavation, to be determined by the building committee.

A well to be dug upon the lot in such location as directed by the building committee, and also to be stoned up so as to leave a diameter of three and a half feet in the clear.

All earth dug out of the cellar and well, to be deposited upon the lot, as may be directed by the building committee.

The foundation walls to be three feet thick at the bottom, $2\frac{1}{2}$ feet thick at the top, and of such height as may be determined by the building committee. The walls to be made of large square block-stone, well faced and bonded.[a]

Underpinning 2 feet wide, and not less than 8 inches thick, to be made of rough split South Ledge stone, with a rough-hammered bevelled wash[b] between the pilasters. The face of the underpinning to project as far forward as the face of the pilasters.

Stone steps, of fine-hammered South Ledge granite, located and of such size as represented on the plan.

All the foundations for piers and partition walls, to be not less than 8 inches thick, and to be placed as represented on the plan of the cellar.

Stone lintels over all the cellar windows, and four stone thresholds of fine-hammered South Ledge granite, to be made of the dimensions marked on the ground plan.

Outside wall of brick to be one foot thick, with pilasters projecting 4 inches beyond the face of the wall, and Corbel-Course[c] and Frieze,[d] as shown on elevation.

Four brick piers in the cellar, each one foot square, and a brick partition 8 inches thick to be carried from the bottom of the cellar to the attic floor, as represented on the plans.

The chimneys to be located and constructed as represented on the ground plan.

Building to be lathed throughout, and plastered with two coats, except the play-room in the basement. Walls not to be plastered underneath the ceiling.

All the windows to have four-course,[e] tooled sandstone caps, and two-course sills. The doors to have five-course caps of the same material.

Twenty ventilating registers, each one foot in diameter, to be furnished and inserted, two in each ventilating flue, one near the floor and the other near the ceiling. All the rooms to be thus ventilated except the play-room.

For the arrangement and sizes of timber in the first three floors, see plan of flooring, as represented on the basement plan.

---

[a] Laid like bricks, so that the joints will not come over each other.

[b] The wash of the stone is the inclined surface for water to run off.

[c] Projections in a wall to sustain the timbers of a floor or roof, are called corbels.

[d] The part of the wall above the pilasters.

[e] Of the thickness of four courses of brick.

Frames to be made entirely of good spruce framing timber. Joists in 4th or attic floor to be 2×9, framed 15 inches between centres. All other timber in roof and observatory to be of the size figured on the plan of roof.

Joists in first three floors to be jointed $\frac{3}{4}$in. crowning[a] to 15 feet in length. All floorings to be bridged with good X bridging where marked on plan of flooring. All large timber to be well and properly secured to the brick walls by suitable anchor irons. The joists to be also secured in a similar manner, as often as once in every ten feet.

Roof to be boarded with suitable $\frac{3}{4}$in. boards, planed, jointed, matched, and suitably tinned. Roof to be bracketed and project as represented on elevation. Observatory to be framed and finished in every respect as represented and figured on plan of observatory, front elevation, and plan of roof.

Lintels $6 \times 7\frac{1}{2}$in. to be furnished for all windows and doors. All floors to be lined with suitable $\frac{3}{4}$in. lining boards laid edge to edge, and nailed with 8d. nails.

The floor to the upper schoolroom to be deafened in the centre of the floor joists with a suitable coat of coarse mortar. All top floors to be of suitable southern hard pine $\frac{7}{8}$in. thick, and not to exceed 6in. in width, well laid and nailed with 12d. floor nails.

Five iron columns to be placed as represented on plans of basement and second story, of size and quality like those in the Pleasant Street Schoolhouse.

Partitions to be arranged as represented on plans. Those in basement and 2d story, to be constructed of 2×5in. partition plank, bridged once with $1\frac{1}{4} \times 5$in. herring-bone bridging. Those in 3d story to be of 2×6in. partition plank, bridged twice with herring-bone bridging as above. All partition plank to be jointed and set edgewise, so as not to exceed one foot between centres. The contractor is also to make all necessary arrangements in the partitions for pipes to convey heated air to all the different apartments, and to do all necessary wood-work preparatory to putting in registers to admit the hot air. The house to be furred throughout with $\frac{3}{4} \times 2\frac{1}{2}$ inch furrings, placed at the distance of one foot between their centres.

Teachers' platforms to be elevated 6 inches above the floor, of the situation and sizes represented on plans.

---

[a] Long timbers are usually made "crowning" in the centre, so as to allow for settling.

All window-frames to be constructed as represented by the drawings, with hard pine pulley styles.[a]

Four cellar window frames, to be made of 2in. chestnut plank, each large enough for a window with 4 lights of 9×12 glass. All sash to be of first quality Eastern pine stock, lip sash, ogee[b] style, 1½in. thick, to be double-hung with suitable weights, cords, and pulleys, and glazed with best quality German glass of such sizes as figured on front elevation.

All windows in 2d and 3d stories to have blinds to slide into the walls upon each side of the window, as shown on detail of window frame.

Four outside doors with side lights, as represented on front elevation, each 8ft. high by 3¼ft. wide, and 2in. thick, 4 panels with bevel joints, to be hung with 3 sets 4in. loose joint butts, and trimmed with suitable mineral knobs. The two front doors to have suitable mortice locks, and the two other doors suitable bolts inside. All other doors to be 3¼×7½ft. 4 panels, 1¾in. thick, with bevel joints, suitably hung with 3 sets of 4in. loose joint butts, and trimmed with suitable mineral knobs and mortice locks.

A flight of cellar stairs, with hard pine treads 1in. thick, placed as represented on cellar plan. All other stairs located as represented on plans, with hard pine risers 1in. thick, and treads 1¼in. thick, and to have cherry newels[c] and hand-rails. All staircases to be ceiled up on the well-room[d] side as high as the hand-rail, and all rooms to be ceiled as high as the window stools, with suitable Eastern pine stock ¾in. thick, not to exceed 6 inches in width, jointed, matched, and beaded.

Cleats to be put up in entries and recitation rooms, sufficient to contain 30 doz. glazed clothes-hooks, placed 6 inches apart, and to be provided with said hooks.

All doors and standing wood-work inside, to be grained in imitation of oak. Jet and brackets to be painted and sanded in imitation of sandstone. Observatory, window frames, and door frames, to be painted white and sanded. All painting to be done with three coats of *pure* white lead and linseed oil, colored as above specified. Tinning upon the roof to be painted with one coat upon the under side, and two upon the upper side, of spruce yellow and boiled linseed oil.

---

[a] The strips in which the window pulleys are placed.

[b] An ogee is a moulding resembling the letter S in its outline.

[c] The posts into which the hand-rail is inserted are called newels.

[d] The space occupied by the stairway is called the well-room.

A piece of raised tin to be placed on each side of each front door to turn the water off outside of the door. A single floor to be laid in the attic, of common white pine boards ⅞in. thick, planed, jointed, matched, and nailed with 8d. nails.[a]

2. Estimate.[b]

Item	Rate	Amount
45 squares of roof	at $12	
33 " ceiling	8	
32 windows	16	
12 "	10	
5 iron columns	12	
4 outside doors with frames and trimmings	16	
6 flights of stairs	50	
30 squares attic flooring	3	
3 hard pine floors		900.00
Iron work for building	45	
Observatory		250.00
40 brackets	2.25	
14 inside doors trimmed	8	
14 door casings	4.75	
35 window casings	1.12½	
2000ft. boards for furring	16	
Labor in furring		75.00
Painting		250.00
32000ft. timber in frame	22	
Deafening $150. Nails $10		
2200yds. plastering	.22	
200000 brick	8.50	
Caps and sills		225.00
175 seats	3.25	
175 "	.80	
125 perch of stone	1.00	
500yd. excavation	.15	
		$7991.62½

[a] The foregoing specification furnishes materials for a great number of useful questions, which the teacher may frame so as to adapt them to the wants of his pupils.

[b] In each item, the expense of labor is included.

3. Estimate of the Materials and Labor required in a Cottage.

From Ranlett's Architect.

		$
296 cubic yds. excavation	@	.09
2538 cubic ft. stone work	@	.10
7 stone sills	@	.50
24 linear ft. steps	@	.14
Cistern work $3.50; 4 hearths	@	3.00
1 marble mantel $50; 2 veined mantels	@	25.00
2 brown stone chimney caps	@	14.00
1451 square yds. plastering	@	.26
36500 brick	@	9.50 per M.
268 linear ft. cornice	@	.24
16196ft. timber in frame	@	2.00 per hund.
474 joist, set in frame and partitions	@	.18
4782 sq. ft. of sheathing and siding	@	.07
2283 " " and iron roof	@	.16½
60 linear ft. 3in. leader, @ 12½c., 114ft.	@	.11
3485 square ft. of interior floor	@	.04
1002 " " veranda "	@	.08
1184 " " " roof	@	.09
172 lin. ft. main cornice	@	.85
83 " " wing "	@	.70
199 " " veranda "	@	.55
15 veranda columns @ $10.50; 3 antæ	@	3.00
164ft. veranda cornice and filling	@	.14
15 steps and rises—back stairs	@	1.00
18 " " " principal "	@	3.50
2 front doors, with side and head lights	@	30.00
11 doors in principal story	@	11.00
10 " second "	@	8.00
10 " wing	@	7.00
7 double windows, first story	@	14.00
8 " " second "	@	10.00

		$	
7 double windows, wing	@	9.00	
8 single "	@	6.50	
6 cellar "	@	2.50	
3 wood mantels @ $4.50; 7 bells	@	3.25	
400 square yds. tight furring	@	.07	
197 linear ft. of blinds	@	.80	
1152 " " of base	@	.04	
12 closets, to shelve and put in hooks	@	4.50	
1000 lb. white lead in oil	@	7.00 per hund.	
43 gallons linseed oil	@	80	
4 " " " boiled	@	.90	
5 " spirits turpentine	@	.50	
½ " varnish	@	4.00	
30 lb. putty, @ 4ct.; 10 lb. litharge	@	.06	
3 lb. glue, @ 20ct.; 2 lb. lampblack in oil	@	.40	
6 lb. chrome yellow in oil	@	.30	
60 days painters' labor	@	1.75	
Hardware,—locks, bolts, window weights, &c.			99.39
			$4550.00

IX. STRENGTH OF MATERIALS.

ALL solid substances may be exposed to four kinds of strains. 1st, they may be torn asunder, as in the case of ropes, tie-beams, king-posts, &c. 2d, they may be crushed, as in the case of columns, posts, &c. 3d, they may be broken across, as in the case of joists, beams, &c. 4th, they may be twisted or wrenched, as in the case of wheel-axles, the screw of a press, the rudder of a vessel, &c. Numerous experiments have been made to determine the strength to which the materials in common use may safely be subjected, and tables have been compiled from the results of those experiments.

35. Table of the Flexibility and Strength of Timber.[a]

Name of wood.	Specific Gravity.	Value of U.[b]	Value of E.[c]	Value of S.[d]	Value of C.[e]
Teak	745	818	9657802	2462	15555
Poon	579	596	6759200	2221	14787
English Oak	969	598	3494730	1181	9836
Do. specimen 2	934	435	5806200	1672	10853
Canadian Oak	872	588	8595864	1766	11428
Dantzic Oak	756	724	4765750	1457	7386
Adriatic Oak	993	610	3885700	1583	8808
Ash	760	395	6580750	2026	17337
Beech	696	615	5417266	1556	9912
Elm	553	509	2799347	1013	5767
Pitch Pine	660	588	4900466	1632	10415
Red Pine	657	605	7359700	1341	10000
New England Fir	553	757	5967400	1102	9947
Riga Fir	753	588	5314570	1108	10707
Do. specimen 2	738		3962800	1051	
Mar Forest Fir	696	588	2581400	1144	9539
Do. specimen 2	693	403	3478328	1262	10691
Larch	531	411	2465433	653	
Do. specimen 2	522	518	3591133	832	
Do. specimen 3	556	518	4210830	1127	7655
Do. specimen 4	560	518	4210830	1149	7352
Norway Spar	577	648	5832000	1474	12180

36. Problems in determining the Strength of Timber.[a]

1. *To find* the strength of direct cohesion of a piece of timber of any given dimensions.

Rule.—Multiply the number of square inches in the transverse section by the value of C in the table, (§ 35,) and the product will be the strength required in pounds.

N. B. If the specific gravity differs from the mean specific gravity of the table, multiply the product by the actual specific gravity, and divide by the tabular specific gravity for the correct strength.

---

[a] Ingram.

[b] Ultimate deflection. [c] Elasticity. [d] Strength. [e] Cohesion.

If b = the breadth in inches, and d = the depth, $b \times d \times C = W$. If either b or d is required, divide W by the product of the remaining factors.

2. *To find* the deflection of a beam fixed at one end, and loaded with any given weight at the other.

Rule.—Divide 32 times the weight multiplied by the cube of the number of inches in the length of the beam, by the continued product of the tabular value of E, the number of inches in the breadth, and the cube of the number of inches in the depth of the beam.

N. B. When the beam is loaded uniformly throughout, multiply the cube of the length by 12 times the weight, instead of 32 times the weight.

3. *To find* the deflection of beams supported at both ends, and loaded in the middle with any given weight.

Rule.—Multiply the cube of the number of inches in the length, by the number of pounds in the given weight, and divide by the continued product of E, the number of inches in the breadth, and the cube of the number of inches in the depth of the beam.

N. B. When the beam is not only supported, but is *fixed* at both ends, the deflection is $\frac{2}{3}$ of that given by the rule.

If the weight is distributed uniformly throughout the length of the beam, the deflection will be $\frac{5}{8}$ of that given by the rule.

4. *To find* the ultimate deflection before their fracture of beams or rods supported at both ends.

Rule.—Multiply U by the number of inches in the depth of the beam, and divide the square of the number of inches in the length by the result.

5. *To find* the ultimate transverse strength of any rectangular beam of timber, fixed at one end and loaded at the other.

Rule.—Find the continued product of S, the number of inches in the breadth, and the square of the number of inches in the depth, and divide the continued product by the number of inches in the length.

If W represents the number of pounds that would produce fracture, l the length in inches, b the breadth in inches, and d the depth in inches, $l = S \times b \times d^2 \div W$; $b = l \times W \div (S \times d^2)$; $d = \sqrt{l \times W \div (S \times b)}$; $S = l \times W \div (b \times d^2)$.

6. *To find* the ultimate transverse strength of any rectangular beam when supported at both ends, and loaded in the centre.

Rule.—Find the continued product of S, 4 times the number of inches in the breadth, and the square of the number of inches in the depth, and divide the product by the number of inches in the length.

$l = 4 \times S \times b \times d^2 \div W$; $b = l \times W \div (4 \times S \times d^2)$; $d = \sqrt{l \times W \div (4 \times S \times b)}$; $S = 4l \times W \div (4 \times b \times d^2)$.

N. B. When the beam is *fixed* at each end and loaded in the middle, the result obtained by the rule must be increased one-half.

When the beam is loaded uniformly throughout its length, the result obtained by the rule must be doubled.

When the beam is *fixed* at both ends, and loaded uniformly throughout, the result obtained by the rule must be trebled.

If the load is to be permanent, it should not exceed $\frac{2}{3}$ of the amount obtained by the rules.

7. *To find* the weight under which a given column will begin to bend, when placed vertically on a horizontal plane.

Rule.—Find the continued product of E, the cube of the number of inches in the least thickness, the number of inches in the greatest thickness, and .2056. Divide this continued product by the square of the number of inches in the length.

If W represents the number of pounds that the column can sustain, d the number of inches in the greatest thickness, b the number of inches in the least thickness, and l the number of inches in the length; $d = W \times l^2 \div (E \times b^3 \times .2056)$; $b = \sqrt[3]{W \times l^2 \div (.2056 \times E \times d)}$; $l = \sqrt{E \times d \times b^3 \times .2056 \div W}$; $E = W \times l^2 \div (d \times b^3 \times .2056)$.

37. Examples for the Pupil.

1. What weights will be required to tear asunder two pieces of beech, each 4in. wide, and 3in. thick, the sp. gr. of the 1st being 696, and that of the 2d 678?

2d *Ans.* 115868 lb.

2. A red pine beam 8½ft. long, 4in. broad, and 6in. deep, is fixed at one end, and loaded with a weight of 500 lb. Required the deflection, when the weight is suspended at its extremity, and also when it is distributed uniformly throughout the length of the beam. 1st *Ans.* 2.51in.

3. A beam of Canadian oak, 5in. broad, 8in. deep, and 25ft. long, is supported at both ends, and loaded with a weight of 3000 lb. Required the deflection when the weight is placed in the centre, and also when it is distributed uniformly throughout the length. 2d *Ans.* 2.23 inches.

4. Required the deflection at the instant of fracture, of an ash beam 30ft. long, 9in. wide, and 6in. deep.

Ans. 54.68 inches.

5. What weight will be required to break a beam of larch, sp. gr. 531, fixed at one end and loaded at the other, the breadth being 3in., depth 6in., and length 6ft?

Ans. 979½ lb.

6. What weight will be required to break a beam of pitch pine, supported at both ends and loaded in the middle, the length being 16ft., the breadth 6in., and the depth 9in.? What weight would be required if the beam were *fixed* at both ends, and loaded uniformly throughout?

1st *Ans.* 16524 lb.

7. What weight will bend a column of New England fir, 8ft. 4in. long, 8in. wide, and 6in. thick, placed vertically on a plane, the weight being applied to its upper extremity?

Ans. 212007.8776 lb.

8. What weight would have been required to break the beam mentioned in Ex. 6, if it had been fixed at each end,

and loaded in the middle? If it were merely supported at each end, and loaded uniformly throughout its length?

38. TABLE SHOWING THE NUMBER OF POUNDS THAT WILL PULL ASUNDER A PRISM ONE INCH SQUARE, OF DIFFERENT MATERIALS, ACCORDING TO THE EXPERIMENTS OF M. MUSCHENBROEK.[a]

Cast gold	22000	Bismuth	2900
Cast silver	41000	Good brass	51000
Anglesea copper	34000	Ivory	16270
Swedish copper	37000	Horn	8750
Cast iron	50500	Whalebone	7500
Bar iron, ordinary	68000		
Ditto, best Swedish	84000	COMPOSITIONS.	
Bar steel, soft	120000	Gold 5, copper 1,	50000
Ditto, razor temper	150000	Silver 5, copper 1,	48500
Cast tin, Eng. block	5200	Swedish copper 6, tin 1,	64000
Ditto, grain	6500	Block tin 3, lead 1,	10200
Cast lead	860	Tin 4, lead 1, zinc 1,	13000
Regulus of antimony	1000	Lead 8, zinc 1,	4500
Zinc	2600		

ACCORDING TO THE EXPERIMENTS OF MR. RENNIE.

	No. of lbs. that would tear asunder a prism 1 in. square.	Length in ft. that would break with its own weight.
Cast steel	134256	39455
Swedish iron	72064	19740
English iron	55872	16938
Cast iron	19096	6110
Cast copper	19072	5092
Yellow brass	17958	5180
Cast tin	4736	1496
Cast lead	1824	384
Good hemp rope	6400	18790
Ditto, 1in. in diameter	5026	18790

39. THE LATERAL STRENGTH OF BARS ONE FOOT LONG, AND ONE INCH SQUARE.[a]

Weight that will break them.	Weight they can bear with safety.	Breaking weight.	Safe weight.
Cast iron 3270 lb.	1090 lb.	Memel fir 390 lb.	130 lb.
Oak 627 lb.	209 lb.	Am. white pine 206 lb.	69 lb.

[a] Ingram.

40. The Cohesive Force of a square inch of Iron of different kinds.[a]

Iron wire	113077 lb.	English iron . . .	65772 lb.
Ditto[b]	93964 "	Welsh iron	64960 "
Swedish iron . .	78850 "	Ditto	55776 "
Ditto	72064 "	German iron . . .	69133 "
Ditto	54960 "	French iron . . .	61000 "
Ditto	53244 "	Russian iron . . .	59472 "
English iron . . .	66000 "	Cast iron	18295 "
Ditto	55000 "	Ditto	19488 "
Ditto	61600 "	Ditto, Welsh . . .	16255 "

41. The Number of Pounds necessary to crush Cubes of 1½ inches.[a]

Aberdeen granite, blue	24536	Craigleith stone, with the strata	15560
White-veined Ital. marble	21738	Ditto across ditto	12346
Very hard freestone . .	21254	Cornish granite	14302
Purbeck limestone . . .	20610	White statuary marble .	13632
Limerick limestone, black	19924	Fine brick	3864
Peterhead granite . . .	18636	Yellow baked brick . . .	2254
Compact limestone . . .	17354	Red brick	1817
Yorkshire paving stone .	15856	Pale red brick	1265
Dundee sandstone	14919	Chalk	1127

One-inch Cubes were crushed by the following weights:

Elm	1284 lb.	English oak . . .	3860 lb.
White deal	1928 lb.	Craigleith stone . .	8688 lb.

Cubes of one-fourth of an inch were crushed by the following weights:

Iron, cast vertically .	11140 lb.	Cast tin	966 lb.
Ditto horizontally .	10110 "	Cast lead	483 "
Cast copper	7318 "		

---

[a] Ingram.

[b] The different numbers represent the different results obtained by the most careful experimenters.

42. Table of Iron and Hempen Cables of equal strength.[a]

Iron Cables. Diameter of Iron Rod.	Hemp Cables. Circumference of Rope.	Resistance.
Inches.	*Inches.*	*Tons.*
$\frac{7}{8}$	9	12
1	10	18
$1\frac{1}{8}$	11	26
$1\frac{1}{4}$	12	32
$1\frac{5}{16}$	13	35
$1\frac{3}{8}$	14 to 15	38
$1\frac{1}{2}$	16	44
$1\frac{5}{8}$	17	52
$1\frac{3}{4}$	18	60
$1\frac{7}{8}$	20	70
2	22 to 24	80

The stress given in the above table is the greatest to which the cables should be exposed, and is about $\frac{1}{2}$ the breaking strain.[b]

A "cable's length," is 120 fathoms.

43. Mean weight of a cubic foot of Stone, and the weight it will sustain with safety.[c]

	Weight.	Pressure.
Sandy Bay granite	168.48 lb.	197000 lb.
Quincy "	167.04 "	156000 "
Concord "	159 "	149000 "
Frankfort "	162 "	148000 "
New York white marble . . .	173 "	85000 "
N. Haven variegated " . . .	175 "	89000 "
Penn. dove marble	170 "	86000 "
Vt. " "	168 "	86000 "
Thomaston blue marble . . .	179 "	90000 "
Connecticut sandstone	164 "	118000 "
North River "	156 "	108000 "
Potomac "	153 "	98000 "

[a] Ure.

[b] A common rope 1ft. long, and 1 inch in circumference, weighs .044 to .046 lb. In a cable, it weighs .027 lb. To find the number of pounds which a rope will sustain, square the number of inches in its girt, and multiply by 200 for common ropes, and by 120 for cables.—*Tredgold.*

[c] Shaw.

44. Problems on the Strength of Iron.

1. *To find* the breadth of a uniform cast-iron beam, to sustain a given weight in the middle.

Rule.—Multiply the number of feet in the length, by the number of pounds to be supported, and divide the product by 850 times the square of the number of inches in the depth. The quotient will give the number of inches in the breadth.

When neither the breadth nor depth is known, but merely the proportion that exists between them; find the continued product of the number of ft. in the length, the number of lb. in the weight, and the ratio of the depth to the breadth; divide the continued product by 850, and extract the cube root of the quotient. The result will be the number of inches in the depth.

If W = the weight to be supported in pounds, l = the length in feet, b = the breadth in inches, and d = the depth in inches, $W = 850 \times b \times d^2 \div l$; $d = \sqrt{(l \times W) \div (850 \times b)}$; $l = 850 \times b \times d^2 \div W$.

2. *To find* the proper breadth and depth of a beam of cast iron supported at both ends, when the load is not in the middle.

Rule.—Measure the number of feet from the point at which the weight is applied, to each support, and find the product of the two numbers; divide 4 times this product by the whole length between the supports, and proceed with the quotient in the same manner as with the length in Problem 1.

N. B. When the load is uniformly distributed over the length of the beam, the depth need be only $\frac{4}{5}$ as great as when it is all placed in the middle.

3. *To find* the proper breadth and depth of an iron beam fixed at one end, and the load applied to the other, or of a beam supported upon a centre of motion.

Rule.—Take the length from the point at which the beam is fixed, or from the centre of motion, to the point where the load is applied, and calculate the strength by the rules in Problem 1, using instead of 850, 212 for cast iron, 238 for wrought iron, or 425 when the weight is uniformly distributed over the length of the beam.

4. *To find* the proper depth of the teeth of wheels.

Rule.—Divide the number of pounds which represents the stress of the wheel at the pitch-circle,[a] by 1500, and extract the square root of the quotient. The result will give the thickness of the teeth in inches.

N. B. The length of the teeth ought not to exceed their thickness. The breadth should be in proportion to the stress upon them, and this stress should not exceed 400 lb. for each inch in breadth.

5. *To find* the proper thickness of the teeth of a wheel, when the power of the first mover is given in pounds, and the velocity per second in feet.

Rule.—Form the continued product of the numbers which represent the power and velocity per second of the first mover, and .073; multiply the number of revolutions the wheel is to make per minute, by the radius the wheel should have if its pitch were two inches; divide the first of these products by the second, and the cube root of the quotient will give the thickness of the teeth in inches.

6. *To find* the proper diameter of a solid cylinder of cast iron to sustain a given weight, when supported at both ends, and the weight applied at the middle of the length.

Rule.—Take the weight in pounds, and the length in feet; multiply the two numbers together, and divide the product by 500; the cube root of the quotient is the number of inches in the diameter.

If W represents the weight in pounds, l the length in feet, and d the diameter in inches, $W = 500 \times d^3 \div l$; $l = 500 \times d^3 \div W$.

---

[a] The *pitch* of the teeth of a wheel, is the distance between the middle points in the bases of two adjacent teeth. It should be at least 2.1 times the thickness of the teeth.

7. *To find* the proper diameter of a solid cylinder of cast iron supported at both ends, to bear a given weight when the strain is not in the middle.

Rule.—Take the weight in pounds, and the distances in feet, from the point where the weight is applied to each of the points of support; divide the continued product of these three numbers by the number of feet in the distance between the points of support, and cut off three figures from the right hand. The cube root of the result will give one half of the diameter of the cylinder in inches.

8. *To find* the proper diameter of a solid cylinder of cast iron when supported at both ends, to sustain a load uniformly distributed over its length.

Rule.—Multiply the number of feet in the length by the number of pounds in the weight, and $\frac{1}{10}$ of the cube root of the product will be the number of inches in the diameter.

$W = 1000 \times d^3 \div l$; $l = 1000 \times d^3 \div W$.

9. *To find* the proper length of a solid cylinder of cast iron, when fixed at one end and loaded at the other; also when the cylinder is supported on a centre of motion.

Rule.—Take the weight in pounds, and the distance of the weight from the point of support in feet, multiply the two numbers together, and $\frac{1}{5}$ of the cube root of the product will be the number of inches in the diameter.

$W = 125 \times d^3 \div l$; $l = 125 \times d^3 \div W$.

10. The strength of direct cohesion, of the materials in tables, § 38 and 40, may be found by Problem 1, § 36, using the numbers in those tables opposite to the given material, instead of the value of C.

11. The lateral strength of iron may be found by the rules for that of timber, using the number opposite to iron in § 39, instead of the value of S in § 35.

12. The strength of a column to resist being crushed, is proportioned to the area of its transverse section. Hence,

to find the weight which will crush any column, multiply the number of inches in the area of its transverse section by the proper number in § 41, and divide the product by $2\frac{1}{4}$; or multiply the number of feet in the area of the transverse section, by the pressure given in § 43. The area of a cylindrical column may be estimated, in making this calculation, at $\frac{7}{9}$ of the square of the diameter. The *true* area of a circle, is found by multiplying the square of the diameter by .7854. The result is the number of pounds.

45. Examples for the Pupil.

1. What is the breadth of a cast-iron beam 30ft. long, and 8in. deep, that will support a weight of 8 tons, placed in the middle? If the length is 30ft. and the breadth 6in., what must be the depth to support 10 tons?

1st *Ans.* 9.88in.
2d " 11.47 "

2. The front of a house is to be broken out to make shops, and the front wall, which is 40ft. long, is to be supported by 2 cast-iron beams, with a prop under the middle of the wall. If there are 4000 c. ft. of wall to be supported, weighing 140 lb. per c. ft., what must be the breadth and depth of each beam, the depth being 5 times the breadth?

Ans. Depth 20.352in.; Breadth 4.0704in.

3. The second story of a building is to project 3ft. over the first. What must be the depth of the fixed iron beams, which are 4 inches broad, supposing the weight supported by each to be 33600 lb.? *Ans.* 7.7in.

4. If the greatest stress at the pitch-circle of a wheel is 6500 lb., what should be the thickness of the teeth?

Ans. 2.08in.

5. If the effective force of the piston of a steam engine is 10000 lb., and its velocity 5 ft. per second, what should

be the thickness for the teeth of a wheel, which is to make 20 revolutions in a minute, and to have 120 teeth?[a]
Ans. 1.46in.

6. What weight will a cast-iron cylinder, supported at both ends, sustain in the middle of its length, the diameter being 8 inches, and the length $16\frac{1}{2}$ ft.? *Ans.* 15515 lb.

7. What must be the diameter of a cast-iron cylinder, which is 20ft. long, to sustain a weight of 33600 lb., the weight being applied at the distance of 5ft. from one end?
Ans. 10.026in.

8. A load of 25000 lb. is to be uniformly distributed over a solid cast-iron cylinder. Required the length of the cylinder, the diameter being 9 inches. *Ans.* 29.16ft.

9. A solid cylinder of cast iron, 8 inches in diameter, is supported in the middle. What may be the length of the arms, to support 10000 lb. at the extremity of each?
Ans. 6.4ft.

10. What weight would pull asunder a hemp rope $2\frac{1}{2}$ inches in diameter? *Ans.* $31412\frac{1}{2}$ lb.

11. What weight distributed uniformly over a cast-iron beam, 20ft. long, 6in. broad, and 8in. deep, will break the beam, it being supported at both ends? *Ans.* 41856 lb.

12. What weight can be sustained with safety by each of the marble pillars of Girard College, estimating their strength as equivalent to that of Pennsylvania Dove marble, the least diameter of the pillars being 5ft?[b]
Ans. 1688610 lb., or, estimating the area at $\frac{7}{9}$ of the square of the diameter, 1672222 lb.

13. The length of each arm of the wrought-iron beam of a balance is 3ft., and the depth is 8 times the breadth.

---

[a] The circumference of a wheel with 120 teeth, and a pitch of 2in., is $120 \times 2 = 240$in. The radius of such a wheel would be $240 \div (2 \times 3.1416) = 38.197$in. See the article on Mensuration.

[b] Architect's Report.

Required the depth and breadth necessary to enable the balance to weigh ½ a ton.

Ans. Depth, $\sqrt[3]{8 \times 3 \times 1120 \div 238} = 4.834$in.
Breadth, .604in.

14. What should be the depth and breadth of a cast-iron beam, 30ft. long, to support a weight of 20 tons, placed 10ft. from one end, the beam being supported at both ends, and the depth being 4 times the breadth?

Ans. Depth, 17.7812in.; Breadth, 4.4453in.

X. SPECIFIC GRAVITY.

THE Specific Gravity of a body, is the ratio of its weight to the weight of an equal volume of some other body assumed as a standard. The standard usually adopted for this purpose is pure distilled water at a given temperature. In England, the temperature of 62° Fahrenheit is generally taken; the French take 32°, or the temperature of melting ice.[a]

46. TABLE OF SPECIFIC GRAVITIES.

Compiled from the Encyclopedia Britannica and other sources.[b]

Substance	Specific gravity	Substance	Specific gravity
ACACIA, inspissated juice	1513	Acid, muriatic	1284.7
Acid, acetic	1007 to 1009.5	nitric	1271.5 to 1583
acetous	1009.5 to 1025.1	phosphoric, liquid	1417
arsenic	3391	solid	2852
boracic, scales	1475	sulphuric	1840.9 to 2125
citric	1034.5	Agate	2348 to 2666.7
fluoric	1500	Air	1.2308

[a] Brande.

[b] The specific gravity of water is fixed at 1000. As a cubic foot of water weighs 1000 ounces avoirdupois, the specific gravity of each article named in the table will represent the weight of 1 cubic foot.

Substance	Specific gravity
Alabaster	2611 to 2876.1
Alcohol, absolute	791
mixed with water	829.3 to 991.9
Alder wood	800
Alum	1750
Amber	1078 to 1085.5
Antimony, fused	6624 to 6860
Apple tree	793
Arsenic, fused	8310
glass of, (arsenic of the shops)	3594.2
Asbestos	2577.9 to 3080.8
Do. mountain cork	680.6 to 993.3
Ash	727[a] to 845
Asphaltum	1070 to 2060
BASALT	2421 to 3000
Beech	696[a] to 852
Beryl, oriental	3549.1
aquamarine	2650 to 2759
Bismuth, molten	9756 to 9822
Blood, human	1054
Bone of an ox	1656
Borax	1740
Boxwood	912 to 1328
Brass, common	7824 to 8395[b]
wiredrawn	8544
Brazil wood	1031
Brick	1557[a] to 2000
Brickwork[c]	1872
Butter	942.3
CADMIUM	8604 to 8694.4
Camphor	988.7
Cannel coal	1270
Caoutchouc	933.5
Castor oil[d]	970
Cedar	457[a] to 561
Chalk[e]	2252 to 2657
Cherry	715
Chestnut[a]	610
Chromium[d]	5900
Citron wood	726.3
Clay[a]	2000
Coal, bituminous[f]	1262 to 1364
anthracite[f]	1500
Cobalt, fused	7645 to 7811
Cocoa wood	1040.3
Coke[a]	744
Copal	1045.2 to 1139.8
Copper,[g] native	7600 to 8508.4
fused	7788 to 8607
wiredrawn	8878
Coral[h]	2680
Cork	240
Cypress	644
DIAMOND	3444.4 to 3550
EARTH, common[a]	1520 to 2000
mean density[i]	5670
Ebony	1209 to 1331
Elm	671
Emerald	2600 to 3155.5
Emery[h]	4000
Ether, acetic	866.4
muriatic	729.6
nitric	908.8
sulphuric	716 to 745
FAT	923.2 to 936.8
Felspar	2438 to 2704
Filbert tree	600
Fir	498 to 553
Flint	2243.1 to 2664.4
GARNET, common	3576 to 3688
precious	4085 to 4352
Gas, ammonia	.73459

[a] Cavallo. [b] 369 c. in. = 1cwt. [c] Benjamin.
[d] Ingram. [e] 13 c. ft. = 1 ton. [f] W. R. Johnson.
[g] A square foot of sheet copper, ¼in. thick, weighs 11 lb. 12oz.
[h] Barlow. [i] Cavendish.

Gas, atmospheric air	1.2308
carbonic acid	1.87
carbonic oxide	1.1777
carburetted hydro.	.73848
chlorine	3.0401
cyanogen	2.2228
fluosilicic acid	4.3984
hydriodic acid	5.4684
hydrogen	.0898484
muriatic acid	1.5353
nitrogen	1.928
nitrous	1.2786
nitrous acid	3.908
nitrous oxide	1.9865
olefiant	1.20377
oxygen	1.3588
phosph. hydrogen	1.0708
steam	.76739
sulph. hydrogen	1.4661
sulphurous acid	2.6097
Glass, bottle	2732.5
crown	2487 to 2520
flint	3000 to 3437
green	2642.3
plate	2520 to 2760
Gold, not hammered	19258.7
hammered	19342
Am. standard	17350[a]
Eng. "	18888
French "	17486
Granite	2613 to 2760.9

Gum ammoniac	1207.1
Arabic	1452.3
guaiacum	1228.9
lac	1139
tragacanth	1316.1
Gunpowder, solid	1745
shaken	932
Gypsum	1872 to 3310.8
HAZEL	606
Hone	2876.3 to 3127.1
Honey	1450
Hornblende	3333 to 3830
ICE[b]	930
Indigo	769
Iodine	4948
Iridium, fused	18680
Iron,[c] bar	7600 to 7788
cast	6953 to 7295
magnetic	4518
meteoric	6480
Ivory	1825 to 1917[b]
JARGON, of Ceylon[d]	4416
Jasper	2358.7 to 2816
Jet	1259
Juniper tree	556
LARD	947.8
Lead	11352 to 11445
Lignum Vitæ	1333
Limestone	1386.4 to 3183
Linden	604
Linseed oil	940.3

---

[a] The specific gravity of the Mint standard gold varies from 17250 to 17500, according to the greater or less proportion of copper used in the alloy. The average specific gravity is about 17350.

[b] Barlow.

[c] A square foot of cast iron, ¼in. thick, weighs 9 lb. 10.6oz ; a sq. ft. of malleable iron of the same thickness, 9 lb. 15.2 oz. ; a bar 1ft. long and 1¼in. square, of cast iron, 9 lb. 8oz. ; a bar of malleable iron, of the same size, 9 lb. 11¼oz. ; a round iron rod, 1ft. long and 1¼in. in diameter, 7 lb. 9.2 oz.

[d] Ingram.

Living men[a] 891
Loadstone 4200 to 4900
Logwood 913
MADDER 765
Magnesia, sulphate of 1797.6
Mahogany 1063
Manganese 6850
Maple 755
Marble, Carrara 2716
Egyptian 2668
various 2516 to 2858
Mercury 13568
Mica 2883
Milk 1020.3 to 1040.9
Molybdena, native 4738.5
Mortar, dry[b] 1384 to 1893
Muriatic acid 1284.7
NAPHTHA 847.5
Nickel, metallic 7421 to 9333.3
forged 8600
Nitre 1900 to 2246
Nitric acid, (aquafortis) 1500[a]
Nitrous acid 1452[a]
OAK[b] 748 to 993
heart of 1170
Obsidian 2348
Oil, of turpentine 870
olive 915.3
whale 923.3
various 857.7 to 1044
Opal 1958 to 2600
Opium 1336.5
Orange tree 705.9
PALLADIUM 11800
Pear tree 661
Pearls 2683
Peat 600 to 1329
Pewter[a] 7471
Phosphorus 1770
Pine 540 to 683
Pitch[a] 1150
Platina 14626 to 22069
Plumbago 1987 to 2267
Poplar 360 to 529.4
Porcelain, China 2384.7
do. European 2145 to 2545
Potash, carbonate 1459.4
Potassium 972.23
Proof spirit 916
Pumice stone 914.5
QUARTZ 2652
Quince tree 705
RHODIUM 11000
Rock crystal 2581 to 2888
Rosin[a] 1100
Ruby 3531 to 4283.3
SAND[b] 1454 to 1886
Sandstone 2142 to 2483.5
Sapphire 3130 to 4283
Scythe stone, fine 2609
Serpentine 2264 to 3000
Silver 10474 to 10510
Slate 26718 to 2752[b]
Soda, sulphate 1439.8
carbonate 1000 to 1500
Sodium 865.07
Spar, heavy[a] 4430
Spermaceti 943.3
Steel 7767 to 7840.4
Stone, common 2000 to 2700
rotten 1981
Sugar, white 1606
Sulphur, native 2033.2
Sulphuric acid 1841
TALLOW 941.9
Tar[a] 1015
Tellurium, native 5700 to 6100
Tin 7063 to 8487

[a] Ingram. [b] Cavallo.

Topaz	3531 to 4061.5	Water, sea	1026.3
Tourmaline	3086 to 3362	well	1001.7
Tungsten	4355 to 6066	Wax, bees'	964.8
Turpentine	991	shoemakers'	897
spirits of	870	Whalebone	1300
ULTRAMARINE	2360	Willow	585
Uranium	7500	Wine, Burgundy	991.5
VAPOR of alcohol	2.58468	Canary	1033
do. absolute	1.985	Champagne	997.9
hydriodic ether	6.7384	Malaga	1022.1
muriatic ether	2.731	Malmsey	1038.2
sulphuric ether	3.182	Port	997
iodine	10.6089	Tokay	1053.8
oil turpentine	6.17	Wolfram	5705 to 7333
sulph. of carbon	3.255	Wootz, hammered[a]	7787
water	.76739	YEW, Dutch	788
Vinegar	1013.5 to 1080	Spanish	807
WALNUT	671	ZINC, common	6862
Water, distilled	1000	pure & compressed	7190.8
Dead Sea	1240.3		

47. PROBLEMS IN SPECIFIC GRAVITY.[a]

I. *To find the magnitude of a body from its weight.*

Find the weight of the body in ounces, and divide by the specific gravity. The quotient will be the number of cubic feet in the contents.

II. *To find the weight of a body from its magnitude.*

Find the number of cubic feet in the body, and multiply by the specific gravity. The product will be the number of ounces in the weight.

III. *To find the specific gravity of a body.*

CASE I. When the body is heavier than water.

Weigh the body both in air and in water. Annex three

---

[a] Ingram.

ciphers to the weight in air, and divide by the difference of the weights. The quotient will be the specific gravity.

CASE II. When the body is lighter than water.

Having weighed the light body in air, and a body heavier than water both in air and water, fasten them together with a slender tie, then weigh the compound in water, and subtract its weight from the weight of the heavy body in water; to the remainder add the weight of the light body in air, and by the sum divide one thousand times the weight of the light body in air. The quotient will be the specific gravity of the light body.

IV. *To find the quantity of each ingredient in a mixture of two substances.*[a]

1. Multiply the specific gravity of the mass by the difference between the specific gravities of the two ingredients, for a *first* product.

2. Multiply the specific gravity of that ingredient whose quantity is desired, by the difference between the specific gravity of the mass, and that of the other ingredient, for a *second* product.

3. Multiply the whole weight of the mass by the *second* product, and divide by the *first* product. The quotient will be the weight of the ingredient sought.

48. EXAMPLES FOR THE PUPIL.

1. How many cubic inches in 1 lb. of white sugar?

Ans. $\frac{8}{803}$ c. ft. = 17.215 c. in.

2. A keg is found to contain 13790 cubic inches. What weight of butter will it hold? *Ans.* 470 lb.

3. A piece of Quincy granite weighs 25 lb. 12½ oz. in air, and 16 lb. 1½ oz. in water. What is its specific gravity?

Ans. 2661.

---

[a] The student will observe that this is a case in Alligation.

4. A piece of copper weighs 18 lb. in air and 16 lb. in water. A piece of elm, which weighs 15 lb. in air, is fastened to the copper, and the compound weighs 6 lb. in water. What is the specific gravity of the elm? *Ans.* 600.

5. What quantity of gold, sp. gr. 19258, and of silver, sp. gr. 10474, must be mixed to form a mass weighing 1cwt. 3qr. 4 lb., and having a specific gravity of 16000?
Ans. 151.44 lb. gold, 48.56 lb. silver.

6. What are the cubical contents of a pillar of Pennsylvania marble, sp. gr. 2720, the weight being 63T. 8cwt. 21 lb.? *Ans.* 835 c. ft. 884 c. in.

7. Each of the pillars of Girard College is 55ft. high, and 6ft. in diameter at the base. What would be the weight of a square block of marble from which a column of the same size could be cut, the specific gravity being 2716?
Ans. 150T. 3qr. 21 lb.

8. It is proposed to float 500 c. ft. of granite on a pine raft 50 ft. long, and 20ft. wide. What must be the depth of the raft in order that it may float at least 6 inches above the water, the specific gravity of the granite being 2620, and the sp. gr. of the pine 600?[a]

9. A raft of elm is 3ft. 6in. thick. To what depth will it sink? *Ans.* 2ft. 4.182in.

10. What must be the depth of a cedar raft, 16ft. long and 10ft. wide, to float 10000 lb. of bricks, the cedar being

[a] All floating bodies sink till they have displaced a quantity of fluid equivalent to their own weight. In this example, the granite would cause the raft to displace 1310 c. ft. of water, to do which, it must sink $1310 \div (50 \times 20) = 1.31$ft. to which the 6in. $=$ .5ft. should be added, making 1.81ft. to represent the buoyancy of the pine. As the sp. gr. of the pine is $\frac{3}{5}$ that of water, it will sink $\frac{3}{5}$ of its depth, leaving $\frac{2}{5}$ for buoyant force. 1.81ft. must therefore be $\frac{2}{5}$ of the depth of the raft, and the depth must be $1.81 \div \frac{2}{5} = 4.525$ft.

of the sp. gr. 550, and the raft floating 3in. out of water? *Ans.* 2ft. 9⅓in.

11. What weight will a raft 30ft. long, 16ft. wide, and 3ft. deep sustain, and float 6 inches above water, the specific gravity of the raft being 580? *Ans.* 22800 lb.

N. B. First find how much above water the raft would float if it were not loaded, and subtract 6 inches to find how much it is sunk by the load. The weight of the load will then be equivalent to the weight of the quantity of water displaced by it.

12. How many inches above water will a raft float, if loaded with 8500 lb., the raft being 12ft. wide, 20ft. long, and 2ft. deep, and of the specific gravity of 625?

Ans. 2.2 inches.

N. B. First find the entire weight of the raft and load, and see what depth of water must be displaced to yield the same weight. Subtract the depth of water displaced, from the depth of the raft, and the remainder will give the part out of water.

13. What is the weight of a sheet of malleable iron, 3ft. 6in. wide, 8ft. long, and $\frac{1}{16}$in. thick?

Ans. 69 lb. 10.4oz.

14. What is the weight of an iron rod, 3in. in diameter, and 16½ft. long?[a] Of a rod 1in. in diameter, and 10¼ft. long? *Ans.* 4cwt. 1qr. 23 lb. 15.2oz.; 1qr. 6lb. 8.1oz.

15. What is the weight of a sheet of copper, 3ft. wide, 6ft. 8in. long, and ⅛in. thick? *Ans.* 1cwt. 5 lb. 8oz.

16. The amount of water displaced by a loaded ship, is found to be 96000 cubic feet. Required the weight of the vessel and cargo, the water being of the specific gravity of 1020. *Ans.* 2732T. 2cwt. 3qr. 12 lb.

---

[a] The weight of iron rods of the same length, is proportioned to the squares of their diameters.

XI. THE ROAD.[a]

49. General Remarks.

When it is proposed to construct a road, the engineer first makes himself acquainted with the face of the country through which the road is to pass, and selects what he considers as the best general route. An instrumental survey is then made of the country along the proposed route, taking levels from point to point, throughout the whole distance, to determine the requisite inclinations of the slopes of the cuttings and embankments, and making borings in all places where excavations are required, to determine the strata through which the cuttings are to be made. A plan and section are then drawn, exhibiting the results of this investigation.

In selecting the route, regard should be had to the supply of materials for constructing the road, and for keeping it in repair. Therefore the position of gravel pits and quarries in the neighborhood of the proposed line, should be well ascertained.

The expense of construction should be proportioned to the traffic expected on the road. If the amount of travel will be great, all steep acclivities should be avoided, either by cutting down the hills and filling up the valleys, or by passing around the base of the hills.

It is recommended by some writers to avoid a dead level, as a moderate inclination of the surface facilitates drainage, and tends to keep the road dry. But, if proper attention is paid to the form of the road, there will be no difficulty in keeping it properly drained, without resorting to any expedient that will be necessarily attended with a loss of power.

---

[a] Brande, Gillespie, Mahan.

The top should be slightly rounded, being made highest in the middle, and gradually sloping to a trench at each side, so that all the water may be carried off. The surface should be as hard and smooth as possible, and, whenever repairs are required, broken stone, pebbles, or hard gravel should be used, if it is possible to obtain either of them.

50. Examples for the Pupil.

1. How many acres per mile will be taken up by a road that is 2 rods wide?—by a road 40ft. wide?—by a road 60ft. wide?

2. In 1678, a contract was made to establish a coach between Edinburgh and Glasgow, a distance of 44 miles. The coach was to be drawn by 6 horses, and the journey between the places, to and from, was engaged to be completed in 6 days.[a] At what average rate did the coach move, if it travelled 9 hours per day?

3. In the year 1763, there was but one line of stage-coaches between Edinburgh and London, which started once a month from each place. It then took a fortnight to perform the journey, which is now completed in less than 48 hours.[a] The number of persons travelling between the two places did not probably exceed 50 per month, but the present intercourse is supposed to amount to at least 300 per day. What has been the rate of increase, both in the rate of travel and in the number of passengers?

4. What will be the cost of a plank road per mile, for a single track 8ft. wide and 3in. thick, with two sills, each 4in. square, at $4.50 per M.,[b] the laying and grading being 75cts. per rod, and superintendence $75 per mile?

---

[a] Brande. Even so recently as the year 1750, the stage-coach from Edinburgh to Glasgow took a day and a half to make the journey.

[b] "Wood is paid for by the cubic foot, unless some one of its dimensions is as small as 4 inches, when board measure is used."—*Gillespie.*

5. Wishing to know my distance from the foot of a steeple which is 125ft. high, I hold a foot rule at arm's length, (which I have found to be equivalent to 2ft. 3in. from the eye,) and find that $1\frac{3}{4}$ inches on the rule intercepts the rays from the top and base of the steeple. What is the distance?[a]

Ans. $1928\frac{4}{7}$ft.

6. The mean velocity of sound through the atmosphere is about 1090ft. per second.[b] What is the distance of a hill on which a cannon is fired, if $8\frac{1}{2}$ seconds elapse between the flash and report?

7. If two supports of a rail that are 3ft. apart, vary $\frac{1}{4}$ of an inch from an exact level, to what elevation per mile would the ascent be equivalent?

8. Estimating the cost of a railroad at $30000 per mile, and the annual repairs and expenses at $2000 per mile, how much might be profitably expended to shorten the road 1 mile, the rate of interest being 6 per cent.?[c]—to shorten it 2m. 6fur. 23r.?

2d *Ans.* $178718.75.

9. An embankment of 27000 c. ft. is to be made. It is estimated that 1 man can loosen 20 c. yd. per day, or load in barrows 25 c. yd., or transport 30 c. yd., or spread and level 80 c. yd. According to this estimate, what will be the cost of the whole, allowing $1.25 per day for wages, and 10 per cent. for shrinkage of the earth, and adding $\frac{1}{20}$ of the amount of wages for tools and superintendence, and

---

[a] The length marked on the rule, is to the distance of the rule from the eye, as the height of the object is to its distance. Distances may also be conveniently measured by pacing, or by noting the time that elapses between the flash and report of a gun.

[b] Pierce.

[c] In laying out a road of any kind, the preference never should be given to the longer of two proposed routes, merely because it can be constructed at a less expense. In order to shorten the distance, the difference in the cost of making, together with a sum, the interest of which would defray the annual repairs and expenses of the road, for the distance saved, may be profitably expended. In the remaining examples of this section, the rate of interest is understood to be 6 per cent.

$\frac{1}{10}$ for contractor's profit, the earth costing 10 cents per c. yd.? *Ans.* \$324.79.

10. The average power of draft of a horse, moving 3 miles per hour for 10 hours a day, being 100 lb., what will be the annual cost of transportation over a road 30 miles long, on which the average friction is $\frac{1}{20}$ of the weight, estimating the amount transported at 50000 tons, and the value of a days' labor of a horse at 75 cents? *Ans.* \$42000.

11. If the road in the preceding example should be improved by macadamizing, or otherwise, so that the friction should be reduced to $\frac{1}{50}$ of the weight, what would be the annual cost of the transportation, and how much might be profitably expended in making the improvement?

2d *Ans.* \$420000.

12. The annual cost of transportation over a road 15 miles long, being estimated at \$35000 per mile, what amount of saving can be effected by expending \$20000 to shorten the road 2 miles, and \$1000000 to reduce the friction to $\frac{1}{2}$ its present amount, the annual cost of repairs being the same in both cases? *Ans.* \$3938333$\frac{1}{3}$.

13. If a hill by friction and gravity, causes 5000 days' work of a horse, at 75 cents per day, which can be avoided by a road along the base of the hill, that will require only 2300 days' work, and if the new road will require an extra annual outlay of \$375 for repairs, how much can be saved by expending \$10000 in making the improvement?

Ans. \$17500.

14. It is calculated that, in locomotives, the evaporation of 1 cubic foot of water per hour, produces a mechanical force of about two horse power, and that a horse on a railway can pull 10 tons with ease.[a] According to this estimate, what load can be drawn by a locomotive which evaporates 175 c. ft. per hour?

---

[a] Chambers.

15. If an engine has sufficient force to draw 92T. 19cwt. 1qr. over level ground, what additional power must be exerted on an ascending grade of 37ft. per mile?[a]

Ans. 13cwt. $3\frac{101}{440}$ lb.

16. Determine the amount of excavation and embankment in the following example, by taking the average of the end areas of each section as the true area of the section,[b] and find the cost of the whole at 10cts. per cubic yard.

Station.	Distance.	End Areas.	Excavation. Cubic feet.	Embankment. Cubic feet.
1	....	0	.....	....
2	561 ft.	1386 sq. ft. excav.	388773	0
3	858 "	1600 " " "	....	0
4	825 "	0 " " "	.....	0
5	820 "	1672 " " emb.	0	685520
6	825 "	528 " " "	0	....
7	330 "	0 " " "	0	....
			2329767	1680140

Cost, $14851.51.

[a] On all inclined planes, the power is to the weight, as the length of the plane is to the height.

[b] This method, which is the one usually employed, gives a result greater than the true contents. Sometimes the calculation is performed by deducing the *middle area* of each section from the arithmetical mean of the heights at the two extremities, but the result thus obtained is too small. The true contents may be found by the

PRISMOIDAL FORMULA.

Find the area of each end of the mass, and also the middle area, corresponding to the arithmetical mean of the heights of the two ends. Add together the area of each end, and four times the middle area. Multiply the sum by the length, and divide the product by six. The quotient will be the true cubic contents required.

The reduction of the contents to cubic yards would be greatly facilitated if the distances of the stations were always some multiple of 54 feet.—*Gillespie.*

XII. THE ENGINEER.

51. The Steam Engine.

A pound of steam at 212° will raise the temperature of a pound of water 970°.[a] But as some of the heat is wasted, the increased temperature may be considered, in practice, as equivalent to 900°. From this fact the following rule is derived:—

"*To find the quantity of steam required to raise a given quantity of water to any required temperature.*—Multiply the number of gallons to be warmed by the number of degrees between the temperature of the cold water and that to which it is to be raised, for a dividend; and to the excess of the temperature of the steam above 212° add 900 for a divisor. The quotient will be the number of gallons formed into steam, required."[b]

A "horse power" was estimated by Boulton and Watt as sufficient to raise 32000 lb. avoirdupois 1 foot high in 1 minute, but in estimating the force of their engines, they used 44000 as a divisor instead of 32000. Desaguliers's estimate was 27500; Smeaton's, 22916; some of the modern English engines are computed at 66000,[c] but the number commonly used is 33000,[d] or 44000 when an allowance of ⅓ is made for friction.

To calculate the power of an engine. Form the continued product of the number of square inches in the area of the cylinder, the number of pounds which represents the effective pressure[e] per square inch, and the number of feet through which the piston moves per minute, and divide by the number of pounds a horse can raise 1 foot per minute.

---

[a] Daniell. [b] Pilkington. [c] Sci. American.

[d] It is customary to consider the friction of the machinery as equivalent to ⅓ of the effect produced.

[e] The effective pressure is the force remaining after making allowance for the waste and friction of the steam.

The following rule is simpler, and in most cases it will be found sufficiently accurate. When the usual estimate of a horse power is employed, and the effective force is 8.4 lb. per square inch, and the distance traversed by the piston 200ft. per minute, the same result will be obtained by either rule.

Square the number of inches in the diameter of the cylinder, and multiply by .04. The product will give the number of horse power.

EXAMPLES.

1. What quantity of water converted into steam at 220° will raise 100 gallons of water at 50° to the boiling point?

Ans. $17\frac{191}{227}$ gallons.

2. What is the power of a steam engine with a cylinder 37 inches in diameter, making the usual estimate of the effective force of the steam and the stroke of the piston?

Ans. 54.76 horse power.

3. Find by each rule, the power of a steam engine that makes 10 strokes per minute, each stroke being 8ft., allowing for friction, $\frac{1}{3}$ of the force, which is 15 lb. per sq. in., the diameter of the cylinder being 40 inches?[a]

Ans. by Rule 1, 60.928 horse power.
" " 2, 64 " "

4. Two steam engines constructed for the island of Ceylon, working 10 hours per day for 300 days in the year, will convert 576000 lb. of paddy[b] into rice worth £116035, while by the common method, the same quantity of paddy converted into rice, would yield only £64799.[c] Allowing £20000 per annum, for interest, repairs, and the extra expense of working the machinery, what is the average daily amount saved by each engine? *Ans.* 104*l.* 2*s.* 4.8*d.*

---

[a] As the piston must move forward and back at each stroke, the distance passed through per minute is 160ft.

[b] Rice in the husks. [c] Partington.

5. An engine at the Wheal Hope mine, in Cornwall, works 3 pumps, the length of stroke of each being 8ft.; their pistons support and lift at each stroke, columns of water, whose joint weights are 27766 lb., and in the month of December, 1826, they made 261890 strokes.[a] Required the velocity per minute, and the total dynamical effect.[b]

Ans. Velocity 46.93ft.; effect 1303058.38 lb.

6. The working effect of 1 bushel of coals, or the number of pounds which could be raised 1ft. high by 1 bushel is called the *Duty* of the engine. Required the duty of the engine in the preceding example, the amount of coal consumed being 1242 bushels. *Ans.* 46838246.

7–10. A cubic foot of water makes 1689 c. ft. of steam, at the temperature of 212°.[c] Estimating the pressure of the atmosphere at 2120 lb.[a] on a square foot, what will be the dynamical effect of 1 lb. of each of the following kinds of fuel?[d]

Fuel.	Spec. gravity.	Weight to evaporate 1 c. ft. of water.	Dynamical effect of 1 lb. of fuel.
Anthracite coal	1500	6.534 lb.	548007 lb. 1ft.
Va. bituminous coal	1364	7.82 "	lb. "
Pa. " "	1262	9.37 "	lb. "
Dry pine wood	336	15.4 "	232512 lb. "

52. The Water-Wheel.

The pressure of water on any surface is equal to the weight of a column of water with the same base as the surface pressed, and of a height equivalent to the depth of the centre of gravity. If the surface is of any regular shape, the centre of gravity corresponds with the centre of the surface.

The actual velocity of water flowing from an orifice, or

---

[a] Moseley.

[b] The dynamical effect is found by multiplying the velocity by the weight.

[c] Daniell.

[d] W. R. Johnson.

falling, will generally be .6 or .7 of the theoretic velocity.[a] To obtain the theoretic velocity per second, find the number of feet in the perpendicular fall of the water, (or in the depth below the surface to the middle of the orifice,) extract the square root, and multiply by 8.018.

The effective force of a water-wheel, may be estimated at $\frac{2}{3}$ of the power applied to it, if the wheel is overshot, and at $\frac{1}{3}$ of the power if the wheel is undershot.[b] To determine the power of a stream of water, measure the breadth and depth of the stream, the height of fall, and the velocity per minute, all in feet; form the continued product of these four numbers, and divide by 792.[c] For undershot wheels take $\frac{1}{2}$ this result.

To calculate the power of machinery or wheelwork, multiply together the effective power, and the lengths of all the driving levers, (or the radii, circumferences, cogs or rounds of the driving wheels,) and divide by the continued product of the lengths of all the leading levers, (or the radii, &c., of the leading wheels.) For the velocity, multiply the velocity of the power by the lengths, radii, circumferences, cogs or rounds of the leading levers or wheels, and divide by the product of the like dimensions of all the driving levers or wheels.

The maximum effect will be produced in machinery of any kind, when the load, or resistance, is $\frac{4}{9}$ of the power, and when the velocity of the machinery at the point of action, is $\frac{1}{3}$ of the greatest velocity of the power.[d]

EXAMPLES.

1. What is the amount of pressure on a dam 75ft. by 12ft., the depth of water being 8ft.?[e]

Ans. 225000 lb.

---

[a] Nicholson. [b] Pilkington.

[c] $\frac{2}{3} \times \frac{D.B.H.V.}{33000} \times 62.5 = \frac{D.B.H.V.}{792}$.

[d] Brande. [e] The depth of the centre of gravity is 4ft.

2. What is the theoretical velocity of water flowing through an orifice, the centre of which is 6ft. 3in. below the surface? *Ans.* 20.045ft. per second.

3. Find the bottom pressure, and the entire pressure upon the bottom and sides of a cube, each side of which measures 8 feet. *Ans.* Bottom pressure, 32000 lb.
Entire " 96000 lb.

4. If the area of an orifice is $2\frac{1}{3}$ sq. ft., and the velocity of the water flowing through it is 15ft. per second, what will be the weight of the water discharged in 1 minute? *Ans.* 131250 lb.

5. A stream 12in. deep, and 22in. broad, moves with a velocity of 88ft. in 15″. Required its effective force, with a fall of 50ft. *Ans.* $40\frac{20}{27}$ horse power.

6. How many cuts are made per minute by the beater of a paper mill, which has 60 teeth, each of which passes by 24 cutters at every revolution, when there are 150 revolutions per minute? *Ans.* 216000 cuts.[a]

7. What force will be exerted at the distance of 2 feet from the centre of a millstone, by a water-wheel with a power of 1000 lb., the diameters of the driving wheels being 8ft., 2ft., and 1ft., and the diameters of the leading wheels being 4ft. and 3ft.? *Ans.* $666\frac{2}{3}$ lb.

8. If the power in the preceding example moved with a velocity of 12ft. per second, what would be the velocity of the millstone? *Ans.* 18ft.

9. What should be the area of the section of a canal, to deliver 90000 c. ft. per hour, the water moving with a velocity of 4ft. per second? *Ans.* $6\frac{1}{4}$ sq. ft.

10. Using 10 c. ft. per second, what time would be necessary to exhaust a pond, the area of which is 10 acres, and

[a] This rapid motion makes a coarse musical note, that can be heard at a great distance from the mill.—*Ure.*

the average depth 3 feet, the pond being supplied by a stream that furnishes 150 c. ft. per minute?

Ans. 48h. 24min.

53. Pumps.[a]

The power employed in working a pump is estimated by multiplying the number of pounds discharged per minute, by the number of feet that the water is raised above the reservoir.

The weight of water in a yard of pipe may be found very nearly by squaring the number of inches in the diameter of the pipe, and increasing the square by $\frac{1}{44}$ of itself.[b] The result will be the weight in pounds avoirdupois.

The number of ale gallons in a yard of pipe may be found very nearly by squaring the number of inches in the diameter, and dividing by 10.

In estimating the power necessary to overcome resistance in pumps, $\frac{1}{5}$ should be added for the friction of the water.

The diameter of the pipes should be at least as great as the diameter of the pump. If it is greater, the friction will be diminished.

EXAMPLES.

1. At the height of 225 ft. above the level of a reservoir, 250 ale gallons are discharged per minute. Required the power of the engine, the weight of one ale gallon of water being $10\frac{1}{5}$ lb.[c] *Ans.* 13.1 horse power.

2. Find the weight of water, and the number of ale gallons, in a pipe 15 rods long, and 3 inches in diameter.

Ans. $759\frac{3}{8}$ lb.; $74\frac{1}{4}$ gallons.

3. On the top of a hill 75 feet high, is a reservoir 40 feet

---

[a] Nicholson, Pilkington, Ferguson.

[b] $\frac{.7854 \times 3 \times 62.5}{144} = \frac{45}{44}$ very nearly.

[c] In all the examples of this section, an allowance of $\frac{1}{4}$ is made for the friction of the machinery. See Sect. 51.

square, and 12ft. deep. What power is necessary to fill the cistern in 45 minutes?

Ans. $\frac{40\times40\times12\times62.5\times75}{45\times44000}\times\frac{6}{5}=54\frac{6}{11}$ horse power.

4. What power is necessary to fill a cistern 30ft. long, 22ft. wide, and 10ft. deep, in 25 minutes, the water being raised 100ft.? *Ans.* 45 horse power.

5. What should be the diameter of the pump in each of the two preceding examples, if there are 40 strokes per minute, the length of the effective[a] stroke being 2ft.?

Ans. Ex. 3: $\frac{40\times40\times12}{45}=426\frac{2}{3}$ c. ft. per minute;

$426\frac{2}{3}\div(2\times40)=5\frac{1}{3}$ sq. ft. area of pump;

$\sqrt{5\frac{1}{3}\div.7854}=2.6$ft. diameter of the pump.

Ex. 4: Diameter 2.05ft.

6. A town of 25000 inhabitants, is to be supplied with water from a river 200 feet below the proposed reservoir. Estimating the average daily consumption at 9 ale gallons for each individual, what must be the power of an engine working 10 hours per day, and what will be the size of the pump, making 30 strokes per minute, the effective stroke being 3ft.? *Ans.* Engine 17.47 horse power.
Area of pump 98 sq. in. nearly.
Diam. of pump 11.17in.

XIII. THE LABORATORY.

54. Chemical Combinations.[b]

In chemical compounds, the following curious facts have been observed.

---

[a] An allowance is made from the stroke of the piston rod for the escape of water through the valves. Pilkington states this allowance at 3 inches.

[b] Draper.

1. The constitution of a compound body is always the same. Thus it has been found that 9 grains of water contain 8 grains of oxygen and 1 grain of hydrogen; and however often the analysis is repeated, this proportion is found to be invariable.

2. The proportions in which bodies are disposed to unite with each other, can always be represented by certain numbers. Thus water is composed of one atom of oxygen and one atom of hydrogen, and as the oxygen atom is 8 times as heavy as that of hydrogen, it follows that in 9 parts by weight, of water, there are 8 parts of oxygen and 1 of hydrogen.

3. If two substances unite with each other in more proportions than one, those proportions bear a simple arithmetical relation to each other; thus 14 grains of nitrogen will successively unite with 8, 16, 24, 32, 40 grains of oxygen, forming five different compounds, which contain respectively 1 atom, 2 atoms, 3 atoms, 4 atoms, 5 atoms of oxygen, and 1 atom of nitrogen.

There are three ways in which the composition of a substance may be expressed: 1, by atom; 2, by weight; 3, by volume. Thus water is composed, by atom, of oxygen 1, and hydrogen 1; by weight, of hydrogen 1, and oxygen 8; and by volume, of hydrogen 2, and oxygen 1.

Elementary bodies are represented in chemistry by letters or *symbols*. A list of the elements and symbols is given in the next section.

A symbolic letter standing alone, represents one atom of the element. Thus C denotes one atom of carbon; O, one atom of oxygen.

To denote more than one atom, we may either repeat the symbol, or a figure may be placed either before or after the symbol: thus OOO, 3O, or O_3 would each represent 3 atoms of oxygen. The latter method is usually adopted. Nitric acid, which is composed of 1 atom of nitrogen and 5 of oxygen, is denoted by NO_5.

To denote a compound formed of several compounds, we employ one or more commas, thus: SO_3,HO, which is the formula of strong oil of vitriol.

The terms, *combining proportion* and *chemical equivalent*, have the same meaning as *atomic weight*.

55. Table of Chemical Equivalents.[a]

Names and Symbols of Elements.		Hydrogen = 1.	Names and Symbols of Elements.		Hydrogen = 1.
Aluminum . .	Al	13.72	Manganese .	Mn	27.72
Antimony . .	Sb[b]	129.24	Mercury . .	Hg[h]	101.43
Arsenic . . .	As	75.34	Molybdenum .	Mo	47.96
Barium . . .	Ba	68.66	Nickel . . .	Nk	29.62
Bismuth . . .	Bi	71.07	Nitrogen . .	N	14.19
Boron . . .	B	10.91	Osmium . .	Os	99.72
Bromine . . .	Br	78.39	Oxygen . . .	O	8.01
Cadmium . .	Cd	55.83	Palladium . .	Pd	53.36
Calcium . . .	Ca	20.52	Phosphorus .	P	31.44
Carbon . . .	C	6.04	Platinum . .	Pt	98.84
Cerium . . .	Ce	46.05	Potassium . .	K[i]	39.26
Chlorine . . .	Cl	35.47	Rhodium . .	R	52.2
Chromium . .	Cr	28.19	Selenium . .	Se	39.63
Cobalt . . .	Co	29.57	Silicon . . .	Si	22.22
Columbium . .	Ta[c]	184.9	Silver . . .	Ag[k]	108.31
Copper . . .	Cu[d]	31.71	Sodium . . .	Na[l]	23.31
Didymium . .	D	?	Strontium . .	Sr	43.85
Erbium . . .	E	?	Sulphur . .	S	16.12
Fluorine . . .	Fl	18.74	Tellurium . .	Te	64.25
Glucinum . .	G	26.54	Terbium . .	Tr	?
Gold	Au[e]	199.2	Thorium . .	Th	59.83
Hydrogen . .	H	1.	Tin	Sn[m]	58.92
Iodine . . .	I	126.57	Titanium . .	Ti	24.33
Iridium . . .	Ir	98.84	Tungsten . .	W[n]	94.8
Iron	Fe[f]	27.18	Vanadium . .	V	68.66
Lantanum . .	La	?	Uranium . .	U	217.2
Lead	Pb[g]	103.73	Yttrium . . .	Y	32.25
Lithium . . .	Li	6.44	Zinc	Zn	32.31
Magnesium . .	Ma	12.89	Zirconium . .	Z	33.67

[a] Compiled from Parnell and Draper. [b] Stibium. [c] Tantalum. [d] Cuprum. [e] Aurum. [f] Ferrum. [g] Plumbum. [h] Hydrargyrum. [i] Kalium. [k] Argentum. [l] Natrium. [m] Stannum. [n] Tungsten or Wolfram.

If compounds are united by a feeble affinity, the sign + is sometimes used. Thus the composition of sulphuric acid may be indicated by SO_3, or by SO_2+O, the latter formula showing that one of the atoms of oxygen is held by a feebler affinity than the other two.

1–58. Find the value of each of the foregoing equivalents, assuming oxygen=100. *Ans.* Al 171.28.
Sb 1613.48, &c. &c.

56. EXAMPLES FOR THE PUPIL.

1–33. Find the atomic weights[a] of each of the following Acids:[b] Acetic, $C_4H_3O_3$; Arsenic, AsO_5; Arsenious, AsO_3; Benzoic, $C_{14}H_5O_3$; Boracic, BO_3; Bromic, BrO_5; Carbonic, CO_2; Chloric, ClO_5; Chromic, CrO_3; Citric, $C_4H_2O_4$; Formic, C_2HO_3; Gallic, C_7HO_3; Hydriodic, HI; Hydrobromic, HBr; Hydrochloric, HCl; Hydrocyanic, $H+C_2N$; Hydrofluoric, HFl; Hydrosulphuric, HS; Hypermanganic, Mn_2O_7; Hyposulphurous, S_2O_2; Hyposulphuric, S_2O_5; Iodic, IO_5; Lactic, $C_6H_5O_5$; Malic, $C_8H_4O_8$; Manganic, MnO_3; Nitric, NO_5; Oxalic, C_2O_3; Phosphoric, PO_5; Silicic, SiO_3; Sulphuric, SO_3; Sulphurous, SO_2; Tannic, $C_{18}H_5O_9$; Tartaric, $C_8H_4O_{10}$.

34–58. Find the atomic weights of each of the following Bases: Alumina, Al_2O_3; Ammonia, NH_3; Oxide of Antimony, SbO_3; Barytes, BaO; Oxide of Chromium, Cr_2O_3; Oxide of Cobalt, CoO; Protoxide of Copper, CuO; Sub-

---

[a] The equivalent of a compound body is always equal to the sum of the equivalents of its constituents.—*Parnell.*

[b] "Compound bodies may, for the most part, be divided into three groups; acids, bases, and salts. By an acid, we mean a body having a sour taste, reddening vegetable blue colors, and neutralizing alkalies; by a base, a body which restores to blue the color reddened by an acid, and possessing the quality of neutralizing the properties of an acid; by a salt, the body arising from the union of an acid and a base. These definitions, however, are to be received with considerable limitation." —*Draper.*

oxide of Copper, Cu_2O; Peroxide of Iron, Fe_2O_3; Protoxide of Iron, FeO; Protoxide of Lead, PbO; Lime, CaO; Magnesia, MaO; Protoxide of Manganese, MnO; Oxide of Mercury, HgO; Suboxide of Mercury, Hg_2O; Oxide of Nickel, NiO; Oxide of Platinum, PtO; Potash, KO; Oxide of Silver, AgO; Soda, NaO; Strontian, SrO; Protoxide of Tin, SnO; Peroxide of Tin, SnO_2; Oxide of Zinc, ZnO.

59. Hydrochloric acid consists of equal volumes of chlorine and hydrogen, united without condensation. Required its specific gravity, the specific gravity of hydrogen being .069, and that of chlorine 2.47.[a] *Ans.* 1.2695.

60. If one volume of carbonic acid gas contains 1 volume of oxygen and 1 volume of carbon vapor, what is the specific gravity of carbon vapor, the specific gravity of carbonic acid being 1.5238, and that of oxygen being 1.1025?
Ans. .4213.

61. The vapor of alcohol is composed of 8 volumes of carbon, 12 volumes of hydrogen, and 2 volumes of oxygen, the whole being condensed into 4 volumes of vapor. Required its specific gravity; the specific gravity of carbon being .4213, that of hydrogen .069, and that of oxygen 1.1025. *Ans.* 1.60085.

62. According to the experiments of Despretz, 1oz. of carbon evolves, during its combustion, as much heat as would raise the temperature of 1oz. of water 14067°. How many pounds of water could be raised from the freezing point to the mean temperature of the human body, (98.3°,) by the 13.9oz. of carbon, which are daily converted into carbonic acid in the body of an adult?[b] *Ans.* 184.3 lb.

---

[a] In determining the specific gravity of gases, air is generally assumed as the standard, = 1.

[b] Liebig.

XIV. GENERAL ANALYSIS.

57. Remarks on the Solution of Questions.

In most treatises on arithmetic, after the fundamental rules have been taught, the various applications of those rules are arranged under different heads, such as Interest, Discount, Practice, Proportion, Profit and Loss, Fellowship, Bankruptcy, &c. This division of the subject is convenient for beginners, but the expert arithmetician must be entirely independent of formal rules; he must be able, by analyzing any question that is proposed to him, to determine what operations are necessary for its solution. Accountants, men of business, and nearly all who are required to make frequent calculations, perform most of their work by analytical processes, and not by the rules that they learned at school.

Mistakes are rarely made in determining the proper mode to be pursued when numbers are to be merely added or subtracted; but when multiplication or division is required, care is often necessary to avoid multiplying when we ought to divide, or dividing when we ought to multiply. Practice and careful attention to the conditions of the question will generally remove all difficulty. Much unnecessary labor may often be avoided, by stating the operations as we proceed, and not performing any of the multiplications or divisions until the statement is complete. Thus, if it were required to multiply 27 by 19, to divide the product by 2 times 171, and to multiply the quotient by $\frac{2}{3}$ of $\frac{7}{8}$ of $\frac{4}{9}$ of 594, the readiest method of arriving at the result, would be to express the operations thus, $\frac{27\times 19}{2\times 171}\times\frac{2\times 7\times 4\times 594}{3\times 8\times 9\times 1}$; then by cancelling, we readily determine the answer, which is 7×33 or 231.

One of the two following methods, will furnish the answer to nearly all questions that admit of an analytical solution.

1st. In the majority of cases, we should endeavor to find from the terms that are given, the value which would correspond to ONE of each of the terms concerning which an answer is required. Having found the answer for ONE, we can easily determine it for the REQUIRED NUMBER in each of the terms of demand.

2d. We are sometimes required to reason from a result to its origin. In such cases it is generally best to *reverse all the operations by which the result was obtained.* Examples illustrating both of these methods, will be found in the two following sections.

Whenever fractions or decimals are involved in the conditions of a question, the same operations must be performed on them, that would be required if their place were occupied by whole numbers. Therefore, if we are ever at a loss whether to multiply or divide by a fraction, all difficulty may be removed by substituting a small integral number, and considering what would then be required.

58. Examples illustrating the First Method.

The pupil should be required to repeat the analysis of all the following examples, and encouraged to give a solution of his own, whenever a different one occurs to him.

1. What is the interest of $635.50 for 3y. 8mo. 24dy. at 6 per cent.?

The interest of ONE dollar for ONE year, is $.06. For 3y. 8mo. 24dy. it would be $3\frac{11}{15}$ times as much, or $\$\frac{56 \times .06}{15}$. The interest of $635.50 will be 635.50 times as much as the interest for ONE dollar, or $\$\frac{635.50 \times 56 \times .06}{15}$[a] $= \$$?

[a] The object of these examples is merely to show how the general principles of analysis can be applied to all classes of questions. All the short processes and contractions that the pupil may have learned, and can retain in his memory, he may be allowed to employ.

2. How much per cent. is gained by selling at 90cts. a pound, tea that cost 75 cents?

A gain of ONE per cent. on a pound, would be \$.0075. The actual gain was 15 cents, which is (.15 ÷ .0075) times ONE per cent., or per cent.?

3. If \$400 at simple interest amounts to \$440.50 in 2y. 3mo., what is the rate per cent.?

The interest is \$40.50. The interest of ONE dollar for ONE year at ONE per cent., would be \$.01. The interest of \$400 for $2\frac{1}{4}$y. at ONE per cent., $= \frac{400}{1} \times \frac{9}{4} \times \frac{.01}{1}$. The NUMBER of per cent. =

$\frac{40.50}{1} \times \frac{1}{400} \times \frac{4}{9} \times \frac{1}{.01} =$ per cent.?

Give an analytical solution of the above question, by first finding the interest for one year.

4. In what time will \$900 amount to \$1044 at 6 per cent., simple interest?

We wish to find in how many years \$900 will yield \$144 interest, at 6 per cent. In ONE year it would yield \$54. The NUMBER of years is therefore 144 ÷ 54 = yr.?

5. If a man's property yields $5\frac{1}{2}$ per cent. simple interest, and his annual income is \$$1093.12\frac{1}{2}$, what is he worth?

We wish to find how many dollars he is worth. If he was worth ONE dollar, his income would be \$$.05\frac{1}{2}$. He is therefore worth as many times ONE dollar, as are equivalent to $1093.12\frac{1}{2} \div .05\frac{1}{2}$, or \$?

6. When money yields but 5 per cent. interest, what is the present worth of \$5316.84 due in 1yr. 7mo. 6dy.?

We wish to find how many dollars will amount to \$5316.84 in $1\frac{3}{5}$ yr. at 5 per cent. ONE dollar would amount to \$1.08. The number of dollars required, is therefore as many times ONE dollar, as are equivalent to \$5316.84 ÷ \$1.08, or \$?

7. What must be the face of a note at 90 days, to be discounted at Bank, at 6 per cent., to yield \$500?

We wish to find how many dollars would yield \$500, and we

first find that ONE dollar would yield \$.9845. The number of dollars required is therefore as many times ONE dollar, as are equivalent to 500 ÷ .9845 = \$?

8. If I gain 25 per cent. on the original cost, by selling merchandise for \$1718.75, how much did it cost me?

How many dollars did it cost? If it had cost ONE dollar, to gain 25 per cent. it must have been sold for \$1.25. It must therefore have cost (1718.75 ÷ 1.25) times ONE dollar, or \$?

9. When exchange on England is at a premium of $8\frac{1}{2}$ per cent., what is the value in sterling money, of \$112?

We cannot so easily find the value of \$1 in English Money, as the value of £1 in Federal Money.[a] If ONE £ = $\$1.08\frac{1}{2} \times \frac{40}{9}$, how many £ = \$112? As many times ONE pound, as there are times $1.08\frac{1}{2} \times \frac{40}{9}$ in 112. $\frac{112}{1} \times \frac{2}{2.17} \times \frac{9}{40} = £$?

10. If 36 men, in $127\frac{1}{2}$ days of $13\frac{1}{2}$ hours, dig a trench $33\frac{3}{4}$ yd. long, $10\frac{1}{2}$ft. deep, and $15\frac{3}{5}$ft. wide, how many men in $7\frac{2}{7}$ days of $12\frac{8}{11}$ hours, will dig a similar trench $82\frac{8}{11}$yd. long, $7\frac{2}{5}$ft. deep, and 10ft. wide?

First, to find how many men in ONE day, of ONE hour, would dig a trench ONE yd. long, ONE ft. deep, and ONE ft. wide, we must multiply 36 by $127\frac{1}{2}$ and by $13\frac{1}{2}$, and divide by $33\frac{3}{4}$, $10\frac{1}{2}$, and $15\frac{3}{5}$, or $\frac{36}{1} \times \frac{255}{2} \times \frac{27}{2} \times \frac{4}{135} \times \frac{2}{21} \times \frac{5}{78}$. Second, in $7\frac{2}{7}$ days it will not require so many men as in one day; working $12\frac{8}{11}$ hours a day, will not require so many men as working one hour a day, &c.; we must therefore divide our first result by $7\frac{2}{7}$ and $12\frac{8}{11}$, and multiply by $82\frac{8}{11}$, $7\frac{2}{5}$, and 10, which will give $\frac{36}{1} \times \frac{255}{2} \times \frac{27}{2} \times \frac{4}{135} \times \frac{2}{21} \times \frac{5}{78} \times \frac{7}{51} \times \frac{11}{140} \times \frac{910}{11} \times \frac{37}{5} \times \frac{10}{1}$ men.

11. How many square yards in a room 28ft. 4in. wide, and 37ft. 6in. long?

The answer is to be in yards, and therefore the dimensions must be reduced to yards. The width is $9\frac{4}{9}$yd., and the length $12\frac{1}{2}$yd. If the room was ONE yard long and ONE yard wide, it would con-

---

[a] Some attention will often be required, to determine which of the quantities should be taken to find the value of ONE.

tain one square yard. But the width is $9\frac{4}{9}$yd., instead of 1yd.; and the length $12\frac{1}{2}$yd., instead of 1yd. We must therefore multiply $1 \times \frac{85}{9} \times \frac{25}{2} =$ sq. yd.?

12. If a man receives $30 for building 8 rods of wall, and he can purchase 3 barrels of flour for $14, and 5cwt. of sugar for 6 barrels of flour, and 126 lb. of tea for 10cwt. of sugar, how many pounds of tea can he purchase by building 24 rods of wall?

We cannot find directly how many pounds of tea he can purchase by ONE rod of wall, but for ONE rod he will receive $\$\frac{30}{8}$. For ONE dollar he can buy $\frac{3}{14}$ of a barrel of flour, and for $\$\frac{30}{8}$ he can buy $\frac{30}{8} \times \frac{3}{14}$ of a barrel. For ONE barrel of flour he can buy $\frac{5}{6}$cwt. of sugar, and for $\frac{30}{8} \times \frac{3}{14}$ bbl., he can buy $\frac{30}{8} \times \frac{3}{14} \times \frac{5}{6}$cwt. For ONE cwt. of sugar he can buy $\frac{126}{10}$ lb. of tea, and for $\frac{30}{8} \times \frac{3}{14} \times \frac{5}{6}$cwt. he can buy $\frac{30}{8} \times \frac{3}{14} \times \frac{5}{6} \times \frac{126}{10}$ lb., which is therefore equivalent to 1 rod of wall, and 24 rods will purchase $\frac{24}{1} \times \frac{30}{8} \times \frac{3}{14} \times \frac{5}{6} \times \frac{126}{10} =$ lb. of tea?

13. A.'s stock in a partnership is $450, B.'s $350, and C.'s $500. How must a loss of $169 be divided between them?

If $1300 loses $169, how much does ONE dollar lose? If ONE dollar loses $\$\frac{169}{1300}$, $450 will lose $\$\frac{450}{1} \times \frac{169}{1300} = \$$, $350 will lose $\$\frac{350}{1} \times \frac{169}{1300} = \$$, and $500 will lose $\$\frac{500}{1} \times \frac{169}{1300} = \$$?

14. Divide 650 into four parts, which shall be to each other in the proportion of $\frac{1}{2}$, $\frac{1}{3}$, $\frac{3}{4}$, and $\frac{7}{12}$.

If one part has $\frac{1}{2}$ of a share, another $\frac{1}{3}$ of a share, another $\frac{3}{4}$ of a share, and the other $\frac{7}{12}$ of a share, then the whole will be $2\frac{1}{6}$ shares. ONE share is therefore $650 \div 2\frac{1}{6}$. Then $\frac{1}{2}$ share $= \frac{1}{2} \times \frac{650}{1} \times \frac{6}{13} =$, $\frac{1}{3}$ share $= \frac{1}{3} \times \frac{650}{1} \times \frac{6}{13} =$, $\frac{3}{4}$ share $= \frac{3}{4} \times \frac{650}{1} \times \frac{6}{13} =$, and $\frac{7}{12}$ share $= \frac{7}{12} \times \frac{650}{1} \times \frac{6}{13} =$?

15. A bankrupt owes $15600, and his property is worth only $10600. How much can he pay on a debt of $450?

On ONE dollar he can pay $\$\frac{10600}{15600} = \$\frac{53}{78}$, and on \$450 he can pay $\$\frac{450}{1} \times \frac{53}{78} = \$$?

16. A. invests a certain sum for a certain time, B. twice as much for $\frac{1}{3}$ of the time, and C. three times as much for $\frac{2}{5}$ of the time. How should they share the gain, which was \$559 ?

Find the value for ONE sum for ONE time, which will be A.'s share. 2 sums for $\frac{1}{3}$ of a time $= \frac{2}{3}$ sum for ONE time = B.'s share. 3 sums for $\frac{2}{5}$ of a time $= 1\frac{1}{5}$ sums for ONE time = C.'s share. Then the whole is equivalent to $1 + \frac{2}{3} + 1\frac{1}{5} = 2\frac{13}{15}$ sums for ONE time. ONE sum for ONE time $= \$559 \div 2\frac{13}{15} = \$$, A.'s share; 2 sums for $\frac{1}{3}$ of a time, or $\frac{2}{3}$ of a sum for ONE time $= \$\frac{2}{3} \times \frac{559}{1} \times \frac{15}{43} =$ \$, B.'s share; 3 sums for $\frac{2}{5}$ time, or $1\frac{1}{5}$ sums for ONE time $=$ $\$\frac{6}{5} \times \frac{559}{1} \times \frac{15}{43} = \$$, C.'s share.

17. A goldsmith mixed 3 lb. of gold 22 carats fine, 5 lb. 20 carats fine, 8 lb. 24 carats fine, and 4 lb. of alloy. What was the fineness of the mixture ?

We wish to find how many carats there are in ONE pound. There are 20 lb. in the whole, containing carats, and therefore in each pound there are $\div 20 =$ carats ?

18. Find the equated time for the payment of \$500 due in 3 months, and \$700 due in 5 months.

The question is, in how many months should \$1200 be paid, and we first find how many dollars could be used for ONE month, to be equivalent to the two debts. If \$5000 could be used ONE month, how many months could \$1200 be used ?

19. A rectangular piece of ground contains 5 acres, and the width is 20 rods. Required the length.

The field contains 800 square rods, and we wish to find how many rods long it is. If it were only ONE rod long and 20 rods wide, it would contain 20 square rods. But as it contains 800 rods, it must be $\frac{800}{20}$ times one rod long, = rods ?

20. A stick of hewn timber 11in. wide and 10in. thick, contains 40 cubic ft. What is its length ?

How many feet long? If it were ONE ft. long, $\frac{11}{12}$ft. wide, and $\frac{5}{6}$ft. thick, it would contain $1 \times \frac{11}{12} \times \frac{5}{6} = \frac{55}{72}$ cubic ft. But as it contains 40 c. ft. it must be $(40 \div \frac{55}{72})$ times ONE ft. long, = ft.?

21. If A. can do $\frac{1}{2}$ of a piece of work in 5 days, B. can do $\frac{1}{3}$ of it in 4 days, and C. can do $\frac{1}{6}$ of it in 2 days, in how many days can they do the whole by working together?

In ONE day they can do $\frac{1}{10} + \frac{1}{12} + \frac{1}{12}$ of the 1 piece of work. Then they can do the 1 piece of work in $1 \div (\frac{1}{10} + \frac{1}{12} + \frac{1}{12}) =$ days?

59. Examples illustrating the Second Method.

1. The greater of two numbers is $5\frac{1}{2}$ times the less, and the sum of the numbers is 52. What are the numbers?

52 is produced by adding the less number to $5\frac{1}{2}$ times the less, and is therefore $6\frac{1}{2}$ times the less.

2. Divide the number 582 into four such parts that the second may be twice the first, the third 21 more than the second, and the fourth 54 more than the first.

582 is produced by adding 21 and 54 to 6 times the first number. Therefore if we subtract 75 from 582, the remainder will be 6 times the first.

3. A farmer bought some horses, cows, and calves for \$1250, giving \$50 apiece for the horses, \$23 apiece for the cows, and \$9 apiece for the calves, and there were three times as many calves as cows, and half as many horses as calves. How many were there of each?

If there had been 2 cows, 6 calves, and 3 horses, they would have cost \$250. But he paid \$1250, then how many times could he repeat the purchase of 2 cows, 6 calves, and 3 horses?

4. If from $5\frac{1}{2}$ times a certain number $8\frac{1}{3}$ be subtracted, $12\frac{1}{2}$ added to the remainder, and the sum divided by $6\frac{1}{4}$, the quotient will be 30. What is the number?

$(30 \times 6\frac{1}{4} - 12\frac{1}{2} + 8\frac{1}{3}) \div 5\frac{1}{2} =$?

5. Five-eighths of a certain number exceeds $\frac{1}{3}$ of it by 21. What is the number?

$\frac{5}{8} - \frac{1}{3} = \frac{7}{24}$. If 21 is $\frac{7}{24} \times$ the number, the number itself is $21 \div \frac{7}{24} =$?

6. What number is that from which if we deduct $\frac{3}{7}$ of itself, and $\frac{2}{9}$ of the remainder, there will be 18 left?

18 is $\frac{7}{9}$ of what is left after deducting $\frac{3}{7}$ of the number. Then the whole number is $\frac{9}{7}$ of $\frac{7}{4}$ of 18 = ?

7. If 30 per cent. is lost by selling shoes at 87½cts. per pair, at what price should they be sold to gain 10 per cent.?

If 30 per cent. is lost, they must be sold at 70 per cent. of the cost. To gain 10 per cent., they should be sold at 110 per cent. of the cost. $\frac{110}{70}$ of $.87\frac{1}{2} =$?

8. B. is 2 years older than A., C.'s age is 4 years more than the sum of A.'s and B.'s, and D.'s age, which is 48, is equal to the sum of the other three. What is the age of each?

48 is the sum of A.'s, B.'s, and C.'s. B.'s is 2 more than A.'s, and C.'s is 6 more than twice A.'s; the three ages are therefore 8 more than 4 times A.'s age.

60. MISCELLANEOUS EXAMPLES IN ANALYSIS.

1. What is the interest of $1872.88 for 7yr. 10mo. 15dy. at 7 per cent.? *Ans.* $1032.43.

2. Bought 18.75yd. of broadcloth, at $4 per yard, and sold the whole for $83.50. What did I gain, and how much per cent.? *Ans.* $8.50 = 11⅓ per cent.

3. At what rate per cent. will $421.50 amount to $674.40 in 6yr. 8mo.?

4. A note of $431 amounted, at its settlement, to $546.29¼. How long had it been on interest, the rate being 6 per cent.? *Ans.* 4y. 5mo. 15dy.

5. What is the face of a note which, at 7 per cent., will yield $111.65 interest in 3y. 8mo.? *Ans.* $435.

6. What is the present worth of $2000, due in 3y. 6mo., interest at 7 per cent.? *Ans.* $1606.43.

7. What must be the face of a note at 60 days, to be discounted at Bank at 7 per cent., to yield $375? *Ans.* $379.65.

8. A commission merchant received $2\frac{1}{2}$ per cent. for the sale of an invoice of merchandise. What was the amount of the invoice, the total amount of the sale and commission being $1666.24? *Ans.* $1625.60.

9. When exchange on England is at a premium of $9\frac{1}{4}$ per cent., what is the value in sterling money, of $137.75? *Ans.* 28*l.* 7*s.* $4\frac{3}{4}$*d.*

10. How many days of $8\frac{1}{2}$ hours, will 42 men require, to build a wall $98\frac{3}{4}$ft. long, $7\frac{1}{2}$ft. high, and $2\frac{3}{4}$ft. thick, if 63 men can build a wall $45\frac{1}{3}$ft. long, $6\frac{7}{12}$ft. high, and $3\frac{1}{4}$ft. thick, in 68 days of $11\frac{1}{3}$ hours? *Ans.* 297 days.

11. Determine from the following table, the degree on each thermometer-scale, that corresponds to 0° of Fahrenheit.

Name of Thermometer.	Where used.	Freezing point.	Boiling point.
Fahrenheit's . . .	Great Britain & United States.	+32°[a]	+212°
Centigrade, or Celsius's	Sweden and France.	0°	+100°
Reaumur's	Germany, Italy, Spain & France.	0°	+80°
Russian, or Delisle's	Russia.	—150°	0°

Ans. Centigrade, $17\frac{7}{9}$° below 0.
Reaumur's, $14\frac{2}{9}$° " 0.
Russian, $176\frac{2}{3}$° " 0.

12. How many square yards in a room 19ft. 6in. wide, and 34ft. 8in. long? *Ans.* $75\frac{1}{9}$ sq. yd.

---

[a] The degrees above zero are indicated by the sign +; the degrees below zero by the sign —.

13. If 33 copecks are equal to 5 English pence, 11 English pence are equal to 3 piasters, 13 piasters are equal to 1 florin, and 5 florins are equal to 29 francs, how many francs are equal to 11000 copecks?

Ans. $202\frac{114}{143}$ francs.

14. A. contributed $380 to an adventure, B. $420, C. $500, and D. $700. What was each man's share of the gain, which was $900? *Ans.* A.'s share $171; B.'s $189; C.'s $225; D.'s $315.

15. Divide 7500 into 5 parts, in the proportions of $\frac{1}{2}$, $\frac{1}{3}$, $\frac{1}{4}$, $\frac{1}{5}$, $\frac{1}{6}$. *Ans.* $2586\frac{18}{87}$; $1724\frac{12}{87}$; $1293\frac{9}{87}$; $1034\frac{42}{87}$; $862\frac{6}{87}$.

16. An echo on the north side of Shipley Church, in Sussex, England, repeats 21 syllables.[a] If the speaker utters 3 syllables a second, at what distance from the echo does he stand, the velocity of sound being 1090ft. per second?

Ans. 3815ft.

17. A man failing in trade owed $75000, to meet which he had property valued at $14500. How much can he pay A., who is a creditor for $10000, B., who is a creditor for $3750, and C., who is a creditor for $12362.50?

Ans. A. 1933\frac{1}{3}$; B. $725; C. $2390.08.

18. The amount contributed by the United States in 1847, for the relief of Ireland and Scotland, has been estimated at $591313.29.[b] In what time would 10 men-of-war consume the same amount, supposing each vessel to have 3 officers, 20 midshipmen, and 1000 sailors; estimating the wages and rations of each officer at $30, of each midshipman at $20, and of each sailor at $16 per month?

Ans. 3mo. 17.57+dy.

19. A farmer mixed 18$\frac{1}{2}$ bushels of wheat, at $1.00 per bushel; 16$\frac{3}{4}$bu. at 1.12\frac{1}{2}$ per bushel; 13$\frac{5}{8}$bu. of barley, at

---

[a] Pierce. [b] Am. Almanac, 1848.

62½cts. per bushel, and 10bu. of oats, at 37½cts. per bushel. What was the mixture worth per peck? *Ans.* \$0.21+.

20. At what temperature does the mercury indicate the same degree on Fahrenheit's and on the Centigrade scale? [See Ex. 11.] *Ans.* —40°.[a]

21. Manson & Hill, of Liverpool, have given to Thomas Morton & Co., of New York, a note of £225 10s., payable in 60 days; one of £196, payable in 60 days; one of £218 7s. 6d., payable in 90 days; and one of £300, payable in 120 days. At what time may the notes all be equitably cancelled by a single payment of £939 17s. 6d.? *Ans.* 86 days.

22. The length of a floor is 37ft. 6in., and the area is 93¾ sq. yd. What is the width? *Ans.* 22ft. 6in.

23. A stick of hewn timber 1ft. 2in. wide and 9in. thick, contains 60⅔ solid feet. What is its length?

Ans. 69ft. 4in.

24. A grain of musk is said to be capable of perfuming for several years, a chamber 12ft. square, without sensible diminution of volume or weight.[b] If the chamber is 8ft. high, and constantly contains an average of 1 particle to every cubic tenth of an inch, how many particles must there be in the grain, supposing it to have lost $\frac{1}{1000}$ of its weight after the air has been changed 5000 times?

Ans. 9953280000000000.

25. A father's age is 6⅗ times his son's age, and the sum of their ages is 34yr. 2mo. 12dy. What is the age of each?

26. Divide \$5000 into four such parts, that the first may be twice the second, the third \$50 less than ½ the second, and the fourth \$100 more than the sum of the first and third.

27. At what temperature does the mercury indicate the

---

[a] If 100° gain 80°, how many degrees will gain 32°?

[b] Moseley.

same degree on Fahrenheit's and on Reaumur's scale? [See Ex. 11.] *Ans.* $-25.6°$.

28. A farmer hired a certain number of boys, and twice as many men, agreeing to pay each man 75 cents a day, and each boy 25 cents. The daily wages of the whole amounted to $5.25. How many were there of each?

Ans. 3 boys; 6 men.

29. A man's age is such that if it be multiplied by 3, and if $\frac{2}{7}$ of the product be tripled, $\frac{2}{9}$ of the result will be 16. Required his age. *Ans.* 28.

30. What is that number, $\frac{3}{7}$ of $\frac{5}{9}$ of which exceeds $\frac{1\frac{4}{7}}{16\frac{1}{2}}$ of itself by 21? *Ans.* 147.

31. At what temperature would the mercury indicate the same degree on the Centigrade and Russian scales? On the Russian and German scales? [See Ex. 11.]

Ans. $+300°$; $+171\frac{3}{7}°$.

32. What number is that, from which if we deduct $\frac{2}{3}$ of $\frac{12}{17}$ of itself, and $\frac{5}{6}$ of $\frac{1}{2}$ of the remainder, there will be $10\frac{1}{2}$ left? *Ans.* 34.

33. If 18 per cent. is lost by selling merchandise at $2050, at what price should it have been sold to gain 10 per cent.?—to gain 25 per cent.?—to lose 10 per cent.?

3d *Ans.* $2250.

34. B. is 5 years older than A.; C.'s age is 5 years more than the sum of A.'s and B.'s; and D.'s age, which is 55, is equal to the sum of the other three. What is the age of each? *Ans.* A. 10; B. 15; C. 30.

35. In examining a piece of charcoal through a microscope, Dr. Hook counted 150 pores in $\frac{1}{16}$ of an inch.[a] At this rate how many pores would there be in one square inch of surface? *Ans.* 5760000.

---

[a] Moseley.

36. A. and B. can do $\frac{11}{15}$ of a piece of work in 1 day; B. and C. can do $\frac{9}{10}$ of it; A. and C. can do $\frac{5}{6}$ of it in the same time. In what time will they all do it working together?

By adding $\frac{11}{15}$, $\frac{9}{10}$, and $\frac{5}{6}$, and dividing the sum by 2, we find the part that they will all do in 1 day.

Ans. $\frac{30}{37}$ of a day.

37. Three men traded in partnership. A. contributed $1500, B. $2250, and C. the remainder. The whole gain was $2700, of which C. received $1200. How much did C. contribute, and what did A. and B. gain?

Ans. C. contributed $3000; A. gained $600; B. gained $900.

38. An estate of $15000 is to be divided among three persons; A. is to receive 5\frac{1}{3}$ as often as B. receives 4\frac{1}{2}$, and B. is to receive 8\frac{1}{2}$ as often as C. receives 4\frac{1}{4}$. What is the share of each? *Ans.* A. $6620.69; B. $5586.21; C. $2793.10.

39. A cistern has three pipes; the first can fill it in $\frac{1}{2}$ an hour, the second can fill it in $\frac{1}{3}$ of an hour, and the third can empty it in an hour. In what time will the cistern be filled, if they all run together? *Ans.* 15 minutes.

40. At what temperature is the mercury as many degrees above zero of Fahrenheit, as it is below zero of the Centigrade?[a] [See Ex. 11.] *Ans.* $11\frac{3}{7}$°.

41. A bathing-tub that holds 147 gallons, is filled by a pipe that brings 14 gallons in 9 minutes, and emptied by a pipe that discharges 40 gallons in 30 minutes. Both pipes having been left open for 3 hours, it is required to find in what time the tub will be filled if the discharging pipe is closed? *Ans.* 1h. $8\frac{11}{14}$ minutes.

42. What sum of money will amount to $1500, in 5 years, at 5 per cent. simple interest? *Ans.* $1200.

---

[a] If the Centigrade moves 100° when the sum of the motions is 280°, how much will it move when the sum of the motions is 32°?

43. At what rate per cent., simple interest, will $700 amount to $1300 in 11 years? *Ans.* $7\frac{61}{77}$ per cent.

44. In what time will $1100 amount to $1750, at 6 per cent. simple interest? *Ans.* $9\frac{28}{33}$ years.

45. Supposing the weight of a molecule of light to be $\frac{1}{1000000}$ of a grain, what should be the velocity of a ball, weighing 1oz. avoirdupois, to have the same momentum.[a] *Ans.* 2.317+ ft. per second.

46. A laborer received $1.50 for every day he worked, and lost 50 cents every day he was idle. He worked twice as many days as he was idle, and at the end of the time he received $42.50. How many days did he work? *Ans.* 34 days.

47. A floor is laid with boards $1\frac{1}{4}$in. thick. How many feet are required, the room being 18ft. 6in. wide, and 20ft. 8in. long? *Ans.* $477\frac{11}{12}$ft.

48. How many clapboards would be required to cover an area of 1376 sq. ft., the clapboards being 4ft. long, and laid with 4 inches to the weather? *Ans.* 1032.

49. A cellar is to be made 40ft. long, and 25ft. wide. How many squares of earth must be removed, the depths at six different points being 8ft., 7ft. 6in., 7ft. 3in., 8ft. 4in., 8ft. 9in., and 5ft. 2in.?[b] *Ans.* $34\frac{13}{18}$ squares.

50. At what temperature is the mercury as many degrees above zero of Fahrenheit, as it is below zero of Reaumur? [See Ex. 11.] *Ans.* $9\frac{11}{13}°$.

51. What is the weight of a stone wall, the solid contents being 2760 c. ft., and the specific gravity 2500? *Ans.* 431250 lb.

---

[a] The momentum of a body is determined by multiplying the weight by the velocity. The velocity of light is 192000 miles per second. Such considerations are supposed to prove that light is without weight.

[b] Assume as the true depth of the cellar, the average of the six measured depths.

52. Wishing to estimate the contents of an irregular pile of wood, I take the dimensions in several places, and find that the average length is 34ft. 6in., the average breadth 27ft. 4in., and the average height 2ft. 3in. If it were piled regularly, I judge that it would occupy only $\frac{2}{3}$ of the space that it now does. Required its estimated contents.

Ans. 11 cords $6\frac{1}{2}$ c. ft.

53. How many cubic feet in a stack of hay, which is estimated to be equivalent to a cylinder 12ft. in diameter and 15ft. high? *Ans.* 1696 c. ft.

54. What must be the area of a roof that would fill a cistern holding 40 hogsheads, with a fall of $\frac{1}{4}$ inch of rain, the roof being of a true pitch?

The roof being of a true pitch, the area of the roof will be $1\frac{1}{2}$ times the area of the building, but the water that falls upon it will be the same as if the roof were flat.

Ans. 24255 sq. ft.

55. A chronometer usually vibrates 4 times in a second. How much must the length of each vibration be increased, in order that it may lose 1 second per day?

Ans. $\frac{1}{345600}$."

56. The area of a cistern is $38\frac{1}{2}$ sq. ft. How many gallons would fill it to the depth of 1ft.?

Ans. 288 gallons.

57. If a package of sugar weighs 6 lb. 2oz. in one scale of a balance, and 8 lb. in the other, what is its true weight?

Ans. 7 lb.[a]

58. A man performed a journey of 135 miles, going twice as far the second day as on the first, and three times as far the third day as on the second. How far did he travel each day?

---

[a] The true weight of any body may be found by a false balance, by weighing the body in each scale, and taking the mean proportional between the two weights. It may also be obtained by first balancing the body with shot or some other article, and then removing the body and placing weights in the scale till the equilibrium is restored.

59. A., B. and C. entered into partnership, contributing in the whole, $4833. B. paid twice as much as A., and C. paid twice as much as A. and B. How much did each contribute?

60. In a certain school of 70 scholars, three times as many study Arithmetic as study Latin, and twice as many learn to read, as study Arithmetic. How many are there in each study?

61. What degree of temperature would be indicated in the same manner on the scales of Fahrenheit and Delisle? [See Ex. 11.] *Ans.* $-1060°$.

62. An estate of $7000 was so divided that the widow received $500 more than the daughter, and the son $1100 more than the widow. What was the share of each?

63. Divide the number 97 into four such parts that the second may be twice the first, the third 7 more than the second, and the fourth 18 more than the first.

64. A thief travels at the rate of 6 miles an hour, and after he has been absent $5\frac{1}{2}$ hours, a constable starts in pursuit, at the rate of 9 miles an hour. In what time will the thief be overtaken?

65. A man when he was married, was three times as old as his wife, but after they had lived together 15 years, he was only twice as old. How old was each at the time of marriage? *Ans.* 45 years; 15 years.

66. At a certain election, the successful candidate had 163 votes more than his opponent, and the whole number of votes polled was 1125. How many did each receive?

67. What sum of money will yield $123.50 in 2 years, at 5 per cent. simple interest?

68. A gentleman distributed $1.95 among 3 beggars, giving the second 25 cents more than the first, and the

third twice as much as the second. How much did each receive?

69. If from three times a certain number 17 be subtracted, the remainder will be 112. What is the number?

70. At what temperature is the mercury as many degrees below zero of Delisle, as it is above zero of Reaumur?—of Celsius?—of Fahrenheit? [See Ex. 11.]

Ans. $52\frac{4}{23}°$; $60°$; $96\frac{4}{11}°$.

71. A merchant owes two of his creditors $1575, and he owes the second but $\frac{4}{5}$ as much as the first. What is the amount of each debt?

72. In a certain school $\frac{1}{2}$ the boys learn to read, $\frac{1}{5}$ learn to write, $\frac{1}{10}$ learn Algebra, $\frac{2}{15}$ learn drawing, and the remaining 4 study Latin. How many are there in the school?

73. One-third of a certain pole is painted green, and $\frac{2}{5}$ of it is painted white, the remainder, which is 8 feet, being in the ground. What is the length of the pole?

74. A man going to market, was met by another, who said: "Good morrow, neighbor, with your hundred geese." He replied: "I have not a hundred; but if I had as many more, and half as many more, and two geese and a half besides, I should have a hundred." How many had he?

75. A man bought 38 pounds of coffee and 95 pounds of sugar; he gave 2 cents per lb. more for the coffee than for the sugar, and the sugar cost twice as much as the coffee. What was the price of each per pound?

Ans. Sugar 8cts.; coffee 10cts.

76. If 4 men can saw 15 cords of oak in the same time that 5 men saw 14 cords of hickory, and if 3 men saw 18 cords of hickory in 3 days, by working 9 hours a day, how many hours a day must 7 men work, to saw 84 cords of oak in 6 days? *Ans.* $6\frac{18}{25}$ hours.

77. There are two such numbers, that if 21 be added to

the first, the sum will be 5 times the second, and if 21 be added to the second, the sum will be 3 times the first. What are the numbers?

$25\frac{1}{5}$ is the difference between $\frac{1}{5}$ of the first, and 3 times the first. The first, therefore, is 9, and the second 6.

78. Two stages are travelling towards each other, one at the rate of $5\frac{2}{7}$ miles an hour, the other $6\frac{1}{9}$ miles an hour. In what time will they meet if they are now $38\frac{3}{4}$ miles apart?

Ans. $3\frac{1149}{2872}$ hours.

79. Two men start from the same place and travel in opposite directions, one at the rate of $4\frac{1}{2}$ miles an hour, and the other $5\frac{3}{11}$ miles an hour. In what time will they be 100 miles apart? *Ans.* $10\frac{10}{43}$ hours.

80. With what velocity must a battering ram, weighing 2000 lb., be moved, to have the same momentum as a cannon ball, weighing 20 lb. and moving 1200ft. per second?

Ans. 12ft. per second.

81. There is a number to which if $\frac{1}{3}$, $\frac{1}{4}$, $\frac{1}{5}$, and $\frac{5}{6}$ of itself be added, the sum will be $\frac{2}{3}$ of $8\frac{7}{9}$ less than 16. What is the number? *Ans.* $3\frac{1241}{1413}$.

82. If A. can do $\frac{1}{2}$ of a piece of work in 5 days, B. can do $\frac{1}{3}$ of it in 4 days, and C. can do $\frac{1}{6}$ of it in 2 days, in what time will they all do $\frac{2}{3}$ of it by working together?

Ans. $2\frac{1}{2}$ days.

83. There are two men of equal ages, but if one was $5\frac{1}{2}$ years older, and the other $9\frac{1}{4}$ years younger, the former would be twice as old as the latter. Required their ages.

Ans. 24yr.

84. If a man can do $\frac{5}{11}$ of a piece of work in 3 days, and a boy can do $\frac{3}{7}$ of it in 5 days, how long will it take them both to do the whole? *Ans.* $4\frac{59}{274}$ days.

85. A hare starts 5 rods before a greyhound, and runs at the rate of 12 miles an hour. After running 48 seconds,

the hound starts in pursuit, and runs 20 miles an hour. In what time will the hare be overtaken?

Ans. 1m. $19\frac{1}{32}$ sec.

86. If 300 tiles that are 9in. long and 6in. wide, will pave a court-yard, how many tiles would be required that are 6in. long and 4in. wide? *Ans.* 675 tiles.

87. How many men will build a wall 240yd. long, 6ft. high, and 3ft. thick, in 8 days of 9 hours, if 7 men can build a wall 40yd. long, 4ft. high, and 2ft. thick, in 32 days of 7 hours? *Ans.* 294 men.

88. A wall which is to be built to the height of 27 feet, has been raised 9 feet in 6 days, by 12 men working 13 hours a day. How many men must be employed to finish it in 2 days, working only 12 hours a day?

Ans. 78 men.

89. Amsterdam exchanges with London, at 34 schillings 4 pfennings per £, and with Lisbon at 56 pfennings for 400 reas. What is the arbitrated exchange between London and Lisbon, by way of Amsterdam?

Ans. £1 $= 2 \oplus 942\frac{6}{7}$.

90. Find the number of vibrations in one second, of each of the rays of light, the velocity of light being 192000 miles per second,[a] and the lengths of the waves being, for red light, .0000256 of an inch.

For orange . .	.0000240	For blue . . .	.0000196
" yellow . .	.0000227	" indigo . .	.0000185
" green . .	.0000211	" violet . .	.0000174[b]

Ans. Red, 475200000000000 vibrations.
Violet, 699144827586207 "
&c. &c. &c.

91. The force available for mechanical purposes in an adult man, is reckoned, in mechanics, equal to $\frac{1}{5}$ of his own weight,

---

[a] Herschel. [b] Draper.

which he can move during 8 hours, with a velocity of $2\frac{1}{2}$ft. per second.[a] What is the momentum[b] for the day's work of a man who weighs 160 lb.? *Ans.* 2304000.

XV. THE COUNTING-HOUSE.

61. PERCENTAGE.

1. THE term per cent. is an abbreviation of the Latin *per centum*, which signifies *by the hundred*. Any number of per cent. of a quantity is therefore equivalent to as many hundredths of that quantity. Thus 7 per cent. is .07; $4\frac{1}{2}$ per cent. is $.04\frac{1}{2}$ or .045; $18\frac{3}{4}$ per cent. is $.18\frac{3}{4}$ or .1875.

2. Per cent. should not be confounded with any of the denominations of Federal Money. Thus, 6 per cent. is not 6 cents, or 6 dollars, but simply $\frac{6}{100}$. 6 per cent. of 125 dollars, is $\frac{6}{100}$ of \$125 = \$7.50; but 6 per cent. of 125 apples is 7.50, or $7\frac{1}{2}$ apples, and 6 per cent. of 125 lb. is 7 lb. 8oz.

3. Any fraction may be reduced to per cent., either by reducing it to a decimal, and stopping the decimal at hundredths' place, or by multiplying the fraction by 100. Thus $\frac{7}{15}$ reduced to a decimal, gives $.46\frac{2}{3}$, or $46\frac{2}{3}$ per cent.; the same fraction reduced to hundredths by multiplying by 100, gives $\frac{700}{15}$ hundredths, or $\frac{700}{15}$ per cent., or $46\frac{2}{3}$ per cent.

4. To determine the value of any quantity when there is a specified gain or loss per cent., we may add or subtract the given percentage from $\frac{100}{100}$ or 1. Thus, if stock is sold

---

[a] Liebig.

[b] The momentum is obtained by multiplying the weight by the distance.

at 25 per cent. advance, it is sold for $\frac{125}{100}$, or 1.25 times its par value; if goods are sold at a loss of 18 per cent., they are sold for $\frac{82}{100}$ of their cost.

62. Problems in Percentage.

I. *To find the gain or loss per cent.*

Make the gain or loss the numerator, and the prime cost the denominator of a fraction, and multiply the resulting fraction by 100. [See Sect. 61, 3.]

a. To find what percentage must be gained on the selling price, to yield any desired profit per cent. on the cost.

Divide the desired percentage of gain, by 100 *plus* that percentage, and reduce the quotient to hundredths.

b. To find the percentage of profit on the cost of merchandise, the percentage gained on the selling price being known.

Divide the percentage gained on the selling price by 100 *minus* that percentage, and reduce the quotient to hundredths.

II. *To determine the value of a quantity, when the value of any percentage is known.*

Divide by the percentage expressed decimally.

Example.—If 2500 is 16 per cent. of a certain number, what is that number? Since .16 *of* a number = .16 *times* that number, 2500 must be .16 times the number sought. The answer, therefore, is 2500 ÷ .16 = 15625.

a. To find the amount which should be added to an insurance, to recover the amount of premium paid.

Divide the premium by the difference between the rate and 100 per cent.

III. *To find the result that any quantity will yield in percentage.*

Multiply the quantity by the result that one will yield.

a. To find the interest, or the amount, of any principal for any given time.

Find the interest, or the amount, of one dollar for the given time, and multiply by the number of dollars in the given principal. This rule will hold good, both for simple and compound interest.

To find the interest of $1 for any given time: To $\frac{1}{2}$ as many cents as there are months, add $\frac{1}{6}$ as many mills as there are remaining days, and the amount will be the interest at 6 per cent. For any other rate, take such part of the interest at 6 per cent. as may be requisite.

EXAMPLE.—What is the interest of $287.75 for 3y. 7mo. 23dy., at $5\frac{1}{2}$ per cent.? The interest of $1 at 6 per cent. $=\frac{1}{2}$ of 43cts. $+\frac{1}{6}$ of 23 mills, or $.218$\frac{5}{6}$. At $5\frac{1}{2}$ per cent. it will be $5\frac{1}{2}$ *sixths*, or $\frac{11}{12}$ as much, or $\frac{11}{12} \times \frac{1.313}{6}$. The interest of $287.75 = $287.75 $\times \frac{11}{12} \times \frac{1.313}{6} =$ $57.72.

b. To find the selling price, to make any proposed gain or loss per cent.

Multiply the prime cost by 1 with the percentage added which is to be gained, or the percentage subtracted which is to be lost.

c. To reduce Sterling to Federal Money.

Multiply the par value of £1, ($$\frac{40}{9}$,) by 1 + the premium. The product will be the exchange value of £1.

Multiply the exchange value of £1, by the number of pounds.

IV. *To find the quantity that will yield any given result in percentage.*

Divide by the result that ONE would yield.

EXAMPLE.—What principal will amount to $500, in 2y. 6mo., at 5 per cent.? $1 would amount to $1.125, and $500 \div 1.125 =$ 444\frac{4}{9}$.

a. To reduce Federal to Sterling Money.

Divide by the exchange value of £1.

b. To find the face of a note to be discounted at bank, in order to obtain any required sum.

Divide the sum required, by the amount that would be received by discounting $1. The quotient will be the number of dollars for which the note should be drawn.

c. To find the amount that a factor can lay out of a sum intrusted to him, and reserve a specified percentage for his commission.

Divide the sum intrusted to him, by 1 + the percentage which is allowed for his commission.

d. To find the prime cost, when the selling price and the gain or loss per cent. are known.

Divide the selling price by the value of 1 with the proposed gain or loss per cent.

e. To find the selling price so as to allow a discount for cash, and gain any proposed rate per cent.

Multiply the prime cost by 1 + the proposed gain per cent., and divide by 1 — the proposed discount per cent.

f. The gain or loss per cent. at any *given* price being known, to find the gain or loss per cent at any *proposed* price.

Multiply the percentage of the prime cost which corresponds to the *given* price, by the *proposed* price, and divide by the *given* price. The quotient, (reduced to hundredths,) will be the percentage of the prime cost which corresponds to the proposed price. The difference between this percentage and 100 per cent., will be the gain or loss per cent.

63. Examples in Percentage.

A. Examples illustrating § 61.

1. Find 25 per cent. of \$13.50; $7\frac{1}{2}$ per cent. of 2cwt. 3qr. 12 lb.; $16\frac{2}{3}$ per cent. of 252 miles; $4\frac{3}{4}$ per cent. of £60; 135 per cent. of 10mo.

2. How many per cent. are equivalent to $\frac{5}{7}$; to $\frac{2}{3}$; $\frac{7}{8}$; $\frac{15}{16}$; $\frac{3}{4}$; $\frac{8}{9}$; $\frac{37}{80}$; $\frac{19}{20}$; $\frac{4}{520}$; $\frac{278}{13}$?

3. What part of the original value is stock worth, when it is $15\frac{1}{2}$ per cent. below par? When it is at a premium of $8\frac{3}{4}$ per cent.? When it is at a discount of $12\frac{1}{2}$ per cent.?

4. How much was received for a farm, bought for \$1350, and sold at an advance of $33\frac{1}{3}$ per cent.? *Ans.* \$1800.

B. Examples under Problem I. § 62.

5. How much per cent. was lost on flour, which was bought at \$5, and sold at $\$4.62\frac{1}{2}$ per barrel?
Ans. $7\frac{1}{2}$ per cent.

6. What percentage must a merchant gain on the total amount of his sales, to be equivalent to a gain of 10 per cent. on the cost? *Ans.* $9\frac{1}{11}$ per cent.

7. A tradesman finds that his profits on a year's business amount to $16\frac{2}{3}$ per cent. of the sales. What percentage of the cost has he gained? *Ans.* 20 per cent

C. Examples under Problem II. § 62.

8. Twenty-seven and a half is 18 per cent. of what number? *Ans.* $152\frac{7}{9}$.

9. What sum should be insured to cover the amount paid for premium and policy, if I wish to insure $1800 on merchandise, at a premium of $\frac{3}{4}$ per cent., the charge for the policy being $1? *Ans.* $1814.61.

D. Examples under Problem III. § 62.

10. Find the interest of £27 7s. 6d. for 1y. 7mo. 18dy. at 5 per cent. *Ans.* £2 4s. $8\frac{1}{2}$d.

11. At what price should I sell broadcloth, which cost $$3\frac{3}{8}$ per yd., in order to gain $11\frac{1}{9}$ per cent.? *Ans.* $3.75.

12. At $9\frac{1}{2}$ per cent. premium, what is the value in U. S. currency, of £27 13s.? *Ans.* $134.56.

E. Examples under Problem IV. § 62.

13. What principal, at 7 per cent. compound interest, would yield an interest of $500 every third year? *Ans.* $2221.80.

14. At 9 per cent. premium, what will be the value in English money, of a Bill of Exchange for $1250.25? *Ans.* £258 1s. 7d.

15. Required the face of a note at 90 days, to yield $500 when discounted at bank. *Ans.* $507.87.

16. A factor receives 5 per cent. commission on the amount that he purchases. If I send him $1000, how much can he lay out, after reserving enough to pay his own commission? *Ans.* $952.38.

17. If I gain 15 per cent. by selling land at $402.50 per acre, what did the land cost per acre? *Ans.* $350.

18. At what price must I sell molasses that cost 25cts. per gallon, in order to gain 10 per cent., after discounting 5 per cent. for cash? *Ans.* $$.28\frac{18}{19}$ per gallon.

19. If $12\frac{1}{2}$ per cent. is gained by selling a house for $3825, what percentage would be gained or lost by selling it for $3230? *Ans.* 5 per cent. lost.

F. Miscellaneous Examples.

20. A bill of goods is purchased on 6 months credit, amounting to $175.75. How much should be paid in cash, at the time of purchase, if the buyer is allowed 5 per cent. for his money? *Ans.* $167.38.

21. At what rate, simple interest, will any principal be doubled in 12y. 6mo.?

22. In what time will $2700 amount to $3132, at 6 per cent. simple interest?

23. What sum of money at 6 per cent. compound interest, will amount to $2750, in 3y. 6mo.? *Ans.* $2241.70.

24. If a merchant receives his usual profit, by selling a quantity of sugar for £46 5s., how much must he raise the price, in order to allow a discount of $7\frac{1}{2}$ per cent.? *Ans.* £3 15s.

64. Percentage on Sterling Money.

In computing interest or discount on English money, for any length of time less than a year, it is customary to omit the shillings and pence in the principal, when they are less than ten shillings; but if they amount to ten shillings or more, they are considered as another pound.

If it is desired to compute the percentage *exactly*, it may be done either by reducing the shillings and pence to the fraction or decimal of a pound,[a] or by multiplying 1 per

[a] Shillings, pence, and farthings may be reduced to the decimal of a pound by inspection, as follows:—*Multiply the number of shillings by* 5, *and call the product hundredths. Reduce the pence and farthings to farthings, increasing their number by* 1, *when it exceeds* 12, *and by* 2, *when it exceeds* 36, *and call the result thousandths. The sum of these two values will be the decimal required.*

To reduce the decimal of a pound to shillings, pence and farthings,

cent. of each denomination, by the number of per cent. required. The latter method is generally the readiest.

EXAMPLE ILLUSTRATING EACH METHOD.

Find $3\frac{1}{2}$ per cent. of £480 10s. 3d.

First Method.	*Second Method.*
10s. 3d. = .5125£	s. d.
480.5125	1 per cent. = £4.80 .10 .03
.035	$3\frac{1}{2} = \frac{7}{2}$ 7
24025625	2)33.60 .70 .21
14415375	16.80 .35 .105
16.8179375	20
20	16.35
16.3587500	12
12	4.305
4.30500	4
4	1.220
1.220	*Ans.* £16 16s. $4\frac{1}{4}$d.
Ans. £16 16s. $4\frac{1}{4}$d.	

The legal rate of interest in England is 5 per cent. For computing interest at this rate, the following rules are convenient. In each case, the principal is supposed to be expressed in pounds, and parts of a pound.

1. Multiply the principal by the number of years, and the product will be the interest in shillings.

2. Multiply the principal by the number of months, and the product will be the interest in pence.

3. Multiply the principal by the number of days, and divide the product by 30; the quotient will be the interest in pence.

For any other rate than 5 per cent., first compute the

Multiply the number of tenths by 2, and the product will be shillings. From the remainder of the decimal subtract $\frac{1}{24}$ of itself, and the figures which stand in the hundredths' and thousandths' places will be farthings.

interest at 5 per cent., and multiply the result by .2 × the number of per cent. For 3 per cent., multiply by .6; for 4½ per cent., multiply by .9; &c. &c.

EXAMPLES.

1. What is the interest of £487 10s. 8d., from March 4th, to Dec. 17, at 6 per cent.?

Mercantile Ans. £23 0s. 4¼d.
Correct Ans. £22 19s. 11d.

2. What per cent. is gained by selling, at £1 10s., velvet that cost £1 2s. 6d., per yard? *Ans.* 33⅓ per cent.

3. How much was received for 25 shares of stock, sold at a premium of 8¾ per cent., the par value being £50 per share? *Ans.* £1359 7s. 6d.

4. What was the interest of six India Bonds, of £100 each, at 3½ per cent., calculating from Sept. 30, 1849; the Bonds having been sold Jan. 15, 1850?

Ans. £6 2s. 6d.

5. What amount of 4 per cent. stock, will yield an income of £150 per annum? *Ans.* £3750.

65. BANKING.

In computing interest at Bank, the time is usually determined in days. When the rate is 6 per cent., the interest is found by multiplying the principal by as many thousandths as are equivalent to the number of days, and dividing the product by 6. Thus the Bank interest of $175.50, for 63 days, is $\$175.50 \times .063 \div 6 = \$1.84$.

For any other rate than 6 per cent., we may first compute the interest at 6 per cent., and add or subtract such part as may be required. For 4 per cent., subtract ⅓ of the interest at 6 per cent.; for 4½ per cent. subtract ¼; for 5 per cent. subtract ⅙; for 7 per cent. add ⅙, and so on.

If a note is given, or a bill drawn for any number of months, calendar months are always understood. A note at 4 months, dated on the 29th, 30th, or 31st of October, would expire on the last day of February, and would be legally due on the 3d of March. The 3d of March is, therefore, a heavy day at bank, as in leap years there are 3 days' payments, and in common years 4 days' payments, which fall due on that day. If either the 3d or 4th of March, in any year except leap year, falls upon Sunday, there will be 5 days' payments falling due on the Saturday previous.

There are two modes of estimating the time that elapses between different dates. The first, is by compound subtraction, which is the method almost invariably adopted in computing interest on notes, payable on demand. The second, is by determining the number of entire calendar months, and then finding how many days are left. This mode is adopted in many counting-houses, and in all banks. Thus, from Oct. 27th, 1850, to March 15th, 1853, would be, according to the 1st method, 2y. 4mo. 18dy.
" " " 2d " 2y. 4mo. 16dy.

From Oct. 31st, 1850, to March 15th, 1853, would be, according to the 1st method, 2y. 4mo. 14dy.
" " " 2d " 2y. 4mo. 15dy.

Bank discount is the same as Bank interest. If a note is discounted at bank, the bank takes off the interest for the time the note has to run, and pays the balance only to the holder of the note.

The number of days which elapse between two given dates, may be found as in the following example:

Required the number of days between March 23d, and Sept. 5th.

We find by adding the number of days in all the intervening time, that Sept. 5th would correspond to March 189th, if the days were all regarded as belonging to March. Between March

23d, and March 189th, would be 166 days, or 23 weeks and 5 days.[a]

If we wish to find on what day of the week any given date will fall, we may proceed in a similar manner. Thus, if March 23d comes on Friday, as there are 23 complete weeks, and 5 additional days between March 23d and Sept. 5th, Sept. 5th will fall 5 days after Friday, or on Wednesday.

March	has	31	days
April	"	30	"
May	"	31	"
June	"	30	"
July	"	31	"
August	"	31	"
Add for Sept.		5	"
		189	
		23	
		166	dy.

A note at 30 days, will fall due in 33 days = 4wk. 5d.; a note at 60 days, in 63 days, = 9wk.; a note at 90 days, in 93 days, = 13wk. 2d.; a note at 120 days, in 123 days, = 17wk. 4d. Therefore a 30 days' note becomes due 5 week days later than the day on which it is given; a 60 days' note, on the same day of the week as the day on which it is given; a 90 days' note, 2 week days later; a 120 days' note, 4 week days later. It is sometimes important to date a note so that it will not mature on Sunday, or on a holiday. This can easily be done by one of the methods above given.

EXAMPLES.

1. How much would an English merchant receive on a note for £600 at 4 months, discounted at 4 per cent.?

Ans. £591 16s.

2. When will a 3 months' note fall due, if dated Aug. 31? —a note at 8 months, dated June 30th?

---

[a] The number of days may also be found by the following table, when the time is less than a year:

TABLE FOR ASCERTAINING THE NUMBER OF DAYS FROM ANY DAY IN THE YEAR, TO ANY OTHER DAY.

1st Mo., Jan.	0	5th Mo., May	120	9th Mo., Sept.	243
2d Mo., Feb.	31	6th Mo., June	151	10th Mo., Oct.	273
3d Mo., Mar.	59	7th Mo., July	181	11th Mo., Nov.	304
4th Mo., Apl.	90	8th Mo., Aug.	212	12th Mo., Dec.	334

RULE.

To the given day of each month, add the tabular number for the month, and subtract the less sum from the greater.

If the two dates are in different years, subtract the result thus found from 365.

In leap years, add 1 to the number after the 28th of February.

3. Determine by each method, the time that elapsed between July 4th, 1776, and June 2d, 1850; between Aug. 18, 1820, and Feb. 29, 1848.

4. How many days were there between Jan. 27, and June 16, 1844?—between Feb. 20, and Oct 8, 1849?

5. New Year's day, A. D. 1850, fell on Tuesday. On what day of the week was Christmas of the same year?

6. Required the avails[a] of a note at 4 months, for $275.50, dated Feb. 12th, 1850, and discounted at a bank in New York, where the legal rate of interest is 7 per cent.

Ans. $268.91.

66. Partial Payments.

When partial payments are made on mercantile accounts which are past due, it is customary to compute interest on the whole debt from the time it became due, and on each payment from the time it was made, until the time of settlement, and to deduct the amount of all the payments, including interest, from the amount of the debt and interest.

The labor of computing interest on each item may be avoided, by multiplying the amount due at first, and the balance due after each payment, by the number of days that they are severally at interest, adding all the products, and dividing the amount by 6000. The quotient will be the interest at 6 per cent.[b]

Example for Illustration.

A debt of $630.25 became due March 15th, on which the following payments were made: April 3d, $170; May 20, $245.30; June 17th, $87.50. How much was due Sept. 5, when the account was settled?

---

[a] The amount received for the note after it is discounted.

[b] It may be readily seen that the same result will be obtained whichever method is adopted, but no accountant who is familiar with both methods, will hesitate to adopt the latter.

	$		Days.		Products.
March 15, Amount due	630.25	×	19	=	11974.75
April 3, 1st payment	170.				
Balance	460.25	×	47	=	21631.75
May 20, 2d payment	245.30				
Balance	214.95	×	28	=	6018.60
June 17, 3d payment	87.50				
Balance	127.45	×	80	=	10196.00
				6000)	49821.10
					8.30351

Ans. $127.45 + $8.30 = $135.75.

The whole debt is on interest from March 15 to April 3, 19 days. The principal is then reduced by a payment, to $460.25, which is on interest from April 3 to May 20, 47 days. In like manner we find that the 2d balance is on interest 28 days, and the 3d, 80 days. The amount due at settlement, is the unpaid balance of the debt, $127.45, together with the interest, $8.30.

EXAMPLES.

1. Sold goods to the value of $650.39, to be paid Jan. 27, 1848. Required the amount due Aug. 19, the following payments having been made on account: Feb. 23, $100; March 15, $150.39; May 20, $200; July 31, $125.

Ans. $86.89.

2. A Georgia merchant bought goods to the amount of $575, and obtained credit till Jan. 31, 1849. But being unable to pay the whole debt at one time, he remitted $100 when it became due, and subsequently paid $75 March 3, $100 March 27, $150 April 17, and the balance June 7. What was the amount of the last payment, the legal rate of interest being 8 per cent.? *Ans.* $158.51.

67. Legal Interest.

The rate of interest varies in different States of the Union. In the examples given in most American works on Arithmetic, 6 per cent. is understood, unless some other rate is specified.

In each of the New England States, in New Jersey, Pennsylvania, Delaware, Maryland, Virginia, North Carolina, Tennessee, Kentucky, Ohio, Indiana, Illinois, Missouri, Arkansas, and the District of Columbia, and on U. S. notes, the rate is 6 per cent. In New York, South Carolina, Michigan, Wisconsin, and Iowa, it is 7 per cent. In Georgia, Alabama, Mississippi, Texas, and Florida, 8 per cent. In Louisiana, 5 per cent., though the bank interest is .06, and conventional interest may be as high as .10. In Maryland, the interest on tobacco contracts is .08. In Mississippi, Missouri, and Arkansas, the interest by agreement may be as high as .10, and in Illinois, Wisconsin, and Iowa, as high as .12.

In the mercantile method of computing interest, compound interest is sometimes charged. But as the courts do not generally allow compound interest, the following rule is recommended in computing interest on notes and bonds:

If any payment exceeds the interest due at the time it is made, deduct it from the AMOUNT,[a] *and compute the subsequent interest on the balance. If the payment is less than the interest, deduct it from the* INTEREST, *reserve the excess of interest to be added to the succeeding interest, and continue the interest on the former principal.*[b]

---

[a] The amount of principal and interest.

[b] "In casting interest on notes, bonds, &c., upon which partial payments have been made, every payment is to be first appropriated to keep down the interest; but the interest is never allowed to form a part of the principal, so as to carry interest. 27 Mass. R. 417; 1 Dall. 378.

"When a partial payment exceeds the amount of interest due when it is made, it is correct to compute the interest to the time of the first payment, add it to the principal, subtract the payment, cast interest on the remainder to the time of the second payment, add it to the remainder, and subtract the second payment, and in like manner from one payment to another, until the time of judgment. 1 Pick. 194; 4 Hen. & Munf. 431; 8 Serg. & Rawle, 458; 2 Wash. C. C. R. 167; see 3 Wash. C. C. R. 350; Ibid. 376.

"When a partial payment is made *before* the debt is due, it cannot

Example for Illustration.

$1000.00 Philadelphia, March 4th, 1841.

For value received, I promise to pay John Smith, or order, one thousand dollars on demand, without defalcation.[a]

William Brown.

December 1st, 1841, received $75.00. July 17th, 1842, received $15.50. August 18th, 1843, received $30.50. December 11th, 1843, received $500.00. January 3d, 1844, received $150.00. What was due on the note, Aug. 18, 1844?

First principal, on interest from March 4, 1841 .		$1000.00
Interest to Dec. 1, 1841 (8mo. 27dy.) . . .		44.50
Amount . .		$1044.50
First payment, exceeding the interest due . .		75.00
Balance for a new principal		969.50
Interest from Dec. 1, 1841, to July 17, 1842 (7mo. 16dy.)	36.52	
Second payment, less than interest due .	15.50	
Excess of interest	21.02	
Interest from July 17, 1842, to Aug. 18, 1843 (13mo. 1dy.)	63.18	
Interest due Aug. 18, 1843 . . .	84.20	
Third payment, less than interest due .	30.50	
Excess of interest	53.70	
Interest from Aug. 18, 1843, to Dec. 11, 1843 (3mo. 23dy.)	18.26	71.96
Amount due Dec. 11, 1843		1041.46
Fourth payment, exceeding the interest due .		500.00
Balance for a new principal		541.46

be apportioned, part to the debt and part to the interest. As if there be a bond for one hundred dollars, payable in one year, and at the expiration of six months fifty dollars be paid in. This payment shall not be apportioned, part to the principal and part to the interest, but at the end of the year interest shall be charged on the whole sum, and the obligor shall receive credit for the interest of fifty dollars for six months. 1 Dall. 124."—*Bouvier's Law Dict.*, vol. 1, p. 717.

[a] The laws of Pennsylvania require the insertion of the words "without defalcation." In other places, they are usually omitted.

New Principal	541.46
Interest from Dec. 11, 1843, to Jan. 3, 1844 (23dy.)	2.07
Amount due Jan. 3, 1844	543.53
Fifth payment, exceeding the interest due .	150.00
Balance for a new principal	393.53
Interest from Jan. 3, 1844, to Aug. 18, 1844 (7mo. 15dy.)	14.76
Amount due Aug. 18, 1844	$408.29

EXAMPLES.[a]

1. Worcester, July 4th, 1840.

For value received, I promise to pay Thomas Jackson, or order, six hundred and thirty-nine dollars, on demand.[b]

$639.00 JOHN WINTER.

Endorsements. Sept. 5, 1840, received $13.25. Jan. 1, 1841, received $1.50. March 17, 1841, received $72.00. Oct. 3, 1841, received $29.50. July 3, 1842, received $9.00. What was due, Jan. 1, 1843? *Ans.* $601.83.

2. Portland, May 13th, 1841.

For value received, I promise to pay George Appleton, or order, nine hundred dollars, on demand.

$900.00 WILLIAM MASON.

Endorsements. Aug. 28, 1843, received $175.00. Dec. 13, 1843, received $10.00. April 13, 1844, received $10.00. May 1, 1844, received $500.00. What was due Sept. 13, 1844? *Ans.* $371.11.

3. New York, Oct. 5th, 1823.

For value received, I promise to pay Brown & Oliver, or order, one thousand dollars on demand.

$1000.00 JAMES THOMAS.

Endorsements. March 4, 1824, received $100.00. July 27, 1825, received $50.00. Oct. 25, 1825, received $100.00. April

[a] For the rate of interest, see p. 220.

[b] In some of the States, a note payable on demand, does not draw interest until demand is made.

13, 1826, received $15.00. Nov. 13, 1826, received $10.00. Dec. 1; 1826, received $500.00. What was due on the note, May 16, 1830? *Ans.* $532.76.

4. Charleston, Aug. 18th, 1840.

For value received, I promise to pay Nathan J. Wilson, or order, four hundred and thirty-one dollars in six months, with interest afterward. EDWARD ELLIS.

$431.00

Endorsements. Feb. 18, 1841, received $31.00. Sept. 15, 1841, received $10.00. Nov. 11, 1841, received $5.00. March 29, 1842, received $100.00. May 13, 1843, received $200.00. Dec. 31, 1843, received $2.50. What was due June 16, 1844? *Ans.* $149.30.

5. Cincinnati, Nov. 1st, 1841.

For value received, we promise to pay Samuel Jones, or order, seven hundred and seventy-five dollars and fifty cents, in two months from date. HENRY THOMPSON & Co.

$775.50

Endorsements. Feb. 27, 1842, received $15.00. August 20, 1842, received $15.00. Dec. 18, 1842, received $15.00. Jan. 1, 1843, received $15.00. April 1, 1843, received $200.00. Feb. 19, 1844, received $21.00. What was the balance due Oct. 1, 1844? *Ans.* $603.64.

68. EQUATION OF PAYMENTS.

The ordinary rule for equation of payments, is founded on the supposition that the interest of the money which is not paid until after it is due, is equal to the discount of that which is paid before it is due. This is not strictly correct, but it corresponds to the usual method of computing discount among merchants.

The labor of equating may be abbreviated by disregarding the cents if they are less than 50, and counting them as an additional dollar if they are more than 50. When the sums are all large, the units of dollars may be disregarded in a similar manner.

To find the equated time for the payment of several debts.

Multiply each charge by the time which has elapsed from the date of the first bill, and divide the sum of the products by the sum of the bills.

To find the equated time for the settlement of an account, in which there are both debits and credits :—

1st. *Find the equated time for each side of the account.*

2d. *Multiply the least side of the account by the time between the dates found for each side, and divide the product by the balance of the account. The quotient will be the time between the date found for the larger side of the account, and the equated time for the settlement of the balance.*

If the date found for the larger amount is the earliest, count back, but if it is the latest, count forward from that date.

EXAMPLE FOR ILLUSTRATION.

Required the time when the balance of the following account became due:

Dr. *Martin Acres in Account with Simon Gray.* *Cr.*

1849				1849			
June 14	To Balance	$875	30	July 10	By Cash	$600	00
Aug. 18	" Mdse.	519	40	Sept. 16	" Mdse.	427	90
Sept. 22	" Bill due	289	37	Nov. 18	" Draft	350	00
Dec. 18	" Do.	318	00	Dec. 20	" Cash	275	00

Drs.		Crs.	
875.30 × 0 =	0	600.00 × 26 =	15600.
519.40 × 65 =	33761.	427.90 × 94 =	40222.60
289.37 × 100 =	28937.	350. × 157 =	54950.
318. × 187 =	59466.	275. × 189 =	51975.
2002.07	122164.	1652.90	162747.60

122164. ÷ 2002.07 = 61. 162747.60 ÷ 1652.90 = 98.

Debits due 61 days after June 14. Credits due 98 days after June 14.

1652.90 × 37 ÷ 349.17 = 175. The balance was therefore due 175 days before the date found for the Dr. side, or 114 days before June 14, which was Feb. 20.

The example may be equated as follows, by disregarding the cents and units of dollars. If the dollars amount to 5 or more, 1 should be added to the number of eagles.

$88 \times 0 = 0$		$60 \times 26 = 1560$	
$52 \times 65 = 3380$		$43 \times 94 = 4042$	
$29 \times 100 = 2900$		$35 \times 157 = 5495$	
$32 \times 187 = 5984$		$28 \times 189 = 5292$	
201	12264	166	16389

$12264 \div 201 = 61.$ $16389 \div 166 = 98.$

$1652.90 \times 37 \div 349.17 = 175$, as before.

Examples for the Pupil.

1. When was the balance of the following account due?

Dr. *James Day in Account with William Knight.* *Cr.*

1848				1848			
Jan. 6	To Mdse.	$427	20	Feb. 20	By Cash	$750	00
Mar. 10	" Do.	316	19	Apl. 16	" Mdse.	95	87
May 18	" Do.	284	72	July 3	" Draft	100	00

Ans. Nov. 29, 1847.

2. Find the equated time for the settlement of the following account:

Dr. North, Harrison & Co. in Acc. with J. E. Oliver. Cr.

1846				1846			
Jan. 1	To Balance	$283	94	Jan. 12	By Cash	$425	00
May 15	" Mdse.	217	33	Apl. 3	" Mdse.	693	13
July 5	" Note	500	00	June 29	" "	37	59
Aug. 30	" Mdse.	894	60				

Ans. Dec. 21, 1846.

The average price of a mixture consisting of several ingredients, is found in nearly the same manner as the average time for the payment of several debts. The method of finding the average price, is usually called Medial Alligation. For Examples, see Sect. 74.

69. ACCOUNTS CURRENT.

An ACCOUNT CURRENT contains a statement of the mercantile transactions of one person with another. On the Dr. side are placed all the payments made, and the amounts of merchandise sold, to the merchant who is furnished with the account; on the Cr. side are entered all sums received, and the amounts of merchandise purchased from the said merchant.

To facilitate the settlement of interest, it is customary to place against each sum the number of days that elapse from the time that each entry becomes due, until the time of rendering the account, and calculate the interest on each item. The balance of interest is entered on the side which has the greatest amount of interest.

EXAMPLE FOR ILLUSTRATION.

Abbott & Clark, Philadelphia, in Account Current with Charles Goodhue & Co., Boston.

DR.

1844		Dol. Cts.	No. days.	Interest.
June 1	To balance due from former acct. . . .	153.50	92	2.354
" 13	To amount due on note for goods . . .	400.00	80	5.333
" 25	To merchandise . .	275.00	68	3.116
July 19	To 90 bbls. flour, at $4.37½	393.75	44	2.887
Sept. 1	To balance of interest .	3.28		
				13.690
		$1225.53		10.412
				3.278

1844	CR.	Dol. Cts.	No. days.	Interest.
June 16	By cash	375.00	77	4.812
" 30	By bill of Arnold & Brown	250.00	63	2.625
July 29	By cash	525.00	34	2.975
Sept. 1	By balance to your debit on a new account .	75.53		10.412
		$1225.53		

1. *Oliver Marriott, Savannah, in Account Current with Daniel Clark, Mobile.*

Dr.

1844		Dol. cts.	No. of days.	Interest
Feb. 16	To balance due from old account,	91.75		
" 25	To merchandise,	163.50		
Apl. 13	To merchandise,	219.25		
May 16	To balance of interest,			

1844	*Cr.*	Dol. cts.	No. of days.	Interest.
Mar. 30	By cash,	400.00		
May 16	By balance to new acct.			

2. *Joseph Mason & Co., Saint Louis, in Account Current with Thompson & Brother, Lexington.*

Dr.

1844		Dol. cts.	No. of days.	Interest.
Jan. 13	To sheeting,	131.50		
Feb. 3	To duck,	87.75		
" 25	To cambric,	240.00		
Mar. 29	To sundries,	300.00		
Apl. 1	To balance of interest,			

1844	*Cr.*	Dol. cts.	No. of days.	Interest.
Jan. 29	By furs,	175.00		
Mar. 3	By bill,	200.00		
Mar. 18	By cash,	380.75		
Apl. 1	By balance to new acct.			

3. *Henry Chatham, Nashville, in Account Current with George Hapgood & Sons, Baltimore.*

Dr.

1844		Dol. cts.	No. of days.	Interest.
Apl. 5	To amount due on note for goods,	500.00		
May 17	To ditto,	119.50		
" 28	To merchandise,	87.25		
June 16	To sundries,	63.00		
July 1	To balance to new acct.			

1844	*Cr.*	Dol. cts.	No. of days.	Interest.
Apl. 9	By cash,	350.00		
" 11	By tobacco,	289.50		
" 23	By bill on Atkins & Jones,	200.00		
July 1	By balance of interest,			

It is more convenient to enter the product of each entry by the number of days, as in ordinary Equation of Payments, instead of computing the interest on each amount. By dividing the balance of products by 6000, we at once obtain the balance of interest at 6 per cent. For any other rate than 6 per cent., the result may be increased or diminished, as in computing Bank Interest.

EXAMPLE.

Wilson, Survilliers & Co. in Account Current with James N. Martin.

DR. CR.

1850		$	ct	dys.		1850		$	ct	dys.	
Jan. 9	Paid for 900*l*. remitted to Brown, Shipley & Co., at 1.09 . . .	4360	00	142	619120	June 1	By proceeds of sales rendered this day. Value due Nov. 6, 1850 .	15141	75	157	23772
" 20	Freight on 30 cases per St. Nicholas . . .	17	10	131	2227						
" "	C. H., charge for warehousing	6	60	"	917						
Feb. 6	Paid duties, and C. H. storage	327	76	114	37392						
" 8	" Marine Insurance	160	05	112	17920						
	1 per cent., (instead of usual charges,) per agreement	151	42								
	5 per cent. commission and guarantee on $15141.75	757	09								
	Numbers from credit side				2377294						
					3054870						
	Amount of products, 3054870 div. by 60 .	509	15								
	Balance due W. S. & Co.	8852	58								
		15141	75			" "	By balance	8852	58		

As the amount of sales in the preceding account is not due, Wilson, Survilliers & Co. should be charged with interest until it becomes due; therefore the product 2377294 is added to the amount of the Dr. products. In forming the products, the cents are disregarded, unless they exceed 50, in which case the number of dollars is increased by 1.

EXAMPLES.

1. A. B. in account with C. D. Debits, 1849, July 3, $5000; July 28, $40; Aug. 16, $800; commission and charges, 6% on $18000. Credits, 1850, March 2, $9000; May 15, $9000. What was the balance of account rendered Jan. 1, 1850. *Ans.* $10617.89.

2. E. F. in acct. with G. H. Debits, 1850, Jan. 20, \$3300; Jan. 31, \$25; Feb. 19, \$400; June 8, \$900; commission and charges, 7% on \$13842.68. Credit, proceeds of sales, due Oct. 29, \$13842.68. Required the balance due July 1. *Ans.* \$7869.86.

The pupil may solve by this method the examples given on page 225.

70. Practice.

In Practice many questions arise that can be solved more readily than by adopting any of the ordinary rules. Many of the operations of business, in which compound numbers are concerned, may be abbreviated by first finding values for the highest denomination, and considering the lower denominations as aliquot parts of the higher.[a]

Table of Aliquot Parts.

Of a dol.		*Of a £.*			*Of a shi.*		*Of a ton.*			*Of a cwt.*			*Of a year.*		
cts.	*\$*	*s.*	*d.*	*£*	*d.*	*s.*	*cwt.*	*qr.*	*ton.*	*qr.*	*lb.*	*cwt.*	*m.*	*d.*	*y.*
50 =	$\frac{1}{2}$	10	=	$\frac{1}{2}$	6 =	$\frac{1}{2}$	10	=	$\frac{1}{2}$	2	=	$\frac{1}{2}$	6	=	$\frac{1}{2}$
$33\frac{1}{3}$ =	$\frac{1}{3}$	6	8 =	$\frac{1}{3}$	4 =	$\frac{1}{3}$	5	=	$\frac{1}{4}$	1	=	$\frac{1}{4}$	4	=	$\frac{1}{3}$
25 =	$\frac{1}{4}$	5	=	$\frac{1}{4}$	3 =	$\frac{1}{4}$	4	=	$\frac{1}{5}$		16 =	$\frac{1}{7}$	3	=	$\frac{1}{4}$
20 =	$\frac{1}{5}$	4	=	$\frac{1}{5}$	2 =	$\frac{1}{6}$	2	2 =	$\frac{1}{8}$		14 =	$\frac{1}{8}$	2	12 =	$\frac{1}{5}$
$16\frac{2}{3}$ =	$\frac{1}{6}$	3	4 =	$\frac{1}{6}$	$1\frac{1}{2}$ =	$\frac{1}{8}$	2	=	$\frac{1}{10}$		8 =	$\frac{1}{14}$	2	=	$\frac{1}{6}$
$12\frac{1}{2}$ =	$\frac{1}{8}$	2	6 =	$\frac{1}{8}$	1 =	$\frac{1}{12}$	1	1 =	$\frac{1}{16}$		7 =	$\frac{1}{16}$	1	15 =	$\frac{1}{8}$
10 =	$\frac{1}{10}$	2	=	$\frac{1}{10}$	$\frac{3}{4}$ =	$\frac{1}{16}$	1	=	$\frac{1}{20}$		4 =	$\frac{1}{28}$	1	10 =	$\frac{1}{9}$
$8\frac{1}{3}$ =	$\frac{1}{12}$	1	=	$\frac{1}{20}$	$\frac{1}{2}$ =	$\frac{1}{24}$		2 =	$\frac{1}{40}$		2 =	$\frac{1}{56}$	1	6 =	$\frac{1}{10}$
$6\frac{1}{4}$ =	$\frac{1}{16}$		10 =	$\frac{1}{24}$	$\frac{1}{4}$ =	$\frac{1}{48}$	8	=	$\frac{2}{5}$	1	4 =	$\frac{2}{7}$	4	24 =	$\frac{2}{5}$
5 =	$\frac{1}{20}$		8 =	$\frac{1}{30}$	9 =	$\frac{3}{4}$	12	=	$\frac{3}{5}$	1	20 =	$\frac{3}{7}$	7	6 =	$\frac{3}{5}$
4 =	$\frac{1}{25}$		4 =	$\frac{1}{60}$	8 =	$\frac{2}{3}$	16	=	$\frac{4}{5}$	2	8 =	$\frac{4}{7}$	9	18 =	$\frac{4}{5}$
$3\frac{1}{3}$ =	$\frac{1}{30}$		3 =	$\frac{1}{80}$	$7\frac{1}{2}$ =	$\frac{5}{8}$	15	=	$\frac{3}{4}$	2	24 =	$\frac{5}{7}$	4	15 =	$\frac{3}{8}$
2 =	$\frac{1}{50}$		2 =	$\frac{1}{120}$	$4\frac{1}{2}$ =	$\frac{3}{8}$	7	2 =	$\frac{3}{8}$	3	12 =	$\frac{6}{7}$	7	15 =	$\frac{5}{8}$

[a] When the number of articles is not very large, and the price consists of several denominations, the answer can generally be obtained most readily by Compound Multiplication.

When the number of articles is large, and the price such as to require but few parts to be taken, the readiest mode of solution is by Practice.

When the quantity and price, consist each of several denominations, the best mode of solution is by Proportion, or by Fractional Analysis.

Similar tables may be made to any required extent, but these are sufficient to show their application.

One of the following rules may be adopted in nearly all questions that admit of abbreviation by Practice:

1. Assume the price at some unit higher than the given price, and take aliquot parts of the assumed price for the answer.

2. Multiply the price by the integers of the highest denomination, and take aliquot parts for the lower denominations.

In the application of either of these rules, there is great room for the exercise of judgment, in determining what parts should be taken to determine the answer most readily.

Examples illustrating the First Rule.

1. At $43\frac{3}{4}$ cents a yard, what is the price of $87\frac{1}{2}$ yards of muslin?

The price at \$1 per yd. would be	\$87.50
at 25cts. $= \$\frac{1}{4}$	21.875
at $12\frac{1}{2}$ " $= \frac{1}{2}$ of $\$\frac{1}{4}$	10.9375
at $6\frac{1}{4}$ " $= \frac{1}{2}$ of $\$\frac{1}{8}$	5.46875
at $43\frac{3}{4}$	\$38.28125

Or, as $\$0.43\frac{3}{4} = \$\frac{7}{16}$, the answer might have been obtained by multiplying \$87.50 by $\frac{7}{16}$.

2. What is the value of $96\frac{1}{2}$ lb. of tea, at 3s. $10\frac{3}{4}$d. per lb.?

	£	s.	d.
The price at £1 per lb. would be	£96	10s.	
at 3s. 4d. $= \frac{1}{6}$£	16	1	8
at 5d. $= \frac{1}{8}$ of 3s. 4d.	2	0	$2\frac{1}{2}$
at $1\frac{1}{4}$d. $= \frac{1}{4}$ of 5d.		10	$0\frac{5}{8}$
at $\frac{1}{2}$d. $= \frac{1}{10}$ of 5d.		4	$\frac{1}{4}$
at 3s. $10\frac{3}{4}$d.	£18	15	$11\frac{3}{8}$d.

Example for illustrating the Second Rule.

What is the value of 11cwt. 3qr. 17lb. of sugar, at £1 3*s*. 6*d*. per cwt.?

	£	*s*.	*d*.	
2qr. = $\frac{1}{2}$cwt.	1	3	6	price of 1cwt.
16lb. = $\frac{1}{7}$cwt.			11	
	12	18	6	price of 11cwt.
1qr. = $\frac{1}{2}$ 2qr.		11	9	" " . 2qr.
		5	10$\frac{1}{2}$	" " . 1qr.
1lb. = $\frac{1}{16}$ 16lb.		3	4$\frac{2}{7}$	" " . . 16lb.
			2$\frac{29}{56}$	" " . . 1lb.
	13	19	8$\frac{1}{4}$[a]	price of 11cwt. 3qr. 17lb.

The following rule is probably as convenient as any that could be given, for computing interest by Practice.

Multiply 1 per cent. of the principal by $\frac{1}{2}$ the even number of months, and if there is an odd month, add 30 to the number of days. Divide the days by 6, and multiply $\frac{1}{10}$ of 1 per cent. by the quotient. If there are any remaining days,[b] take as many 60ths of 1 per cent. Add the numbers so obtained, and their sum will be the interest at 6 per cent. For any other rate, increase or diminish the result, as in the Bank Rule.

Example for Illustration.

What is the interest of $9763.25 for 2yr. 9mo. 8dy.?

1 per cent., or the interest for 2 months, is $97.6325. Multiplying by 16, we obtain the interest for 2yr. 8mo. The 1mo. remaining, added to the 8 days, gives 38 days, in which 6 is contained 6 times, with 2 remainder. As the interest

1 per cent. is	97.6325
$\frac{1}{2}$ of 2yr. 8mo. is	16
	585.7950
	976.325
36dy. is .6 of 2mo.	58.5795
2dy. is $\frac{1}{30}$ of 2mo.	3.2544
	$1623.9539

[a] The fraction is $\frac{17}{56}$, but as nothing is reckoned less than $\frac{1}{4}$*d*., it is called $\frac{1}{4}$.

[b] When the remainder is 4, it is more convenient to diminish the quotient by 1, and call the remainder 10, taking $\frac{10}{60}$ or $\frac{1}{6}$ of 1 per cent for the interest.

for 6 days is .1 of 1 per cent., or $9.763+, the interest for 36 days is 6 times as much. The two remaining days are $\frac{1}{30}$ of 60 days, we therefore add $\frac{1}{30}$ of $97.63+.

There are a variety of other contractions that may frequently be adopted in practice. A few are given below, which will often be found useful.

(1.) When the multiplier consists of any number of 9's, increase it by 1, and subtract the multiplicand from the product. Thus, 18473 × 9999 = 184730000 — 18473 = 184711527.

(2.) To multiply by 5, divide the multiplicand by .2. Thus, 187 × 5 = 187 ÷ .2 = 935. To divide by 5, multiply the dividend by .2.

(3.) To multiply by 25, divide the multiplicand by .04. Thus, 1289 × 25 = 1289 ÷ .04 = 32225. To divide by 25, multiply the dividend by .04.

(4.) To multiply by 75, multiply by 100, and subtract $\frac{1}{4}$ of the product. Thus, 18645 × 75 = 1864500 — 466125 = 1398375. To divide by 75, divide by 100, and add $\frac{1}{3}$ of the quotient.

(5.) To multiply by 125, divide the multiplicand by .008. Thus, 1641 × 125 = 1641 ÷ .008 = 205125. To divide by 125, multiply the dividend by .008.

(6.) To multiply by 375, divide by .008, and multiply the quotient by 3. Thus, 294 × 375 = 294 ÷ .008 × 3 = 110250. To divide by 375, multiply by .008 and divide by 3.

(7.) To multiply by 625, divide the multiplicand by .0016. Thus, 4812 × 625 = 4812 ÷ .0016 = 3007500. To divide by 625, multiply the dividend by .0016.

(8.) To multiply by 875, multiply by 1000, and subtract $\frac{1}{8}$ of the product. Thus, 735 × 875 = 735000 — 91875 = 643125. To divide by 875, divide by 1000 and add $\frac{1}{7}$ of the quotient.

(9.) To multiply by any number within 12 of 100, 1000, &c., annex to the multiplicand as many zeros as there are figures in the multiplier, and subtract as many times the multiplicand as are equivalent to the excess of 100, 1000, &c., over the multiplier. Thus, 24796 × 99989 = 2479600000 — (11 × 24796) = 2479327244.

(10.) To square a number ending in 5, multiply the number of tens by one more than itself, and place 25 at the right of the product. Thus, 3 × 4 = 12, and 35 × 35 = 1225; 12 × 13 = 156, and 125 × 125 = 15625; 6 × 7 = 42, and 65 × 65 = 4225.

(11.) When the tens in two numbers are alike, and the sum of the units is 10, to obtain the product multiply the number of tens by one more than itself for the hundreds, and place the product of the units at the right of this product, for the tens and units. Thus, $4 \times 5 = 20$, and $43 \times 47 = 2021$; $42 \times 48 = 2016$; $44 \times 46 = 2024$; $7 \times 8 = 56$, and $72 \times 78 = 5616$; $71 \times 79 = 5609$, &c.

(12.) The sum of two numbers multiplied by their difference, is equal to the difference of their squares. Hence we may readily find the product of two numbers, one of which is as much above as the other is below, a certain number of tens. Thus, $87 \times 73 = (80 + 7) \times (80 - 7) = 80^2 - 7^2 = 6400 - 49 = 6351$.

(13.) To square any number between 50 and 60, add the units of the given number to 25 for the hundreds, and annex the square of the units for the tens and units. Thus, for the square of 51; $25 + 1 = 26$ *hundreds*, and $1 \times 1 = 1$; hence $51 \times 51 = 2601$. In like manner $53 \times 53 = 2809$; $59 \times 59 = 3481$.

(14.) When one figure of the multiplier is an aliquot part of one or more of the remaining figures, the work may be abbreviated as in the following example:—

Multiply 489.137 by 7261.8.

We see at once that 18 is a multiple of 6, and 72 is a multiple of 18. Therefore, multiplying first by 6, we take 3 times the product for the product by 18, and 4 times the product by 18, for the product by 72.

```
      489.137
       7261.8
     --------
    2934822      = prod. by 6.
       8804466   = prod. by 3 × 6 = prod. by 18.
   35217864      = prod. by 4 × 18 = prod. by 72.
   ------------
   3552015.0666
```

(15.) In the ordinary mode of determining the greatest common divisor of two numbers, any prime factor or square number, contained in one number but not in the other, or any prime factor or square number in a remainder that is not in the preceding divisor, may be rejected, and the work thus abbreviated. For example, let the greatest common measure of 689 and 2279 be required.

Here the square number 4 is a factor of 212, and not of 689. We therefore divide 212 by 4, and immediately obtain the greatest common measure. In the application of this principle to the reduction of fractions, we observe that 53 divides 689 13 times, and it divides 212, 4 times. It therefore divides $3 \times 689 + 212$ or 2279, $3 \times 13 + 4$ or 43 times. Therefore $\frac{689}{2279} = \frac{13}{43}$.

```
689 ) 2279 ( 3
      2067
      ----
   4 ) 212
      ----
G. C. Meas. 53 ) 689 ( 13
                 53
                 ---
                 159
                 159
                 ---
```

Reduce $\frac{457}{563}$ to its lowest terms.

```
457 ) 563 ( 1
      457
      ---
      106 = 2 × 53
```

Neither 2 nor 53 being factors of 457, the fraction is already in its lowest terms.

EXAMPLES.

1. Find the cost of 15lb. 10oz. of tea, at \$.37½ per lb.; at \$.25; at \$.31¼; at \$.43¾; at \$.50; at \$.56¼; at \$.66⅔; at \$.87½.

2. What is the price of 89⅞yds. of broadcloth, at \$4.75 per yd.? *Ans.* \$426.31¼.

3. What is the value of 49A. 3R. 15r. of land, at \$125 per acre? *Ans.* \$6230.47.

4. What is the value of 96yds. 3qr. 3na. of broadcloth, at £1 2s. 6d. per yard? *Ans.* £109 1s. 1¼d.

5. What will 17T. 11cwt. 2qr. 21lb. of iron cost, at \$19.75 per ton? *Ans.* \$347.29.

6. What is the cost of 163A. 2R. 25r. of land, at \$15.75 per acre? *Ans.* \$2577.59.

7. What is the value of 364yds. 3qr. 1na. of sheeting, at 12½cts. a yard? *Ans.* \$45.60.

8. Bought 76bu. 3pk. of potatoes, at 37½cts. a bushel; 19bu. 2pk. of wheat, at \$1.10 a bushel; 37bu. 1pk. of

barley, at 62½cts. a bushel; and 10T. 15cwt. of hay, at $16.00 a ton. What was the amount of the whole?

Ans. $245.51.

9. What is the interest of $2375 for 5yr. 11mo. 23dy., at .055 per year? *Ans.* $781.21.

10. What is the interest of $4814.25 for 3yr. 7mo. 14dy., at .06 per year? At .075 per year?

Ans. $1046.297; $1307.87.

11. Multiply 576.3 by 99; by 999000.

12. Multiply 7.894 by 5; by 25; by 7500.

13. Multiply 48.302 by 1250; by 375000.

14. Divide 1879.4 by 5; by 250; by 75.

15. Divide 4449.17 by 125; by 375.

16. Multiply 3.0872 by 525125, by adding three partial products.

17. Multiply 41909 by 999625125, in the most expeditious manner.

18. Multiply 89443 by 625; by 875.

19. Divide 141.982 by 625; by 875.

20. Multiply 89443 by 625875.

21. Multiply 283172 by 9992; by 991.

22. What is the square of 15? of 85? of 115?

23. What is the product of 73 × 77? 12 × 18? 44 × 46?

24. What is the product of 81 × 89? 75 × 75? 34 × 36?

25. What is the product of 16 × 24? 19 × 21? 35 × 45? 89 × 71? 67 × 53? 78 × 82? 96 × 84? 113 × 107? 112 × 128?

26. What is the square of 58? of 56? 52? 55? 57? 54?

27. Find the greatest common divisor of 804 and 938; of 741 and 1083; of 1343 and 1817.

28. Reduce each of the following fractions to its lowest terms: $\frac{301}{473}$, $\frac{781}{1207}$, $\frac{667}{899}$, $\frac{1147}{1333}$, $\frac{941}{1711}$.

29. Multiply 476384 by 9995125625.

30. Required the interest of $1729.50 for 3yr. 7mo. 16dy., at 7 per cent.

71. CAUSE AND EFFECT.

In all questions which can be solved by ratio or proportion, if the multiplying terms are written in one column, and the dividing terms in another, the factors common to both columns may be cancelled, and the answer obtained by dividing the product of the remaining factors in the column of multipliers, by the product of the remaining factors in the column of divisors.

The terms of a proportion may be distinguished into *causes* and *effects*, and the alternate products of either cause by the other effect, are equal. For example, if 8 men build 6 rods of wall in a day, 4 men will build 3 rods in the same time. Stating the proportion, we have

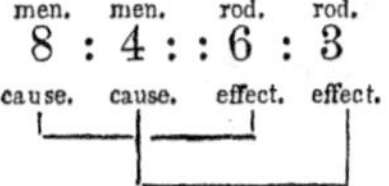

The product of the extremes, and the product of the means, give us the product of each cause by the effect of the other cause. If then we write each effect opposite to its cause, our multipliers and divisors will be obtained without difficulty.

Men, animals, and times, are evidently causes, because the increase of either of them, will increase the effect produced.

In questions of freight, we may regard distances and bulk as causes, producing money for their effect.

The principal of a sum of money at interest, is a cause, and the interest is an effect.

A little practice will give great facility in distinguishing between causes and effects, in all cases of common occurrence, and by acquiring readiness in making this distinction, a vast amount of labor may be saved.

In arranging the terms according to this rule, when we reach the required term, or *term of demand*, its place may be supplied with a dash. Then, as the product of all the factors on each side must be equal, the missing term may be found by cancelling, and dividing the product of all the numbers on the side opposite to the dash, by the product of all the numbers on the side in which the dash is included.

Examples for Illustration.

1. If 18 men, in 6 days of 8 hours, build a wall 150 feet long, 2 feet wide, and 4 feet high, in how many days of 12 hours will 24 men build a wall 200 feet long, 3 feet wide, and 6 feet high?

causes.	men 18	24 men.
	days 6	— days.
	hours 8	12 hours
effects.	long 200	150 long.
	wide 3	2 wide.
	high 3 6	4 high.
		9 days *Ans.*

Commencing our statements, we write 18 men, 6 days of 8 hours, as cause, and the effect, which is a wall 150ft. long, 2ft. wide, and 4 ft. high, on the opposite side. Opposite to each of these terms, we write — days, 12 hours, 24 men, as cause, and 200 long, 3 wide, 6 high, as effect. Cancelling the like factors, we have but 3 × 3 on the side of the multipliers, and 1 on that of the divisors. 3 × 3 ÷ 1 is therefore the missing term, or number of days required. If either term were fractional, the denominator, representing a divisor, should be transposed to the opposite side. By proceeding in this manner, a statement may be made as soon as the question can be proposed.

2. If $\$27\frac{1}{2}$ buy $4\frac{3}{4}$ yards of cloth, that is $\frac{5}{8}$yd. wide, how

many yards of like quality, that is $\frac{5}{4}$yd. wide, may be bought for \$$13\frac{3}{4}$?

STATEMENT.

$27\frac{1}{2}$	$13\frac{3}{4}$
—	$4\frac{3}{4}$
$\frac{5}{4}$	$\frac{5}{8}$

Reducing the mixed numbers to improper fractions, we transpose the denominators, writing them above the causes, — then cancel and divide as before.

Ans. $\frac{19}{16} = 1\frac{3}{16}$yd.

8		
4̸	4̸	denoms.
2 4̸	2̸	
5̸5̸	5̸5̸	
—	19	
5̸	5̸	

EXAMPLES FOR THE PUPIL.

1. How many men in 18 months, will build a wall that 108 men can build in 16 months? [a]

N. B. The effect in this example is 1 wall.

2. How many bushels of meal will serve 54 persons 12 months, if 15 persons consume 12 bushels in 2 months? [b]

3. If 27 men build a cistern 30ft. long, 16ft. wide and 10ft. high, in 4 weeks, by working 5 days in a week, and 9 hours a day, how many men, working 6 days in a week, and 12 hours a day, will build a cistern 48ft. long, 12ft. wide, and 20ft. high, in 9 weeks? [c]

4. What is the interest of \$360 for 3y. 4mo., at 6 per cent.? [d]

5. If \$16 gain \$3 in 5 mo., how much ought \$24 to gain in 10mo.? [e]

6. If the freight of 1800 lb. for 56 miles, is \$1.50, how far may 1T. 4cwt. 12 lb. be carried for \$6.75? [f]

7. How much wheat, at \$1.20 a bushel, must be

---

a

—	108
18	16
1	1

b

54	15
12	2
12	—

c

27	—
4	6
5	12
9	9
48	30
12	16
20	10

d

1	360
12	40
—	.06

e

16	24
5	10
—	3

f

1800	2700
56	—
6.75	1.50

given in exchange for 90 barrels of flour, at $4.75 per barrel? [a]

8. If the rent of 19A. 3R. of land is £4 10s., what will be the rent of $73\frac{1}{2}$A.? [b]

9. If the expenses of a family of 8 persons are $40 in 10 weeks, how many persons can be supported $12\frac{1}{2}$ weeks for $100? [c]

10. The shadow of a stick that is 5ft. 6in. high, measures 3ft. 4in. What is the height of a tree whose shadow measures 75ft. at the same time? [d]

11. If $29\frac{1}{3}$bu. of wheat yield 1760bu. in 5 years, how much will $45\frac{1}{2}$bu. yield in 6 years at the same rate?

12. If 10 compositors, in 2 days of 10 hours, set $66\frac{2}{3}$ pages of types, each page containing 45 lines of 50 letters, how many compositors will set $94\frac{2}{7}$ pages, each page containing 35 lines of 40 letters, in $2\frac{3}{4}$ days of 8 hours?

13. If 19 men, in $71\frac{3}{4}$ days of $10\frac{3}{4}$ hours, dig a trench $41\frac{1}{2}$yd. long, $5\frac{7}{8}$ft. deep, and $7\frac{4}{5}$ft. wide, how long a trench, that is $8\frac{2}{5}$ft. deep, and $4\frac{7}{10}$ft. wide, will 11 men dig in $291\frac{1}{2}$ days of $4\frac{7}{9}$ hours? *Ans.* $50\frac{244235}{687078}$yd.

14. If 18 men in $9\frac{1}{2}$ months consume flour worth $78.75, when wheat is $1.12½ per bushel, how many months will $145 supply 35 men with flour, when wheat is $1.00 per bushel? *Ans.* $10\frac{59}{490}$mo.

15. If 2500 slates, each 8 inches long and 5 inches wide, will cover a roof, how many will be required that are 6 inches long and 4 inches wide? *Ans.* $4166\frac{2}{3}$.

16. A pile of wood 60ft. long, 10ft. high, and 8ft. thick,

---

[a]

—	90
1.20	4.75

[b]

2	4
79	147
—	99

[c]

2	
8	—
10	25
100	40

[d]

3	2
11	—
75	10

was sold for \$238$\frac{1}{3}$. What would be the price of a pile 20ft. long, 8ft. high, and 4ft. thick, at the same rate?

Ans. \$31$\frac{7}{9}$.

17. If 5 men, by working 8 hours a day for 12 days, can build a wall 40 rods long, 2 feet thick, and 6 feet high, how many men, working 9 hours a day, will build a similar wall, 30 rods long, 3 feet thick, and 8 feet high, in 4 days?

Ans. 20 men.

18. If 7 men, in 8$\frac{1}{2}$ days, by working 9 hours a day, can build $\frac{1}{3}$ of a wall that is to be raised 12 feet, how many days must 11 men work, when the days are 8 hours long, to raise the same wall 5 feet? *Ans.* 7$\frac{427}{704}$ days.

19. If \$1000 gain \$11$\frac{1}{9}$ in 80 days, how much will \$2500 gain in 120 days, at the same rate?

20. If a man, by walking 3 miles an hour, for 6 hours a day, can accomplish a journey in 12 days, in how many days would a man walk the same distance, at the rate of 2$\frac{1}{2}$ miles an hour, for 9 hours a day?

21. If 42$\frac{1}{2}$ bushels of corn, that weighs 51$\frac{1}{4}$ pounds a bushel, can be bought with 23 bushels of wheat, that weighs 56$\frac{3}{4}$ pounds a bushel, how much corn, weighing 60 pounds a bushel, would be equivalent to 100 bushels of wheat that weighs 54 pounds a bushel? *Ans.* 150$\frac{975}{5221}$bu.

22. If a man travels 240 miles in 8 days, when the days are 12 hours long, how many miles will he travel in 24 days, when the days are 16 hours long?

23. If the freight of 2T. 6cwt. for 28 miles, is \$14.50, what will be the freight of 9T. 4cwt. for 96 miles?

24. If 4 men, in 3 days of 8 hours, build 40 rods of wall, how many rods will 18 men build in 5 days of 9 hours?

Ans. 337$\frac{1}{2}$ rods.

25. How many men, in 24 days of 16 hours, will do three times as much work as 18 men can perform in 32 days of 12 hours?

26. If 14 men, in $5\frac{1}{2}$ days, by working 8 hours a day, reap $38\frac{1}{2}$ acres of grain, how many men will reap $37\frac{1}{2}$ acres in $6\frac{2}{3}$ days, by working 9 hours a day?

27. How much wheat, that weighs 60 lb. per bushel, would be required to supply a garrison of 1400 men 9 months, if 2800 bushels, weighing 58 lb. per bushel, supply 800 men $3\frac{1}{2}$ months? *Ans.* 12180bu.

28. How many hours a day must 15 men work, to dig a trench 400ft. long, 6ft. wide, and 3ft. deep, in $187\frac{1}{2}$ days, if 72 men can dig a trench 250ft. long, 8ft. wide, and 4ft. deep, in $31\frac{1}{4}$ days, by working 7 hours a day?

29. How many men can be furnished with 4 suits each, by 1140 yards of cloth that is $1\frac{1}{2}$yd. wide, if 2016 yards, $\frac{3}{4}$yd. wide, furnish 112 men 3 suits apiece?

30. If 16 compositors set 150 pages of types, each page containing 48 lines, and each line 50 letters, in 3 days of 10 hours, how many compositors will be required to set 500 pages of 72 lines each, and 45 letters in a line, in 6 days of 8 hours? *Ans.* 45.

31. If $1700, at 6 per cent., yield an interest of $350 in a given time, what will be the interest of $3900 at 7 per cent. for one half the time?

32. If 30 reams are required for 1500 pamphlets of 10 sheets each, how many reams will be required for 740 pamphlets of $12\frac{1}{2}$ sheets each?

33. If 17 yards of serge, that is $\frac{3}{4}$ wide, are required to line a cloak containing 9.75 yards of cloth that is $5\frac{1}{3}$ quarters wide, how many yards of a yard wide, would line a cloak containing 10.5 yards of cloth $6\frac{1}{8}$ quarters wide?

Ans. $15\frac{2559}{3328}$yd.

34. If a cistern discharges $83\frac{1}{3}$ gallons of water in 1.3 hours, how much will it discharge in $6\frac{5}{7}$ hours?

35. When exchange on London is at a premium of $9\frac{1}{2}$ per cent., what is the value of $1863.50, in English money, the par of exchange being £1 = $$4.44\frac{4}{9}$?

72. EXCHANGE.

The term EXCHANGE, in commerce, is generally employed to designate that species of mercantile transactions, by which the debts of individuals residing at a distance from their creditors, are cancelled without the transmission of money.

A BILL OF EXCHANGE is an order addressed to some person at a distance, directing him to pay a certain sum to the person in whose favor the bill is drawn, or to his order. The person who draws the bill is called the *drawer;* the person in whose favor it is drawn, the *remitter* or *payee;* the person on whom it is drawn, the *drawee.* The drawee is also called the *acceptor*, when he has accepted, or engaged to pay the bills.

Though bills of Exchange are originally drawn by creditors on their debtors, they are very rarely transmitted directly, but pass from hand to hand like any other circulating medium, and are bought and sold in the market. When the remitter disposes of a bill, he writes his name on the back, and is termed the *endorser*. If he endorses in favor of any particular individual, he gives a *special endorsement,* and such *endorsee* must also endorse the bill if he negotiates it. But if the endorsement is blank, the bill may be passed at pleasure from hand to hand. Every endorser, as well as the acceptor, is held responsible for the payment of the bill, and may be sued for its recovery.

INLAND, or DOMESTIC EXCHANGE, includes the commercial transactions within the limits of one country. FOREIGN EXCHANGE relates to the transactions of one country with another.

The TRUE PAR OF EXCHANGE is the value of the currency of one country estimated in the currency of another, by comparing the quantity of gold and silver in their respective coins. The exchange with England apparently furnishes an exception to this rule, the nominal par being 4.44\frac{4}{9}$ per £, while the actual value of the pound sterling, which is the real par, is about $4.87. Hence, exchange on England is generally said to be from 8 to 10 per cent. above par.

The COURSE OF EXCHANGE, or the fluctuation above or below par, depends generally on the amounts due between different countries. Thus, when the debts and credits between two countries are equal, the real exchange is at par. But if New York owes London more than London owes New York, there will be a greater demand for bills on London, and this demand will produce a rise in the price, or cause the bills to be at a *premium.* The premium, however, can never exceed the cost of transporting specie; for if it did, all debts would be paid in money or merchandise, instead of bills of exchange. The *nominal* premium, however, may exceed the cost of remitting coin, when the nominal par is above the real par.

The operation of Bills of Exchange, may be explained by a single example.

If A. of Boston owes B. of Paris, and C. of Paris owes D. of Boston, A. purchases in the market a *bill* upon Paris; that is, he buys of D. an order on his debtor C., to pay A. or his order the amount desired. A. endorses the bill, and sends it to B., who receives payment from C. Thus the two debts are cancelled by a single remittance; the inconvenience of exporting and re-importing coin is removed, and all danger of loss is obviated by sending three bills (called the First, Second, and Third of Exchange), either of which being paid, the others are void.

An ACCEPTANCE is an engagement to pay the amount of the bill, and may be either *absolute* or *qualified.* An absolute acceptance binds the drawee when the bill becomes due, and in making it the drawee usually writes "Accepted," and subscribes his name at the bottom, or across the body of the bill. A qualified acceptance implies some condition, as the sale of merchandise, &c., and does not bind the acceptor until the condition is complied with. If a bill is made payable at a certain time after sight, the acceptance should be dated.

A bill should be presented for payment during the regular hours of business, on the day it becomes due.

When acceptance or payment has been refused, the holder should give immediate notice to all the parties whom he intends to hold responsible for the payment of the bill. This notice is usually accompanied with a PROTEST, which is an instrument prepared by a public notary, stating that acceptance or payment has been demanded and refused, and that the holder of the bill intends to recover any damages which he may sustain in consequence.

In some places on the continent of Europe, banks of deposit are established, and exchanges are frequently made by transferring the amounts credited on the books of the bank, from one person to another. The deposits on which these credits are based, are called *banco,* and they usually bear a premium above the ordinary currency of the country. This premium is called the *agio.*

The comparative market value of gold and silver is constantly varying, and the mint value is differently estimated by different governments. Thus, in England the relative worth of the two metals is as 1 to 14.29; in France as 1 to 15.52, and in the United States as 1 to 15.99. In England, silver is so much overvalued, that it would banish the gold coins from circulation, were there not a statute providing that *only gold shall be legal tender in all payments of more than* 40 *shillings.* The relative value of the precious metals should always be considered, in estimating the true par of exchange with any country.

Domestic Exchange.

Inland Exchange is usually effected by checks or Drafts, similar in form to the following:—

$1275.25 Philadelphia, June 3, 1850.

Sixty days from date, pay to James N. Lewis, or order, Twelve Hundred and Seventy-Five Dollars and Twenty-Five Cents, and charge the same to my account.

William Morris.

To Markham & Jones,
Merchants, Cincinnati.

The premium or discount on drafts, may be owing either to a difference in the value of the circulating medium, or to fluctuations in the demand.

The English denominations of shillings and pence, are still retained in this country to some extent. At the formation of the Constitution, the continental currency had suffered a greater depreciation in some of the colonies than in others. Thus, while a pound in New England was worth $3.33⅓, in Pennsylvania it was but $2.66⅔, and in New York but $2.50. The value in Federal Money, of the old currencies of the different States, is as follows:—

A shilling of New England, Virginia, Kentucky, or Tennessee, is 16⅔ cents.

A shilling of New York or North Carolina, is 12½ cents.

A shilling of New Jersey, Pennsylvania, Delaware, or Maryland, is 13⅓ cents.

A shilling of South Carolina or Georgia, is $21\frac{3}{7}$ cents.

We cannot remind teachers too often of the signal benefits they may confer upon their pupils, by communicating collateral know-

ledge to them;[a] that is, such knowledge as is directly connected with the subject of their lessons, though rarely, if ever, found in a text-book. This practice should be commenced with a child the first day he enters the school-room, and should never be discontinued until the day when, for the last time, he leaves it. If teachers would make themselves familiar with such books as Miss Mayo's Lessons on Objects; Mrs. Hamilton's Questions; the first fifty pages of Wilmsen's Children's Friend, and similar works, it would be impossible for them to keep school, or even to hear a recitation, without overflowing with information, both instructive and delightful. The school-room would then cease to be a place so far out of the world; and the gulf which has so long separated it from actual life would be bridged over. When it was our fortune to be a teacher of the Greek and Latin classics, we used to think it as much a part of our daily duty to be prepared with illustrative anecdotes and historical facts, drawn from the manners and customs of other nations and times, in order to render each lesson more useful and interesting, as to be prepared for translation or syntax. The whole business of the school-room, from morning till night, should, in this way, be made attractive and profitable. Children do love information which is adapted to their capacities, and they will desire to go where it can be found, as naturally as bees to flowers. An absurd objection is sometimes urged against such a course; namely, that it will only amuse children, turn what should be toil into pastime, and create a disrelish for close, pains-taking, solitary application. This objection is theoretic merely. It is never made by those who have tried the experiment. It is urged only by such as are too ignorant or too indolent to make the necessary preparation. Not only reason, but experience, proves that it is the best possible means of kindling a desire for knowledge in the bosoms of the young; and when this desire is once kindled, the teacher has only to direct the car, instead of dragging it.

We propose, on the present occasion, to give a specimen of the kind of instruction we mean; to show, by an example, how collateral knowledge may be appropriately introduced to illustrate and enrich the matters contained in the text-book. And we may remark, in passing, that it is strange how any teacher can ever use the term *text-book,* without being reminded that it is only a collection of *texts,* which it is his duty to explain and illustrate.

In our cities, every merchant and most business men have

[a] The remainder of this Section was originally published in the December numbers of the Common School Journal, for 1847.

much to do with bills of exchange and promissory notes. In the country, too, almost every man has something to do with notes of hand, either as promisor or payee, endorser or endorsee. If a man borrows money, he makes these instruments; if he lends money, he receives them. Every respectable man is liable to be on a jury, where questions respecting this class of securities are tried; and no man is so poor, so ignorant, or so far outside of all society, as not to hear conversations about them.

Suppose, then, a class of advanced scholars, whose minds have been previously awakened by a proper course of instruction, to be asked in what way they suppose that commercial transactions between the merchants of different nations are carried on. The citizens of the United States, for instance, send abroad their productions to different quarters of the world, to the amount of a hundred millions of dollars or more annually. In what manner do they receive their pay? In money, or otherwise? Should any one say *in money;* then explain to him the immense trouble, risk, and expense, of bringing a hundred millions of dollars from other countries, across the ocean to this, which amount must soon be sent abroad again to pay for foreign productions which we want.

Here the historical fact may be stated, that we learn from the Pandects,[a] that, when a Roman capitalist had lent money to a foreigner, the common mode of collecting the debt was, to send a trusty slave to the foreign country to receive the debt and its interest, and to bring them home. But this was necessarily both expensive and perilous.

Some may suppose that money may not be remitted for the settlement of each transaction, but that the traffic may be carried on by barter. One merchant may send flour, and receive his pay in cutlery; another beef, and receive broadcloth, &c. It will be easy to answer any suggestions of this kind by showing, that, on such a plan, each man would have to trade in everything, or, at least, in a great variety of things, and with a great number of men. But the same man could never trade in books, leather, jewelry, iron-ware, silks, pork, tea, fish, fruits, logwood, flour, cotton, rice, oil, feathers, coal, hemp, molasses, indigo, otter skins, &c. &c.; or if, by any possibility, one man could trade in all these things, he never could trade with all parts of the world from which they come. *Barter*, therefore, must always be confined to a small number of articles, and to the same place.

[a] Pandects, the digest of the civil law published by Justinian. Explain who Justinian was, and when and where he lived.

Here the subject may be dismissed, for the first day, and the children sent away *to devise* or *to ascertain* in what way commercial transactions between different nations may be made expeditious, safe, and cheap. We suggest the suspension of the subject at this point, because we deem the proper course, in regard to all instruction, to be, first, to awaken the child's mind to a sense of the necessity or desirableness of knowledge, and to put it into an inquiring or receptive state; and then, secondly, to rectify the views which his unaided judgment may suggest, or to impart, when necessary, the precise knowledge he needs.

At subsequent recitations, let the subject be taken up again, and either the pupils or the teacher will explain it as follows:—

A., in Boston, is about to ship flour for the Liverpool market. B., in Liverpool, is a corn[a] merchant, and will buy A.'s flour. At the time of this transaction, C., also a merchant in Boston, wants cloths from D., a manufacturer in Manchester. When A. ships his flour, he draws a bill of exchange on B., in Liverpool, in which he requests B. to pay to himself, or to some other person named on the face of the bill, and *to the order* of whoever is so named, a sum stated, supposed to be the value of the flour when it shall reach Liverpool. But C., in Boston, who wants cloths from D., in Manchester, has no money in England to pay for them; C. therefore buys A.'s bill on B., and pays for it in our money. If the bill be made payable to A.'s order, A. *endorses* it to C.; and C. then *endorses* it to D.,—or he endorses it *in blank*, as it is called, which phrase the teacher must explain,—and sends it to his agent or correspondent in England. When the bill arrives in England, C.'s agent or correspondent presents it to B.; and, the flour having arrived, B. *accepts*, that is, promises to pay it, according to its terms. It is then taken to the manufacturer D., and, the cloths having been bought, it is delivered to him. D. therefore becomes B.'s creditor, and receives payment from him in English money, as A. had received his pay from C. in Boston money. Thus the transaction is completed without the trouble, expense, or risk of sending a cent of money across the ocean, to be sunk by storms, or plundered by pirates.

But suppose, in the above case, that C., after having bought

[a] Let it be explained that the word "corn," in England, never has the same signification as with us. Here, it is commonly used for *Indian corn* or *maize;* but in England, it is a *generic* term, and means wheat, barley, and other cereal grains.

A.'s bill of exchange, does not wish to use it for three months; but his neighbor E. wants it, or one like it, to be used immediately. Is there any way by which C. can transfer this bill to E., receive his money, and so have the use of it for the three months,—or must he go back to A., have the bargain rescinded, the bill cancelled, and a new one drawn in favor of E.? There is such a way. When a bill or note is drawn payable *to the order* of any one, it is payable to whomsoever that one shall order it to be paid. In the case supposed, therefore, C. has only to write his name on the bill, with the words, "Pay to E.," and E. receives C.'s whole interest in it. If he says, "Pay to E., or his order," then E. may order it to be paid to any one else; and so on. It is this transferability, or quality of being transferable from hand to hand, that makes bills of exchange and promissory notes *negotiable* instruments. This word *negotiable* is an important one, and the meaning of it should be precisely understood.

By the civil law of the European continent, bills of exchange and promissory notes were early recognised as mercantile instruments, and, from their nature, negotiable. But in England there was a strong prejudice against the assignment or transfer of debts, because of the abuses liable to be practised, if one man could buy up a debt against another, and sue and imprison him on it. There are still laws, both in England and in this country, against buying up debts for purposes of oppression. Anciently, the common law of England forbade the assignment or negotiation of promissory notes. But the statutes of 3 and 4 Anne gave negotiability to notes, placing them, as mercantile contracts, on the same footing as inland bills of exchange. These English statutes have been generally adopted in the United States, as a part of our common law.

At the present time, therefore, bills of exchange and promissory notes, by their quality of negotiability, are the means by which debts and credits are transferred from one person to another, with safety, despatch, and economy. They afford means of circulation for all the property they represent, and thus they enlarge in every country its stock of circulating wealth, or its means of trade.

Promissory notes are of two kinds;—*negotiable*, and *not negotiable*.

A *negotiable* note expresses on its face that it is payable, not only to the person named in it, but to any other person who shall acquire the legal interest in it. If it be made payable to John Stiles *or order*, it is then negotiable by *endorsement*; if to John Stiles *or bearer*, it is then negotiable *by delivery*.

A note *not negotiable*, expresses on its face that it is payable to the particular person named in it,—as to John Stiles. Such a note is payable only to the party named.

All valid promissory notes *import* a valuable consideration; that is, an action at law may be sustained upon them without specially setting forth or proving a consideration for the note. In this they differ from other unsealed contracts.

In a promissory note, there are two original parties,—the maker of the note, who is called the *promisor;* and the party to whom it is made payable, who is called the *promisee* or *payee.*

A valid *negotiable* promissory note is a written promise for the payment of *money, at all events.*

The promise must be *in writing*, but it may be in ink or pencil; and all but the signature may be in printed letters. The signature gives efficacy to the note, and must be in the handwriting of the promisor or of his authorized agent.

The form of words used is not material, provided the note contains a written *promise* to pay. A mere acknowledgment of indebtedness is not sufficient; Thus, "I owe you $300," though in writing, is only a due bill; it is not a promissory note.

The note must be for the payment of *money.* Therefore, a written promise to pay in goods or labor, is not a negotiable promissory note, although put into the form of a note, and payable "to order" or "bearer."

A negotiable note must, on its face, fix and make certain the *amount* of money to be paid, either in words or figures. Hence, a written promise to pay "all that shall be due on final settlement," or "all that shall be realized from the growing crop," or "all that shall be received from John Stiles," is not a negotiable promissory note. Even though a part of the sum to be paid should be made certain, on the face of the note, it is yet not a negotiable promissory note, even for the part which is so made certain.

The money, by the note itself, must be made payable *at all events*, and *independently of any contingency.* Therefore, a written promise to pay "when certain goods are sold," or "when a certain ship arrives," is not a negotiable promissory note. So, if the note is made payable out of a particular fund, as "my next month's wages," it is not a negotiable promissory note.

And the promise must be to pay money on a day certain,—a y fixed by the note itself. Therefore, a promise to pay $100, "when A. shall come of age," is not a negotiable promissory note; for A. may never come of age. But a note promising to pay $100 when A. shall die, is valid; for A.'s death on some day is certain; and the note, by its own terms, fixes that day for payment.

A note must contain no uncertainty as to the person to whom it is payable. Hence, a written promise to pay to "A. or B.," is not a negotiable promissory note. But a written promise to pay to "A. or bearer," is good; for, in legal effect, such a note is payable to "bearer;" and any person who has legal possession of the note, and presents it for payment, is the "bearer" intended.

A note may be issued with a blank for the payee's name; and, in such case, any *bonâ fide* holder may fill up the blank with his own or any other name, and the note will then be treated as though it had been valid in all respects from its date.

It is indispensable that the maker's name should appear on the note as promisor. The name, however, may be written in ink or pencil, and at the top, or bottom, or in the margin of the paper.

It is not indispensable to the validity of a note that it should be dated, because it is allowable to show the time when it was made, by evidence extrinsic to the note itself; but this is always expensive, often difficult, and sometimes impossible. If a note be postdated or antedated, the time of its actual issue may always be shown, when required for substantial justice.

It is not necessary that a note should specify any place of payment; but when it is the intent of the parties that it shall be paid at a particular place, the place must be specified in the body of the note. A memorandum at the bottom of the note, or on its margin, is not sufficient.

Nor is it essential that a note should be attested. An attestation, however, in Massachusetts, takes a note out of the statute of limitations, as to the payee, his executor or administrator.

A promissory note may be made by one person, or by two or more persons. When made by two or more persons, it may be joint, or joint and several.

When two or more persons sign a note written thus: "We promise to pay," &c., it is a joint note only. If they sign a note written thus: "I promise to pay," &c., it is a joint and several note. When a note is joint, all the promisors must be jointly sued; if joint and several, either promisor may be sued alone.

When a note written thus, "We promise," &c., is signed thus,

"A. B. as principal, and C. D. as surety," it is still the *joint* note of A. B. and C. D. Had it been written, "I promise," &c., and signed in the same way, it would be a joint and several note. The words "principal" and "surety" only show the relation of the makers to each other; they do not affect other parties.

By the phrase "negotiable promissory note," is meant an instrument, negotiable and possessing all the privileges of a promissory note in commerce. A note not negotiable is nevertheless a binding contract between the parties to it.

It is in reference to the transfer of a note from hand to hand,—like a bank bill or a Bank of England note,—that the question of its negotiability becomes material.

THE TRANSFER OF NOTES.

A note may be transferred by delivery, or by endorsement.

AS TO TRANSFER BY DELIVERY.—The rule is, that no person whose name is not on the note, as a party thereto, is liable *on the note.*

Therefore, when a note payable to bearer, or endorsed in blank, is transferred by the holder, *by delivery only*, the party transferring it is not liable upon it.

By not endorsing it, he is understood to mean that he will not be responsible on it; and such, therefore, is the contract between him and the party receiving it.

But if, in such case, the note is received, by the party to whom it is delivered, as a conditional payment of a debt previously due him, or as a conditional satisfaction of any other valuable consideration then given, the party transferring it, if the note is dishonored, (that is, if not paid,) on legal presentment and notice, will be responsible for the debt, or consideration, though not directly suable on the note.

And though a party transferring a note by delivery only, is not liable *on the note*, he is not exempt from *all* obligations or responsibilities.

In the first place, by legal implication, he warrants his own title to the note, and his right to transfer it by delivery.

Then he warrants that the note is genuine, and not forged or fictitious.

And he warrants, moreover, that he has no knowledge of any facts which make the note worthless; for instance, if the note be a bank note, and the party transferring it knows the bank has failed, and conceals this knowledge, his act is a fraud, and the consideration he received may be recovered back. The fraud makes void the contract. And even if the failure of the bank, at the time of the transfer, was unknown to either of the parties to it, it is the better opinion that the transferrer must bear the loss, because it is implied in the transaction that the note would be paid on due presentment.

As to Transfer by Endorsement.—When a note is payable to a person, or his order, it is properly transferable only by endorsement, as nothing else will give to the holder a legal title, so that he can, at law, hold the parties to the note directly liable to him.

By a mere *assignment* of a negotiable note, the holder acquires only the same rights that the assignment would give him, if the note were *not* negotiable.

No particular form of words is required to make an endorsement legal; generally it is enough if the signature of the endorser is on the note, without any words at all; and this is the usual mode of endorsing notes.

The endorsement may be on either side, or any part of the note, or on a paper annexed to it, and in ink or in pencil.

A note transferable by delivery only, may be endorsed; and then the endorser incurs the same obligations and liabilities as if the note had been originally made transferable by endorsement only.

The time of endorsing a note may be material, for if a person, (not the payee of a negotiable note,) endorses it *when it is made*, he will be liable at all events, not as endorser, but as guarantor. If he endorses it afterwards, (not being a regular endorser,) he will be liable if his act is founded on any legal consideration, but not otherwise.

Every endorser, by his endorsement, contracts with every *subsequent* holder of the note,—

1. That the instrument itself, and the signatures antecedent to his, are genuine. 2. That he, (the endorser,) has a good title to the note. 3. That he is competent to bind himself as endorser by his endorsement. 4. That the maker is competent to bind himself as maker, and will, on presentment, pay the note. 5. That if, when duly presented, it is not paid by the maker, the endorser, on due notice, will pay it.

An endorsement may be in "*blank*," or "*in full*," or "*restrictive*," or "*general*," or "*qualified*," or "*conditional*."

A "*blank*" endorsement is merely the name of the endorser written on the note.

After such an endorsement, a note may be transferred by delivery only, and be circulated like a bank note; and any holder may write out, over the endorser's name, the contract *implied by law* on the part of the endorser, and sue upon it.

An endorsement is said to be "*in full,*" when it mentions the name of the person in whose favor it is made, and then the endorsee can transfer his interest in it only by writing his own endorsement on it.

In order to make an endorsement "*restrictive,*" there must be express words, showing that intent; as, "Pay to John Stiles *only.*"

An endorsement is said to be "*general,*" when it is in blank, or payable to the endorsee or *order.*

A "*qualified*" endorsement is one which affects the liability of the endorser, but not the negotiability of the note; as when to the endorsement is added, "*without recourse,*" or "at endorsee's own risk," &c.

A "*conditional*" endorsement limits the validity of the endorsement to some future event, and may be either precedent or subsequent; as, 1st, "Pay John Stiles the within on my marriage;" or, 2d, "Pay John Stiles, or order, the within in six months, unless he sooner receives it from my agent."

Whoever receives an endorsed note, contracts with the endorser, (and if there are many, with each of them,) that the note shall be presented to the promisor for payment at the proper time; that no extra time for payment shall be allowed; and that notice of non-payment shall be immediately given to the endorser; and a default in any of these particulars discharges the endorser.

Due presentment for payment requires that the note should be presented *as soon as it becomes due.* If the holder could delay a day, he might two days, or a year; but any delay may injuriously affect the endorser, and his remedy against other persons. Therefore, if the holder of the note does not present it to the promisor *on the day* it becomes due, the endorsers are discharged.

And the rule is so, although the holder received the note so near the time of its maturity as to make the demand in legal time impossible.

And such demand for payment is required though it is known that the maker is dead or an insolvent.

Where a note is made payable on demand, the time at which payment must be demanded, depends on the circumstances of the case, the rule being that payment must be demanded *in reasonable time.*

And in Massachusetts, by statute, the endorser is excused, if the demand for payment on the maker is not made within sixty days from the date of the note.

If a note is payable generally, that is, without any place being designated, it may be presented at the maker's counting-house or dwelling-house. If it is presented at the counting-house, it must be within the hours in which, by the usage of the city or place, counting-houses are kept open; if at the dwelling-house, then at hours while the family are up, and the maker may be presumed not to have gone to bed.

And where a note is made payable at a particular place, the demand must be made at the place fixed, as well as at the proper time; otherwise the endorser is discharged.

Where a note is payable by a partnership, presentment to either of the partners is sufficient. Where the promisors are only joint contractors, and not partners, demand must be made on each.

The demand must be made *with the note;* and if any particular bank or place is fixed for payment, the note must be there, in order to make the demand valid.

On the failure of the maker to pay, the holder must give due notice of it to each party liable to him; and if he fails to do so to any party, such party is discharged.

And when the endorser lives in the same place with the holder, notice may be given on the day when the demand was made, or the day after, but not later.

When the endorser and holder live in different towns, the notice may be by mail, by special messenger, or by private hand.

And notice by the mail on the day, or the day after, is good, but not later.

Where there are numerous endorsers, each is entitled to notice, and each is to give notice to all parties prior to himself; and each endorser has *the next day after receiving notice*, in which to give notice to any prior party whom he seeks to hold liable to himself.

73. Arbitration of Exchange.

Merchants often find an advantage in remitting bills circuitously, rather than directly to the place where they are due. The determination of the value of such remittances is called ARBITRATION OF EXCHANGE, and is best determined by the CHAIN RULE.

EXAMPLE FOR ILLUSTRATION.

A French merchant wishes to pay in London a bill of £1500. How many francs must he pay to procure remittances through Russia, Hamburg, and Spain, allowing £13 = 75 roubles, 5 roubles = 9 marcs of Hamburg; 3 marcs = 1 Spanish dollar; and 9 dollars = 50 francs.

We write the quantities which are equivalent to each other, as antecedent and consequent, *making each consequent of the same denomination as the next antecedent.* The like factors on opposite sides are cancelled, and the products divided as in § 71, to obtain the answer.

	£13	= 75	roubles.
rou.	5	= 9	marcs.
marcs	3	= 1	dollar.
dol.	9	= 50	francs.
francs	—	= 1500 £	
	13)	375000	
		28846fr. $15\frac{5}{13}$c.	

The question may be otherwise stated in the following manner: If £13 produce 75 roubles, 5 roubles produce 9 marcs, 3 marcs produce $1.00, and $9.00 produce 50 francs, how many francs will £1500 produce? The second set, or set of demand, contains but a single cause and effect. The first, or given set, contains a number of causes and effects, but they are so connected, that all the terms may be multiplied together, as a single compound term. Thus, if £13 produce 75 roubles, and 5 roubles produce 9 marcs, £13 will produce $\frac{75}{5}$ of 9 marcs, and £13 × 5 will produce 75 × 9 marcs. In the same way, it may be shown that £13 × 5 × 3 = $75 × 9 × 1, and £13 × 5 × 3 × 9 = 75 × 9 × 1 × 50 fr. Then how many francs will £1500 produce?

	£13	1500£
rou.	5	
mar.	3	
$	9	
		75 rou
		9 mar.
		1 $
fr.	—	50 fr.

EXAMPLES FOR THE PUPIL.

1. A London merchant wishing to pay 1000 milrees in Lisbon, remits as follows: To Amsterdam, at 36 schillings 7 groats per £; thence to Cadiz, at 17 groats for 2 rials of

plate; thence to Leghorn, at 17 pezze for 100 rials; thence to Lisbon, at 1497 rees for 2 pezze. How many pounds did he remit? *Ans.* £152 3s. 3½d.

2. If a merchant of New York remits $5000 to Havre, at 5fr. 35c. for $1; thence to London, at 49fr. for £2; thence to Hamburg, at 1 marc for 1s. 6d.; and thence to St. Petersburg, at 8 roubles for 17 marcs, how many roubles can he pay with his remittance?[a]

Ans. 6850rou. 74cop.

3. If 33 copecks are equal to 5 English pence, 11 English pence are equal to 3 piastres, 13 piastres are equal to 1 florin, and 5 florins are equal to 29 francs, how many francs are equal to 9000 copecks? *Ans.* 165f. 92c.

4. If a man receives $30 for building 8 rods of wall, and he can purchase 3 barrels of flour for $14, and 3cwt. of sugar for 4 barrels of flour, and 21lb. of tea for 2cwt. of sugar, how many pounds of tea could he purchase by building 17 rods of wall? *Ans.* 107lb. 9¼oz.

5. If 13 days' work will purchase 1 hogshead of molasses, and 2 hogsheads of molasses are worth 5 tons of hay, and 3 tons of hay are worth 4 bags of coffee, how many bags of coffee can be bought with 39 days' labor?

Ans. 10 bags.

6. If 70 braces of Venice are equal to 75 braces of Leghorn, and 7 braces of Leghorn are equal to 4 yards, how many yards are there in 79.375 braces of Venice?

Ans. $48\frac{117}{196}$yd.

7. A merchant in New York orders £500 sterling, due him in London, to be sent by the following circuit: To Hamburg, at 15 marcs banco per £; thence to Copenhagen, at 100 marcs banco for 33 rix-dollars; thence to Bordeaux,

---

[a] The pupil must carefully observe the rule, and make each consequent of the *same denomination* as the next antecedent.

at 3 rix-dollars for 18 francs; thence to Lisbon, at 125 francs for 18 milrees; and thence to New York at $1.25 per milree. What was the arbitrated value of a dollar by this remittance?[a] *Ans.* 3s. 8.89 + d.

8. Amsterdam exchanges with London, at 34 schillings 4 pfennings per £, and with Lisbon at 52 pfennings for 400 reas. What is the arbitrated exchange between London and Lisbon, by way of Amsterdam?

Ans. £1 = 3 ⊕ $169\frac{3}{13}$.

9. The exchange between New York and London is $4.84 per pound; between London and Amsterdam, 35 schillings per pound; between Amsterdam and Paris, 58 groats for 6 francs; between Paris and Venice, 10 francs per ducat; and between Venice and Cadiz, 360 maravedis per ducat. How many maravedis will be equivalent to $4500, by this circuitous remittance? *Ans.* $1454260\frac{1660}{3509}$ mar.

10. If 100 lb. of Amsterdam are equal to 105 lb. of Antwerp, 100 lb. of Antwerp to 142 lb. of Genoa, 100 lb. of Genoa to 70 lb. of Leipsic, and 100 lb. of Leipsic to 104 lb. of America, how many lb. of Amsterdam are equal to 1491 lb. of America? *Ans.* $1373\frac{57}{91}$ lb.

74. ALLIGATION.

When it is desired to make a mixture of a given value, with a variety of ingredients, the following method is usually adopted:—

1. Having written the values of the ingredients in a perpendicular column, connect by a line each value that is less than the required average with one or more that is greater, and each value that is greater with one or more that is less.

2. Write the difference between each value and the average,

[a] After finding the number of dollars which are equivalent to £500, the arbitrated value of $1 is found by dividing £500 by that number.

opposite the ingredient with which that value is connected, and the difference, (or the sum of the differences, if there be more than one,) opposite each ingredient, will be the quantity of that ingredient required.

Examples for Illustration.

1. How much sugar at 5cts., 7cts., 8cts., 10cts., and 12cts., must be mixed together, that the mixture may be worth 9cts. a pound?

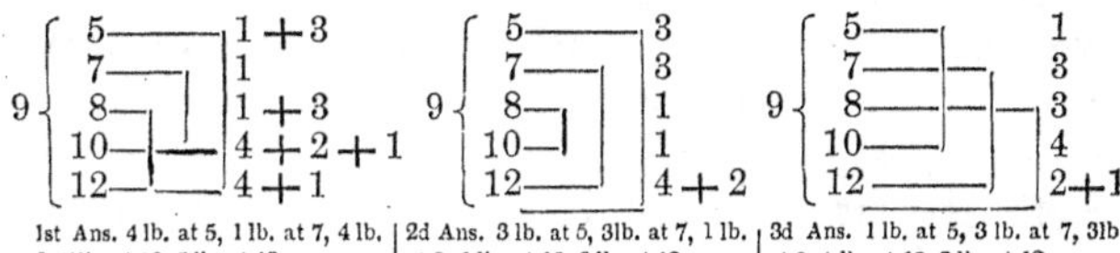

1st Ans. 4 lb. at 5, 1 lb. at 7, 4 lb. at 8, 7 lb. at 10, 5 lb. at 12. | 2d Ans. 3 lb. at 5, 3lb. at 7, 1 lb. at 8, 1 lb. at 10, 6 lb. at 12. | 3d Ans. 1 lb. at 5, 3 lb. at 7, 3lb at 8, 4 lb. at 10, 3 lb. at 12.

We may obtain as many answers as there are different ways of connecting the numbers *above*, with those *below* the average.

To prove the rule correct, let us examine the second of the above answers. If we were mixing sugars at 5 and 12cts. to sell the mixture at 9cts., we should gain 4cts. on every pound of the former, and lose 3cts. on every pound of the latter. Then, on 3 lb. of the former we should gain 12cts. and on 4 lb. of the latter we should lose 12cts.; therefore, if we mix these quantities, we shall neither gain nor lose by selling the mixture at 9cts. In the same way it may be shown that 3 lb. at 7cts. and 2 lb. at 12cts., 1 lb. at 8cts. and 1 lb. at 10cts. may be sold at the average of 9cts., and the same reasoning will prove the truth of each of the other answers.

2. A farmer wishes to mix 10 bushels of barley at 50cts., 4 bushels of oats at 40cts., and 16 bushels of rye at 75cts. with wheat at $1.25, and corn at 90cts. a bushel, so that the mixture may be worth $1.00 per bushel.

We may regard the limited quantities as a single ingredient of 30 bushels, worth 62cts. a bushel. Proceeding in the usual way, we find that 25 bushels at 62cts., 25 at 90cts., and 48 at $1 25, would give us a mixture of the desired average value.

1.00 { 62 — 25; 90 — 25; 1.25 — 38 + 10

But as we have 30 bushels at 62cts., we must take $\frac{30}{25}$ or $\frac{6}{5}$ of these proportionate quantities, and we have 30 bushels at 90cts., and $57\frac{3}{5}$bu. at $1.25, for the answer.

In most questions in Alligation, an infinite number of answers may be obtained, but it will readily be perceived that the preceding method gives only a few of those answers. The following rule is not only more general, but also more analytical in its character.

Assume any quantity you please of each ingredient, find the cost of the whole, and also the cost of the same quantity at the mean rate proposed. If the assumed quantities cost TOO MUCH, take such additional quantities of the lower priced, or such diminished quantities of the higher priced ingredients, as will exactly counterbalance the excess. If TOO LITTLE, take such additional quantities of the higher priced, or such diminished quantities of the lower priced ingredients, as will exactly counterbalance the deficiency.

EXAMPLE FOR ILLUSTRATION.

In what proportion should I mix sugars at 5cts., 7cts., 8cts., 10cts., and 12cts., in order that the mixture may be worth 9cts. a pound?

3 lb.	at 5cts.	=	.15
1 lb.	at 7cts.	=	.07
2 lb.	at 8cts.	=	.16
4 lb.	at 10cts.	=	.40
5 lb.	at 12cts.	=	.60
15 lb.	cost		1.38
15 lb.	at 9cts.	=	1.35
Excess			.03

We may commence by taking any quantity we please of each ingredient, and finding the cost of the whole. If, for example, we take 3 lb. at 5cts., 1 lb. at 7cts., 2 lb. at 8cts., 4 lb. at 10cts., and 5 lb. at 12cts., (making 15 lb. in all,) the whole cost will be $1.38. But 15 lb. at 9cts. would cost only $1.35, therefore our estimated quantities give an excess of 3 cents above the required cost. To balance this excess, we must either add some of the sugar that costs less than the average price proposed, or take out some of that which costs more than the average. For every pound that we add at 5 cents, there will be a deficiency of 4 cents from the mean rate; for every pound at 7cts., a deficiency of 2cts.; for every pound at 8cts., a deficiency of 1ct. Then we may either add $\frac{3}{4}$ of a lb. more at 5cts., or $1\frac{1}{2}$ lb. more at 7cts., or 3 lb. more at 8cts., or 1 lb. more at 8cts. and 1 lb. at 7cts., or any other quantity that will make a deficiency of 3cts.

If the quantities first assumed had given a *deficiency* instead of an *excess*, we should have been obliged to take some additional quantity of one or more of the ingredients whose value is *greater* than the proposed average, or to take out some of the ingredients whose value is less than the average. The cost of the quantities assumed in the margin, would be only 91cts., but 12 lb. at 9cts. would cost $1.08. There is, therefore, a deficiency of 17cts., which may be balanced by taking enough of the sugar that costs more than 9cts., to make an excess of 17cts. Thus, 5 lb. additional at 12cts., and 2 lb. at 10cts., would answer the required conditions. The pupil may determine other values for himself.

4 lb.	at 5cts.	=	.20
3 lb.	at 7cts.	=	.21
2 lb.	at 8cts.	=	.16
1 lb.	at 10cts.	=	.10
2 lb.	at 12cts.	=	.24
12 lb.	cost		.91
12 lb.	at 9cts.	=	1.08
	Deficiency		.17

The following are therefore some of the answers to the proposed question:—(1.) $3\frac{3}{4}$ lb. at 5cts., 1 lb. at 7cts., 2 lb. at 8cts., 4 lb. at 10cts., and 5 lb. at 12cts. (2.) 3 lb. at 5cts., $2\frac{1}{2}$ lb. at 7cts., 2 lb. at 8cts., 4 lb. at 10cts., and 5 lb. at 12cts. (3.) 3 lb. at 5cts., 1 lb. at 7cts., 5 lb. at 8cts., 4 lb. at 10cts., and 5 lb. at 12cts. (4.) 3 lb. at 5cts., 2 lb. at 7cts., 3 lb. at 8cts., 4 lb. at 10cts., 5 lb. at 12cts. (5.) 4 lb. at 5cts., 3 lb. at 7cts., 2 lb. at 8cts., 3 lb. at 10cts., 7 lb. at 12cts.

Let the pupil perform the above examples by *taking out* some of the quantities originally assumed.

Examples for the Pupil.

1. A mixture is made of 24bu. of grain at $1.10 per bu., 28bu. at $.90, and 44bu. at $.60. How must the mixture be sold, in order to gain $\$\frac{3}{16}$ per bushel?[a]

Ans. $1 per bushel.

2. What is the fineness of a composition, consisting of 3 lb. 6oz. of gold, 23 carats fine, 4 lb. 8oz. of 21 carats, 3 lb. 9oz. of 20 carats, and 2 lb. 2oz. of alloy?

Ans. 18 carats.

3. A grocer mixes 65 lb. of sugar at $.08, 30 lb. at $.07,

[a] The first three examples are not, strictly speaking, examples in Alligation, but, in accordance with general custom, they are inserted in this place. See Remark at the close of Section 68.

25 lb. at $.05, and 20 lb. at $.06¼. How must the mixture be sold, in order to gain 25 per cent.?

Ans. 8¾cts. per lb.

4. In what proportions may I mix teas at 60cts., 70cts., $1.10, and $1.20, per pound, so as to gain 12½ per cent. by selling at $.90 per pound?[a] *Ans.* 4 lb. at 60; 3 lb. at 70; 1 lb. at $1.10; 2 lb. at $1.20.

5. A corn merchant bought wheat at $1.20, at $1.10, at $.90, and at $.70 per bushel; but the markets having fallen, he is desirous to sell at $.80 per bushel, and is willing to lose 20 per cent. In what proportions may a mixture be made, to answer the conditions of the question?

Ans. 3bu. at $1.20, 1bu. at $1.10, 1bu. at $.90, 2bu. at $.70.

6. A cask contains 50 gallons of acid, worth 70 cents per gallon; how much will it contain after the value of the liquor has been reduced to 60 cents per gallon, by pouring in water? *Ans.* 58⅓ gallons.

7. A silversmith mixes 20oz. of silver, at 5s. 9d. per ounce, with two other kinds, at 5s. 6d., and 5s. 3d., and as much alloy as reduces the mass to the value of 4s. 10d. per ounce. Required the weight of the whole composition.

Ans. 68$\frac{8}{29}$oz.

8. A refiner melts 14 lb. of gold, 22 carats fine, with 16 lb. of 20 carats fine. How much alloy must be added, in order to make the composition 18 carats fine?

Ans. 4$\frac{8}{9}$ lb.

9. A New York merchant shipped 8000 bushels of grain, consisting of wheat, barley, and rye, which he sold in London at 7s. 4d. per bushel. The prime cost of the white wheat was 6s. 6d., the red wheat 5s. 9d., the barley 3s. 4d., and the rye 4s. 3d. per bushel. The port charges and other

---

[a] An infinite number of answers can be obtained to some of the remaining examples in this section, besides the ones that are given.

incidental expenses, amounted to £58 6s. 8d.; the freight was 10d. per bushel, and he gained 9¼d. per bushel by the transaction. How many bushels of each kind of grain would answer the conditions of the question?

Ans. $2285\frac{5}{7}$bu. white wheat; $3857\frac{1}{7}$bu. red wheat; $285\frac{5}{7}$bu. barley; $1571\frac{3}{7}$bu. rye.

75. General Average.[a]

Whenever any sacrifice of property is made, or any expense necessarily incurred, for the preservation of a ship and cargo, the loss is divided among all the parties interested, either as owners of the vessel, or of the property on board.

"Thus, where the goods of a particular merchant are thrown overboard in a storm, to save the ship from sinking; or where the masts, cables, anchors, or other furniture of the ship are cut away or destroyed for the preservation of the whole; or money or goods are given as a composition to pirates to save the rest; or an expense is incurred in reclaiming the ship, or defending a suit in a foreign court of admiralty, and obtaining her discharge from an unjust capture or detention; in these and the like cases, where any sacrifice is deliberately and voluntarily made, or any expense fairly and *bonâ fide* incurred, to prevent a total loss, such sacrifice or expense is the proper subject of a general contribution, and ought to be ratably borne by the owners of the ship, freight, and cargo, so that the loss may fall equally on all, according to the equitable maxim of the civil law:—No one ought to be enriched by another's loss."[b]

The loss is distributed in such cases, by General Average. But if any sacrifice is made for the sake of the ship only, or of the cargo only, the loss must be borne by the parties immediately interested, and is consequently defrayed by a *particular* average.

In New York, only ½ of the value of the freight is contributory; in the other United States ports, ⅔ of the value is taken. The remainder of the freight is reserved for seamen's wages, because if the seamen were laid under this

[a] McCulloch, Hunt's Mer. Mag. [b] Mr. Serjeant Marshall.

obligation, they might be tempted to oppose a sacrifice necessary for the general safety.

When the loss of masts, or ship furniture, is compensated by general average, as the new articles are supposed to be of more value than those that were lost, only $\frac{2}{3}$ of the value is contributed.

EXAMPLES.

1. It became necessary, in the Downs, to cut the cable of a ship destined for Hull; the ship afterwards struck on a bar, and the captain was compelled to cut away the mast, and throw overboard part of the cargo; in which operation another part was injured. The ship, being cleared from the sands, was forced to take refuge in Ramsgate harbor, to avoid further injury from the storm. Required, from the following statement, the proper adjustment of the loss.

AMOUNT OF LOSSES.		CONTRIBUTORY VALUES.	
Goods of A., cast overboard . . .	$2000	Goods of A., cast overboard . . .	$2000
Damage of B.'s goods .	800	Sound value of B.'s goods	4000
Freight of A.'s goods .	400	Goods of C. . . .	2000
$\frac{2}{3}$ of new cable, anchor, and mast . . .	800	" " D. . . .	8000
Pilotage, port-duties, and expenses of bringing ship off the sands .	700	" " E. . . .	20000
Adjusting average .	16	Value of the ship . .	8000
Postage . . .	4	$\frac{2}{3}$ of freight . . .	3200
Total loss . . .	$4720	Tot. contributory values	$47200

Ans. The owners of the vessel receive $800; C. pays $200.
A. receives . . . $1800; D. " $800.
B. " . . . $400; E. " $2000.

2. In a storm, goods belonging to A., worth $500, were thrown overboard, and the losses of the owners of the vessel amounted to $1500. Adjust the general average, the total

value of A.'s goods being $800; B.'s goods, $1200; C.'s goods, $3000; value of the ship, $10000; total freight, $1500. *Ans.* A. receives $400; B. pays $150.
Owners receive $125; C. pays $375.

XVI. STATISTICS.

76. PRODUCT OF MINES IN THE UNITED STATES, ACCORDING TO THE CENSUS OF 1840.[a]

States and Territories.	Iron, Cast & Bar. Tons.	Lead. Pounds.	Gold. Value.	Other Metals. Value.	Coal. Tons of 28 bushels.	Salt. Bushels.	Granite, Marble, &c. Value.
Maine........	6122	....	....	$1600	....	50000	$107506
N. Hampshire..	1445	1000	....	10300	1069	1200	16038
Massachusetts.	15336	....	....	2500	....	376596	790855
Rhode Island..	4126	....	....	....	1000	....	17800
Connecticut ...	10118	....	....	....	1357	1500	313469
Vermont	7398	....	....	70500	....	....	33855
New York.....	82781	670000	....	84564	....	2867884	1541480
New Jersey ...	18285	....	....	39550	....	500	35721
Pennsylvania..	185639	....	....	100200	1274709	549478	238831
Delaware......	466	....	....	....	....	1160	16000
Maryland......	16776	....	....	28800	7929	1200	22750
Virginia.......	24697	878648	$51758	....	379569	1745618	84489
N. Carolina....	1931	10000	255618	1000	53	4493	3350
South Carolina	2415	....	37418	....	....	2250	3000
Georgia	494	....	121881	....	....	....	51990
Alabama	105	....	61230	....	845	....	13700
Mississippi	....	....	....	....	....	....	....
Louisiana	2766	....	....	....	....	....	....
Tennessee.....	25802	....	1500	....	498	....	30100
Kentucky......	32843	....	....	....	23131	219695	19592
Ohio	42702	....	....	16000	125775	297350	195831
Indiana........	830	....	....	....	8644	6400	35021
Illinois........	158	8755000	200	....	15282	20000	74228
Missouri.......	298	5295455	....	15600	8904	13150	28110
Arkansas......	....	....	....	....	196	8700	15500
Michigan......	601	....	....	....	....	....	2700
Florida........	....	....	....	....	....	12000	2650
Wisconsin.....	3	15129350	....	....	....	....	968
Iowa..........	....	500000	....	....	357	....	350
Total,							

EXAMPLES.

1. Find the total products embraced in the above table.

2. Estimating the average value of the iron at $22 per ton, lead at 4½ cents per lb., coal at $3.75 per ton, and salt at $.40 per bushel, what was the entire value of the above products?

---

[a] Tucker's Progress.

77. TABLE OF THE AGRICULTURAL PRODUCTS OF THE UNITED STATES, ACCORDING TO THE CENSUS OF 1840.[a]

States and Territories.	Bushels of Wheat.	Bushels of Barley.	Bushels of Oats.	Bushels of Rye.	Bushels of Buckwheat	Bushels of Ind. Corn.	Bushels of Potatoes.
Maine........	848166	355161	1076409	137941	51543	950528	10392280
N. Hampshire..	422124	121899	1296114	308148	105103	1162572	6206606
Massachusetts.	157923	165319	1319680	536014	87000	1809192	5385652
Rhode Island..	3098	66490	171517	34521	2979	450498	911973
Connecticut...	87009	33759	1453262	737424	303043	1500441	3414238
Vermont......	495800	54781	2222584	230993	228416	1119678	8869751
New York....	12286418	2520068	20675847	2979323	2287885	10972286	30123614
New Jersey...	774203	12501	3083524	1665820	856117	4361975	2072069
Pennsylvania..	13213077	209893	20641819	6613873	2113742	14240022	9535663
Delaware.....	315165	5260	927405	33546	11299	2099359	200712
Maryland.....	3345783	3594	3534211	723577	73606	8233086	1036433
Virginia.......	10109716	87430	13451062	1482799	243822	34577591	2944660
N. Carolina....	1960855	3574	3193941	213971	15391	23893763	2609239
S. Carolina....	968354	3967	1486208	44738	72	14722805	2698313
Georgia.......	1801830	12979	1610030	60693	141	20905122	1291366
Alabama......	828052	7692	1406353	51008	58	20947004	1708356
Mississippi....	196626	1654	668624	11444	61	13161237	1630100
Louisiana.....	60	....	107353	1812	....	5952912	834341
Tennessee.....	4569692	4809	7035678	304320	17118	44986188	1904370
Kentucky.....	4803152	17491	7155974	1321373	8169	39847120	1055085
Ohio..........	16571661	212440	14393103	814205	633139	33668144	5805021
Indiana........	4049375	28015	5981605	129621	49019	28155887	1525794
Illinois........	3335393	82251	4988008	88197	57884	22634211	2025520
Missouri......	1037386	9801	2234947	68608	15318	17332524	783768
Arkansas......	105878	760	189553	6219	88	4846632	293608
Michigan......	2157108	127802	2114051	34236	113592	2277039	2109205
Florida........	412	30	13829	305	....	898974	264617
Wisconsin.....	212116	11062	406514	1965	10654	379359	419608
Iowa..........	154693	728	216385	3792	6212	1406241	234063
Dist. of Col....	12147	294	15751	5081	272	39485	12035
Total........							

EXAMPLES.

1. Find the total products of each article embraced in the above table.

2. Find the value of the entire product of the wheat, at $1.10 per bushel; of the barley, at $.65 per bu.; of the oats, at $.40 per bu.; of the rye, at $.70 per bu.; of the buckwheat, at $.75 per bu.; of the corn, at $.80 per bu.; of the potatoes, at $.37½ per bu.

3. What percentage of the entire wheat harvest was produced in 1840, by the four principal wheat-growing States?

[a] Tucker's Progress.

78. TABLE OF AGRICULTURAL PRODUCTS.—CONTINUED.[a]

States, &c.	Hay. Tons.	Tobacco. Pounds.	Rice. Pounds.	Cotton. Pounds.	Wool. Pounds.	Sugar. Pounds.	Dairy Products.
Maine......	691358	30	....	....	1465551	257464	$1496902
N. Hamp....	496107	115	....	....	1260517	1162368	1638543
Mass.......	569395	64955	....	....	941906	579227	2373299
R. Island...	63449	317	....	....	183830	50	223229
Connecticut	426704	471657	....	....	889870	51764	1376534
Vermont....	836739	585	....	....	3699235	4647934	2008737
New York .	3127047	744	....	....	9845295	10048109	10496021
New Jersey	334861	1922	....	....	397207	56	1328032
Pennsylv'a.	1311643	325018	....	....	3048564	2265755	3187292
Delaware ..	22483	272	....	334	64404	....	113828
Maryland...	106687½	24816012	....	5673	488201	36266	457466
Virginia....	364708½	75347106	2956	3494483	2538374	1541833	1480488
N. Carolina	101369	16772359	2820388	51926190	625044	7163	674349
S. Carolina.	24618	51519	60590861	61710274	299170	30000	577810
Georgia.....	16969¾	162894	12384732	163392396	371303	329744	605172
Alabama....	12718	273302	149019	117138823	220353	10143	265200
Mississippi.	171	83471	777195	193401577	175196	77	359585
Louisiana...	24651	119824	3604534	152555368	49283	119947720	153069
Tennessee..	31233	29550432	7977	27701277	1060332	258073	472141
Kentucky...	88306	53436909	16376	691456	1786847	1377835	931363
Ohio........	1022037	5942275	....	....	3685315	6363386	1848869
Indiana.....	178029	1820306	....	180	1237919	3727795	742269
Illinois.....	164932	564326	460	200947	650007	399813	428175
Missouri ...	49083	9067913	50	121122	562265	274853	100432
Arkansas..	586	148439	5454	6028642	64943	1542	59205
Michigan....	130805	1602	....	....	153375	1329784	301052
Florida.....	1197	75274	481420	12110533	7285	275317	23094
Wisconsin..	30938	115	....	....	6777	135288	35677
Iowa.......	17953	8076	....	....	23039	41450	23609
Dist. of Col.	1331	55550	....	....	707	....	5566
Total							

EXAMPLES.

1. Which of the States yielded the heaviest product of each of the foregoing articles, and what percentage of the entire product of each article, was the heaviest product?

2. Find the average amount yielded by each of the producing States, of each article.

3. Find the total product of each article, and the value of the cotton, at 12½ cents per pound.

4. Find what percentage of each of the foregoing articles was produced by the New England States; by the Middle States; by the Southern States; by the South-western States; by the North-western States.

[a] Tucker's Progress.

79. TABLE OF OCCUPATIONS IN THE UNITED STATES, ACCORDING TO THE CENSUS OF 1840.[a]

States and Territories.	Mining.	Agriculture.	Commerce.	Manufactures.	Ocean Navig.	Internal Navig.	Learned Professions.	Total.
Maine.........	36	101630	2921	21879	10091	539	1889	
N. Hampshire..	13	77949	1379	17826	452	198	1640	
Vermont.......	77	73150	1303	13174	41	146	1563	
Massachusetts.	499	87837	8063	85176	27153	372	3804	
Rhode Island..	35	16617	1348	21271	1717	228	457	
Connecticut....	151	56955	2743	27932	2700	431	1697	
N. Eng. States	......	....	....	....	......	......	....	675082
New York.....	1898	455954	28468	173193	5511	10167	14111	
New Jersey....	266	56701	2283	27004	1143	1625	1627	
Pennsylvania..	4603	207533	15338	105883	1815	3951	6706	
Delaware......	5	16015	467	4060	401	235	199	
Maryland......	320	72046	3281	21529	717	1528	1666	
Dist. of Col....	......	384	240	2278	126	80	203	
Middle States..	......	....	....	....	......	......	....	1251560
Virginia.......	1995	318771	6361	54147	582	2952	3866	
N. Carolina....	589	217095	1734	14322	327	379	1086	
South Carolina,	51	198363	1958	10325	381	348	1481	
Georgia........	574	209383	2428	7984	262	352	1250	
Florida........	1	12117	481	1177	435	118	204	
South. States..	......	....	....	....	......	......	....	1073879
Alabama	96	177439	2212	7195	256	758	1514	
Mississippi....	14	139724	1303	4151	33	100	1506	
Louisiana......	1	79289	8549	7565	1322	662	1018	
Arkansas......	41	26355	215	1173	3	39	301	
Tennessee.....	103	227739	2217	17815	55	302	2042	
Southw. States	......	....	....	....	......	......	....	713107
Missouri.......	742	92408	2522	11100	39	1885	1469	
Kentucky......	331	197738	3448	23217	44	968	2487	
Ohio	704	272579	9201	66265	212	3323	5663	
Indiana........	233	148806	3076	20590	89	627	2257	
Illinois........	782	105337	2506	13185	63	310	2021	
Michigan......	40	56521	728	6890	24	166	904	
Wisconsin	794	7047	479	1814	14	209	259	
Iowa..........	217	10469	355	1629	13	78	365	
Northw. States	......	....	....	....	......	......	....	1085242
Total.......	15211	3719951	117607	791749	56021	33076	65255	

EXAMPLES.

1. Fill all the blanks in the above table.

2. Find the average number employed in each occupation, in each of the New England States; in each of the Middle States; in each of the Southern States; in each of the South-western States; in each of the North-western States; in the entire Union.

---

[a] Tucker's Progress.

80. TABLE SHOWING THE STATE OF EDUCATION IN THE UNITED STATES, AND THE NUMBER OF WHITE PERSONS OVER 20 YEARS OF AGE WHO COULD NOT READ OR WRITE, ACCORDING TO THE CENSUS OF 1840.[a]

States and Territories.	Colleges and Univ.	Students.	Academies and Gram. Schools.	Scholars.	Primary Schools.	Scholars.	Scholars at Public Charge.	Illiterate.
Maine	4	266	86	8477	3385	164477	60212	3241
New Hampshire	2	433	68	5799	2127	83632	7715	942
Vermont	3	233	46	4113	2402	82817	14701	2276
Massachusetts	4	769	251	16746	3362	160257	158351	4448
Rhode Island	2	324	52	3664	434	17355	10749	1614
Connecticut	4	832	127	4865	1619	65739	10912	526
New England States								
New York	12	1285	505	34715	10593	502367	27075	44452
New Jersey	3	443	66	3027	1207	52583	7128	6385
Pennsylvania	20	2034	290	15970	4978	179989	73908	33940
Delaware	1	23	20	764	152	6924	1571	4832
Maryland	12	813	133	4289	565	16851	6624	11817
District of Columbia	2	224	26	1389	29	851	482	1033
Middle States								
Virginia	13	1097	382	11083	1561	35331	9791	58787
North Carolina	2	158	141	4398	632	14937	124	56609
South Carolina	1	168	117	4326	566	12520	3524	20615
Georgia	11	622	176	7878	601	15561	1333	30717
Florida	....	......	18	732	51	925	14	1303
Southern States								
Alabama	2	152	114	5018	639	16243	3213	22592
Mississippi	7	454	71	2553	382	8236	107	8360
Louisiana	12	989	52	1995	179	3573	1190	4861
Arkansas	....	......	8	300	113	2614	....	6567
Tennessee	8	492	152	5539	983	25090	6907	58531
South-west'n States								
Missouri	6	495	47	1926	642	16788	526	19457
Kentucky	10	1419	116	4906	952	24641	429	40018
Ohio	18	1717	73	4310	5186	218609	51812	35394
Indiana	4	322	54	2946	1521	48189	6929	38100
Illinois	5	311	42	1967	1241	34876	1683	27502
Michigan	5	158	12	485	975	29701	998	2173
Wisconsin	....	......	2	65	77	1937	315	1701
Iowa	....	......	1	25	63	1500	....	1118
North-west'n States								
Total								

EXAMPLE.

1. Fill the blanks in the above table, and find the percentage of the entire population[b] embraced in each class, in each section of the Union.

---

[a] Tucker's Progress.

[b] See Table of Population, page 36.

81. SUMMARY OF THE ANNUAL PRODUCTS OF INDUSTRY IN THE SEVERAL STATES, AS ESTIMATED IN THE CENSUS OF 1840.[a]

States and Territories.	VALUE OF ANNUAL PRODUCTS FROM						
	Agriculture.	Manufactures.	Commerce.	Mining.	Forest.	Fisheries.	Total.
Maine...	$15856270	$5615303	$1505380	$327376	$1877663	$1280713	
N. Hamp.	11377752	6545811	1001533	88373	449861	92811	
Vermont	17879155	5685425	758899	389488	430224	....	
Mass....	16065627	43518057	7004691	2020572	377354	6483996	
R. Island	2199309	8640626	1294956	162410	44610	659312	
Connec't	11371776	12778963	1963281	820419	181575	907723	
N. E. S.	74749889	....	....	....	....	....	187657294
N. York..	108275281	47454514	24311715	7408070	5040781	1316072	
N. Jersey	16209853	10696257	1206929	1073921	361326	124140	
Penns'a..	68180924	33354279	10593368	17666146	1203578	35360	
Delaw'e.	3198440	1538879	266257	54555	13119	181285	
Maryland	17586720	6212677	3499087	1056210	241194	225773	
D. of Col.	176942	904526	802725	....	....	87400	
Middle S.	....	....	....	....	....	....	390558303
Virginia.	59085821	8349218	5299451	3321629	617760	95173	
N. Caro'a	26975831	2053697	1322284	372486	1446108	251792	
S. Caro'a	21553691	2248915	2632421	187608	549626	1275	
Georgia..	31468271	1953950	2248488	191631	117439	584	
Florida..	1834237	434544	464637	2700	27350	213219	
S. States	....	....	....	....	....	....	175321836
Alabama	24696513	1732770	2273267	81310	177465	....	
Mississ'i	26494565	1585790	1453686	....	205297	....	
Louisi'a.	22851375	4087655	7868898	165280	71751	....	
Arkans's.	5086757	1145309	420635	18225	217469	....	
Tenness..	31660180	2477193	2239478	1371331	225179	....	
S. W. S.	....	....	....	....	....	....	138607378
Missouri.	10484263	2360708	2349245	187669	448559	....	
Kentuc'y.	29226545	5092353	2580575	1539919	184799	....	
Ohio.....	37802001	14588091	8050316	2442682	1013063	10525	
Indiana..	17247743	3676705	1866155	660836	80000	1192	
Illinois..	13701466	3213981	1493125	293272	249841	....	
Michigan	4502889	1376249	622822	56790	467540	....	
Wiscon.	568105	304692	189957	384603	430580	27663	
Iowa...	769295	179087	136525	13250	83949	....	
N. W. S.	....	....	....	....	....	....	170989925
Total	....	....	....	....	....	....	1063134736

EXAMPLE.

1. Fill all the blanks in the above table, and find what per centage of the total products was derived from each source. Find also the proportion to each inhabitant of the total products of industry in each State.

---

a Tucker's Progress.

XVII. PERMUTATION AND COMBINATION.

82. Permutation.

Permutation shows the number of changes that can be made in the order of a given number of things.

PROBLEM I.

To find the number of changes that can be made of any given number of things, all different from each other.

How many changes may be made in the position of 4 persons at table?

If there were but two persons, *a* and *b*, they could sit in but two positions, *ab* and *ba*. If there were three, the third could sit at the head, in the middle, or at the foot, in each of the two changes, and there could then be $1 \times 2 \times 3 = 6$ changes. If there were 4, the fourth could sit as the 1st, 2d, 3d, or 4th, in each of these 6 changes, and there would then be $1 \times 2 \times 3 \times 4 = 24$ changes.

RULE.

Multiply together the series of numbers 1, 2, 3, &c., up to the given number, and the product will be the number sought.

1. How many variations may be made in the order of the 9 digits? *Ans.* 362880.

2. How many changes may be made in the position of the letters of the alphabet?
Ans. 403291461126605635584000000.

3. How long a time will be required for 8 persons to seat themselves at table in every possible order, if they eat 3 meals a day?

PROBLEM II.

Any number of different things being given, to find how many changes can be made out of them by taking a given number of the things at a time.

If we have five things, each one of the 5 may be placed before each of the others, and we thus have 5×4 permutations of 2 out of 5. If we take 3 at a time, the third thing may be placed as 1st, 2d, and 3d, in each of these permutations, and we have $5 \times 4 \times 3$ permutations of 3 out of 5. For a similar reason we have $5 \times 4 \times 3 \times 2$ permutations of 4 out of 5, &c.

RULE.

Take a series of numbers, commencing with the number of things given, and decreasing by 1, until the number of terms is equal to the number of things to be taken at a time. The product of all the terms will be the answer required.

4. How many changes can be rung with 8 bells, taking 5 at a time? *Ans.* 6720.

5. How many numbers of 4 different figures each, can be expressed by the 9 digits?

6. In how many different ways may 10 letters of the alphabet be arranged? *Ans.* 19275223968000.

PROBLEM III.

To find the number of permutations in any given number of things, among which there are several of a kind.

How many permutations can be made of the letters in the word *terrier?*

If the letters were all different, the permutations, according to Problem I., would be $1 \times 2 \times 3 \times 4 \times 5 \times 6 \times 7 = 5040$. But the permutations of the three r's would, if they were all different, be $1 \times 2 \times 3$, which could be combined with each of the other changes; the number must therefore be divided by $1 \times 2 \times 3$. For the same reason, it must also be divided by 1×2, on account of the 2 e's. Then the true number sought is $\frac{1 \times 2 \times 3 \times 4 \times 5 \times 6 \times 7}{1 \times 2 \times 3 \times 1 \times 2} = 420$.

RULE.

Take the natural series, from 1 up to the number of things of the first kind, and the same series up to the number of things of each succeeding kind, and form the continued product of all the series.

By the continued product, divide the number of permutations

of which the given things would be capable, if they were all different, and the quotient will be the number sought.

7. How many changes can be made in the order of the letters, in the word Philadelphia? *Ans.* 14968800.

8. How many different numbers can be made, that will employ all the figures in the number 119089907343?

9. How many permutations can be made with the letters in the word Cincinnati? *Ans.* 201600.

83. Combination.

Combination shows in how many ways a less number of things may be chosen from a greater.

If we have ten articles, each may be combined with every one of the nine remaining ones, and therefore we may have 10×9 permutations of 2 out of 10. But each combination will evidently be repeated; thus, we have *ab* and *ba*, *ac* and *ca*, &c. Therefore, the number of combinations will be $\frac{10 \times 9}{2}$.

If now we add an eleventh article, each of the eleven may be joined to each of the combinations of the remaining ten, and we shall have $\frac{11 \times 10 \times 9}{1 \times 2}$ permutations. But each combination will be three times repeated; thus we shall have *abc*, *bac*, and *cab*; *abd*, *bad*, and *dab*, &c. The number of combinations of 3 out of 11 will therefore be $\frac{11 \times 10 \times 9}{1 \times 2 \times 3}$. Hence we obtain the following

RULE.

Write for a numerator the descending series, commencing with the number from which the combinations are to be made, and for a denominator the ascending series, commencing with 1, giving to each series as many terms as are equivalent to the number in one combination.

Cancel the like factors in the numerator and denominator, and divide.

10. How many combinations of 4 letters, can be made from the alphabet? *Ans.* 14950.

11. How many combinations of 7 can be made from 18 apples? *Ans.* 31824.

12. How many ranks of 10 men, may be made in a company of 80?

13. How many locks of different wards, may be unlocked with a key of 6 wards? [Find the number of combinations of 1, 2, 3, 4, 5, and 6 in 6, and the sum of all the combinations will be the number required.] *Ans.* 63.

XVIII. INVOLUTION AND EVOLUTION.

84. INVOLUTION.

INVOLUTION is the repeated multiplication of a number by itself.

The product obtained by Involution is called a *power*. The *root* is the number involved, or the *first power*. If the root be multiplied by itself, or employed twice as a factor, the product is the second power. If the root is employed three times as a factor, it is raised to the 3d power; if 5 times, to the 5th power, &c. Thus, 2 is the 1st power of 2 or 2^1. 2×2 or 4, is the 2d power of 2, or 2^2. $2 \times 2 \times 2$ or 8, is the 3d power of 2, or 2^3. $2 \times 2 \times 2 \times 2 \times 2$ or 32, is the 5th power of 2, or 2^5. The power is usually denoted by a small figure over the right of the root, called the *exponent*, or *index*. When there is no exponent, the number is regarded as the 1st power.

The second power is often called the square, because the number of square feet in any square surface, is obtained by multiplying the number of feet in one side by itself.

The third power is often called the cube, because the number of cubic feet in any cubical block, may be obtained by raising the number of feet in one side to the 3d power.

The 4th power is sometimes called the bi-quadrate, or the square squared; the 5th power, the first sursolid; the 6th power, the square cubed, or the cube squared; the 7th power, the second sursolid; the 8th power, the bi-quadrate squared; the 9th power, the cube cubed; the 10th power, the 1st sursolid squared, &c.

If the exponents of any two powers of the same number be added, we shall obtain the exponent of their product. Thus $6^3 \times 6^5 = 6 \times 6 \times 6 \times 6 \times 6 \times 6 \times 6 \times 6 = 6^8$; $4^2 \times 4^3 = 4 \times 4 \times 4 \times 4 \times 4 = 4^5$.

In any two powers of the same number, if we subtract the smaller exponent from the larger, we shall obtain the exponent of their quotient. Thus $6^8 \div 6^5 = \frac{6 \times 6 \times 6 \times 6 \times 6 \times 6 \times 6 \times 6}{6 \times 6 \times 6 \times 6 \times 6} = 6 \times 6 \times 6 = 6^3$.

We may represent any power of a number by multiplying its exponent. Thus, the 7th power of 5 is 5^7; the 3d power of 2^2 is 2^6, because $2^2 \times 2^2 \times 2^2 = 2^6$. These properties form the basis of the system of Logarithms.

1. What is the 2d power of 6? the 3d power?

2. Find the value of $.9^4$; 12^3; $(\frac{1}{2})^5$; 2^9.

3. Find the value of 16^4; 1.6^4; $.16^4$; $(\frac{11}{13})^3$.

4. What is the square of 13.68? of $9\frac{2}{5}$?

5. What is the difference between 3^4 and 4^3?

6. What is the value of 1^{17}; 3^7; $2^3 \times 2^2$?

7. What power of 9 is equivalent to $9^5 \times 9^3$; $9^2 \times 9^{10}$; $9^4 \times 9^6$; $9 \times 9^7 \times 9^8$?

8. Multiply 127^9 by 127^7, and divide the product by 127^{15}.

9. Divide 31^9 by 31^9; 17^8 by 17^5; 42^7 by 42^6.

10. What is the sixth power of $4\frac{1}{10}$?

11. What is the 9th power of 5^3? the 12th power of 18^5? the 24th power of 17^2?

85. EVOLUTION.

EVOLUTION is the process by which we discover the root of any given power. Thus, 3 is the 2d root of 9, the 3d root of 27, the 5th root of 243, because $9 = 3^2$; $27 = 3^3$; $243 = 3^5$. So the 2d or square root of 49 is 7; the 3d or cube root of 125 is 5; the 4th root of 16 is 2; the 5th root of 1024 is 4, &c. We may denote a root by a *radical sign*, or by a *fractional exponent*.

The radical sign is $\sqrt{}$, and when employed by itself denotes the square root. If we wish to denote the 3d, 5th, 7th, &c. root, the index of the root is written above the radical sign, thus, $\sqrt[3]{}$, $\sqrt[5]{}$, &c. In fractional exponents, the numerator expresses the power of the number, and the denominator expresses the root. Thus, $(27)^{\frac{1}{3}} = \sqrt[3]{27}$; $(16)^{\frac{3}{4}} = \sqrt[4]{16^3}$; $(32)^{\frac{4}{5}} = \sqrt[5]{32^4}$, &c.

The product, or the quotient, of two second, third, or other roots, is the 2d, 3d, &c., root of the product or quotient. Thus, $\sqrt[3]{27} \times \sqrt[3]{125} = \sqrt[3]{27 \times 125}$ or $\sqrt[3]{3375}$. For $27 = 3^3 = 3 \times 3 \times 3$, and $125 = 5 \times 5 \times 5$. Then $27 \times 125 = 3 \times 3 \times 3 \times 5 \times 5 \times 5 = 3 \times 5 \times 3 \times 5 \times 3 \times 5 = 15^3$. Therefore $\sqrt[3]{27 \times 125} = 15$. In a similar manner it may be shown that $\sqrt[3]{3375} \div \sqrt[3]{125} = \sqrt[3]{27}$.

The power of any root may be obtained by multiplying the fractional exponent. Thus, the 4th power of $27^{\frac{2}{3}} = 27^{\frac{8}{3}}$. For by the last proposition, $\sqrt[3]{27^2} \times \sqrt[3]{27^2} \times \sqrt[3]{27^2} \times \sqrt[3]{27^2} = \sqrt[3]{27^8} = 27^{\frac{8}{3}}$.

The root of any power or root may be obtained by dividing the exponent by the index of the desired root. Thus, $\sqrt[3]{3^{\frac{1}{5}}} = 3^{\frac{1}{5} \div 3} = 3^{\frac{1}{15}}$.

This is the converse of the last proposition. For if the 3d power of $3^{\frac{1}{15}}$ is $3^{\frac{3}{15}}$, or $3^{\frac{1}{5}}$, the 3d root of $3^{\frac{1}{5}}$ must be $3^{\frac{1}{15}}$.

If the numerator and denominator of fractional indices be multiplied or divided by the same number, the value of the quantity is not altered. Thus, $3^{\frac{4}{6}} = 3^{\frac{8}{12}} = 3^{\frac{2}{3}}$. For the multiplication of the numerator involves the number to a certain power, and the multiplication of the denominator extracts the corresponding root. Then the 3d root of the 3d power, the 5th root of the 5th power, &c., is the 1st power.

We may multiply or divide any two roots of the same number, by adding or subtracting the fractional exponents. Thus, $\sqrt[3]{2} \times \sqrt{2} = 2^{\frac{1}{3} + \frac{1}{2}} = 2^{\frac{5}{6}}$; $\sqrt[3]{5} \div \sqrt[4]{5} = 5^{\frac{1}{3} - \frac{1}{4}} = 5^{\frac{1}{12}}$. For by the last proposition we have $\sqrt[3]{2} \times \sqrt{2} = \sqrt[6]{2^2} \times \sqrt[6]{2^3}$, which is equivalent to $\sqrt[6]{2^5}$ or $2^{\frac{5}{6}}$. Also $\sqrt[3]{5} \div \sqrt[4]{5} = \sqrt[12]{5^4} \div \sqrt[12]{5^3} = \sqrt[12]{5}$ or $5^{\frac{1}{12}}$.

When the exact root of a number can be obtained, it is called a *rational number*. An *irrational number*, or *surd*, is one whose exact root cannot be obtained. Thus, $\sqrt{16}$, $\sqrt[3]{27}$, $\sqrt[3]{64}$, $\sqrt[4]{81}$, are rational numbers, equivalent to 4, 3, 4, 3, respectively. But $\sqrt{5}$, $\sqrt[3]{19}$, $\sqrt[5]{16}$, are all surds, and their roots can only be obtained approximately.

A number which has a rational root, is called a perfect power. Thus, 16 is a perfect 2d power, and a perfect 4th power, but an imperfect power of any other degree. But 5, 7, 12, &c., are imperfect powers of any degree.

1. What is the square root of 9? the cube root of 8?

2. What is the 4th root of 81? the 5th root of 32?

3. What is the value of $\sqrt[9]{1}$; $25^{\frac{1}{2}}$; $\sqrt[6]{64}$; $64^{\frac{8}{16}}$?

4. What is the product of $\sqrt[4]{8}$ by $\sqrt[4]{12}$; of $\sqrt[7]{9}$ by $7^{\frac{1}{7}}$?

5. Multiply $\sqrt[4]{3}$ by $\sqrt[6]{3}$; $7^{\frac{1}{2}}$ by $\sqrt[9]{7}$; $4^{\frac{2}{3}}$ by $\sqrt[4]{4^2}$.

6. Divide $6^{\frac{1}{3}}$ by $6^{\frac{1}{5}}$; $\sqrt[4]{5}$ by $\sqrt[5]{5}$; $17^{\frac{1}{3}}$ by $\sqrt[9]{17}$.

7. Find the 4th power of $\sqrt{9}$; the 6th power of $8^{\frac{4}{5}}$.

8. What is the cube root of 7^6; the 5th root of $11^{\frac{4}{7}}$?

86. Roots of all Powers.[a]

When the exponent of a power can be resolved into two or more factors, by successively extracting the roots denoted by those factors, we may obtain the root desired. Thus, as $12 = 3 \times 2 \times 2$, the cube root of the square root of the square root of a number, is equal to the 12th root. So the square root of the square root is the 4th root; the cube root of the cube root is the 9th root; the cube root of the square root is the 6th root, &c.

The following rule will, however, be generally found more convenient for determining the roots of all powers greater than the cube.

---

[a] It does not seem necessary to give the ordinary rules for the extraction of the square and cube roots, as the pupil is supposed to be already familiar with them. But the following formulas will probably be found useful, in explaining the usual method of finding the trial and complete divisors.

The square of any number consisting of tens and units =
The square of the tens +
(2 × the tens + the units) × the units.

The cube of any number consisting of tens and units =
The cube of the tens +
{3 × the square of the tens +
3 × the tens × the units +
the square of the units.} × the units.

The entire portion of the root which has been found, at any step, may be considered as the tens, and the next root figure will then represent the units.

GENERAL RULE.

At the left of the number whose root is required, arrange as many columns as are equal to the index of the root, writing 1 at the head of the first or left hand column, and zero at the head of each of the others.

Divide the number into periods of as many figures as the index of the root requires. Write the root of the left hand period as the first figure of the true root.

Multiply the number in the first column by the root figure, and add the product to the second column; add the product of this sum by the root figure to the third column, and so proceed, *subtracting* the product of the last column from the given number.

Repeat this process, stopping at the last column, and thus proceed, stopping one column sooner each time, until the last sum falls in the second column.

To determine the second root figure, consider the number in the last column as a trial divisor, and proceed with the second root figure thus obtained,[a] precisely as with the first.

Continue the operation until the root is completed, or the approximation carried as far as is desired.

In order to avoid error, observe carefully the value of each root figure and each product. Thus, if the first root figure is hundreds, the number in the second column will be hundreds,—in the third, ten thousands,—in the fourth, millions, &c.

Examples for Illustration.

1. What is the third root of 205692449327?

---

[a] If the root figure thus found proves too large, erase it and try a smaller number.

1	0	0	205692449327(5903
	5 thous.	25 mill.	125 bill.
	5	25 c. d.	806,92
	5	50	80379 mill.
	10	(1) 75 t. d.	3134,49327
	5	1431 ten thous.	313449327 un.
	(1) 159 hun.	8931 c. d.	
	9	1512	
	168	(2) 10443 t. d.	
	9	53109 un.	
	(2) 17703 un.	104483109 c. d.	

The complete divisors are marked c. d., the trial divisors, t. d. The figures at which the new additions commence are marked (1), (2). The partial dividends by which each root figure is determined, are distinguished by a comma. They always terminate with the first figure of the period that is annexed. The abbreviations, thous., mill., &c., show the value of the figures against which they are placed.

2. Extract the 5th root of 858533232.56832.

1	0	0	0	0	858533232.56832(61.2
	6 tens.	36 hund.	216 thous.	1296 ten thou.	7776 hund. thous.
	6	36	216	1296 c. d.	8093,3232
	6	72	648	5184	66996301
	12	108	864	(1) 6480 t. d.	139369315,6832
	6	108	1296	2196301 un.	1393693156832
	18	216	(1) 2160	66996301 c. d.	
	6	144	36301 un.	2232904	
	24	(1) 360	2196301	(2) 69229205 t. d.	
	6	301 un	36603	4554528416 ten thous.	
	(1) 301 un.	36301	2232904	696846578416 c. d.	
	1	302	36906		
	302	36603	(2) 2269810		
	1	303	7454208 thous.		
	303	36906	2277264208		
	1	304			
	304	(2) 37210			
	1	6104 hund.			
	(2) 3052 tenths	3727104			

The additions to the left hand column may be made mentally, and thus shorten the labor. There are other abbreviations, for which the student is referred to the Chapter on Approximations.

TABLE OF ROOTS AND POWERS.

1st power,	1	2	3	4	5	6	7	8	9
2d power,	1	4	9	16	25	36	49	64	81
3d power,	1	8	27	64	125	216	343	512	729
4th power,	1	16	81	256	625	1296	2401	4096	6561
5th power,	1	32	243	1024	3125	7776	16807	32768	59049
6th power,	1	64	729	4096	15625	46656	117649	262144	531441
7th power,	1	128	2187	16384	78125	279936	823543	2097152	4782969
8th power,	1	256	6561	65536	390625	1679616	5764801	16777216	43046721
9th power,	1	512	19683	262144	1953125	10077696	40353607	134217728	387420489
10th power,	1	1024	59049	1048576	9765625	60466176	282475249	1073741824	3486784401
11th power,	1	2048	177147	4194304	48828125	362797056	1977326743	8589934592	31381059609
12th power,	1	4096	531441	16777216	244140625	2176782336	13841287201	68719476736	282429536481
13th power,	1	8192	1594323	67108864	1220703125	13060694016	96889010407	549755813888	2541865828329
14th power,	1	16384	4782969	268435456	6103515625	78364164096	678223072849	4398046511104	22876792454961
15th power,	1	32768	14348907	1073741824	30517578125	470184984576	4747561509943	35184372088832	205891132094649

The first root figure in each of the following examples may be found by the table of Powers and Roots.

1. Extract the square root of 350026681.

2. Extract the square root of 3; 5; 6.5.

3. Extract the cube root of 2924207.

4. Extract the cube root of 13; 12.5.

5. Extract the fifth root of 65.7748550151.

6. Extract the 7th root of 1.246688292353624506368.

87. APPLICATION OF THE SQUARE ROOT.

The areas of any similar figures are proportioned to the squares of their like dimensions.

The area of any circle is equal to the square of its diameter multiplied by .7854.

The circumference of a circle is equal to its diameter multiplied by 3.1416.[a]

The area of a triangle is equal to the base multiplied by half the height.

In any right-angled triangle, the square of the longest side is equal to the sum of the squares of the other two sides.

The distance through which bodies fall, when falling freely, are as the squares of the times. In a vacuum, a body would fall $16\frac{1}{12}$ft. in 1 second. Then we have the proportion, letting n represent any number of seconds,

$$\begin{array}{llll} \textit{sec.} & \textit{sec.} & \textit{ft.} & \textit{ft.} \\ (1)^2 : & n^2 :: & 16\frac{1}{12} : & \text{distance in } n \text{ seconds.} \end{array}$$

Any three terms of this proportion being given, the fourth may be readily found. But it should be remarked, that in consequence of the resistance of the air, the space

[a] The more exact ratio is, 3.14159265358979323846264338328.

actually fallen through is somewhat less than that given by the formula.

If b represents the base of a right-angled triangle, p the perpendicular, and h the hypothenuse, $h=\sqrt{p^2+b^2}$; $b=\sqrt{h^2-p^2}$; $p=\sqrt{h^2-b^2}$.

The square root of the area of any surface will give the side of a square, equal in area to the given surface.

EXAMPLES.

1. What is the diameter of a circle that is 16 times as large as one whose diameter is 13 feet? *Ans.* 52ft.

2. The area of a circle is 7632 feet; what is the diameter? *Ans.* 98.5ft.

3. A horse is fastened to a post in the centre of a field. What is the length of a rope that will allow him to graze an acre? *Ans.* 7.136 rods.

4. A ladder 75 feet long, rests against the trunk of a tree at a point 50 feet from the ground. How far is the foot of the ladder from the root of the tree? *Ans.* 55.9ft.

5. The length of a room is 18 feet, and the width 12 feet. What is the distance between the opposite corners? What length of rope would reach from an upper corner to the opposite lower corner, the height being 10 feet? *Ans.* 21.6ft.; 23.832ft.

6. The circumference of a circle is 29 rods. What is the side of a square having an equal area? *Ans.* 8.18 rods.

7. Two ships left the same port; one sailed 125 miles north, the other 100 miles east. How far were they then apart? *Ans.* 160m. 24.96r.

8. A kite accidentally lodged in the top of a tree, but the line breaking, I measure its length, which is 210 feet.

What is the height of the tree, the foot being 189 feet from my standing place? *Ans.* 91.53ft.

9. Desiring to know the height of a precipice, I drop a stone from the summit, and observe by my watch that it strikes the ground in $3\frac{1}{2}$ seconds. What is the height?[a] *Ans.* 197.02ft.

10. A bag of sand is dropped from a balloon $1\frac{1}{4}$ miles above the surface of the earth. How long will it be in falling?[a] *Ans.* 20.25sec.

When one number bears the same ratio to a second as the second does to a third, the second number is called a *mean proportional* between the other two. Thus, in the proportion 3 : 6 : : 6 : 12, 6 is a mean proportional between 3 and 12.

The mean proportional between any two numbers is equal to the square root of their product.

11. Find a mean proportional between 7 and 252.

12. Find a mean proportional between .75 and 12.

13. Find a mean proportional between $\frac{1}{5}$ and $\frac{1}{405}$.

14. Find a mean proportional between $\frac{7}{200}$ and .875.

15. Find mean proportionals between $\frac{1}{16}$ and 16; 5 and 6; 25 and 13; $\frac{2}{7}$ and $\frac{5}{9}$.

88. APPLICATION OF THE CUBE ROOT.

The solid contents and the weights of similar bodies are to each other as the cubes of their diameters, or of their similar sides.

The solid contents of a sphere may be found by multiplying the cube of the diameter by .5236.

The cube root of the solid contents of any body, will give the side of a cube, equal in solidity to the given body.

---

[a] No allowance is made for resistance of the air, in the answers that are given.

EXAMPLES.

1. What are the solid contents of the earth, supposing it a perfect sphere, whose diameter is 7920 miles?
Ans. 260120860876.8 cubic miles.

2. If a ball 2 inches in diameter, weighs $1\frac{1}{2}$ pounds, what would be the weight of a similar ball 6 inches in diameter? *Ans.* $40\frac{1}{2}$ lb.

3. What is the side of a cubical box that will hold 1 bushel? *Ans.* 12.907in.

4. What is the side of a cubical pile that contains 256 cords of wood? *Ans.* 32ft.

5. If a tree 1 foot in diameter, yields 2 cords of wood, how much wood is there in a similar tree that is 3ft. 6in. in diameter? *Ans.* $85\frac{3}{4}$ cords.

6. If a pound avoirdupois of gold is worth $200, and a cubic inch weighs $11\frac{1}{4}$ oz., what would be the value of a gold ball 1 foot in diameter? *Ans.* $243000.

7. What is the diameter of a ball that weighs 27 times as much as one 3 feet 6 inches in diameter?
Ans. 10ft. 6in.

8. If a hollow sphere 3 feet in diameter and $2\frac{1}{3}$ inches thick, weighs 12 tons, what would be the dimensions of a similar sphere that would weigh 324 tons?
Ans. Diameter 9ft.; thickness 7 inches.

9. What is the side of a cubical block of wood, that weighs as much as a sphere of the same material, 15 inches in diameter? *Ans.* 12.09 inches.

10. The length of a ship's keel is 70ft., the breadth of beam 25ft., and the depth of the hold $12\frac{1}{2}$ft. Required the dimensions of another vessel, built on the same model, but of twice the tonnage.

11. If a ship whose keel measures 90 feet, carries 420 tons, what will be the tonnage of a similar vessel with a keel 60ft. long?

XIX. PROGRESSION, OR SERIES.

89. ARITHMETICAL AND GEOMETRICAL PROGRESSION.

LET a represent the less extreme of a series, l the greater extreme, n the number of terms, s the sum of all the terms in an arithmetical series, p the product of all the terms in a geometrical series, d the arithmetical difference, and r the geometrical ratio. Then

In Arithmetical Progression.		*In Geometrical Progression.*	
$l = a + (n-1)d$	(1)	$l = a \times r^{n-1}$	(5)
$s = n\dfrac{(a+l)}{2}$	(2)	$p = \sqrt{(a \times l)^n}$	(6)
$a = l - (n-1)d$	(3)	$a = \dfrac{l}{r^{n-1}}$	(7)
$d = \dfrac{l-a}{n-1}$	(4)	$r = \sqrt[n-1]{\dfrac{l}{a}}$	(8)

In comparing these tables, we see that

addition corresponds to *multiplication;*
subtraction " " *division;*
multiplication " " *involution;*
division " " *evolution.*

If, therefore, we had a series of numbers bearing the same ratio to the natural series, as an Arithmetical to a Geometrical Progression, the labor of multiplication would be reduced to that of simple addition, and involution to simple multiplication. Such a series constitutes a TABLE OF LOGARITHMS.

Examples.

1. Determine the value of n, when a, d, and l are given, by the 2d method of analysis, stated in Section 57.

By formula (1) we perceive that if 1 be subtracted from n, the remainder multiplied by d, and a added to the product, the sum will be l. Reversing the operation, if we subtract a from l, divide the remainder by d, and add 1 to the quotient, the sum will be n.

Ans. $n=\frac{l-a}{d}+1.$

2. From formula (2) find the value of l, when a, n, and s are given, and the value of a, when l, n, and s are given.

Ans. $l=\frac{2s}{n}-a$; $a=\frac{2s}{n}-l.$

3. Determine the value of n, when the values of a, l, and s are known.

Ans. $n=\frac{2s}{a+l}.$

4. From formulas (1) and (2) determine the value of a, when d, n, and s are known.

Substituting for l in formula (2), its value in formula (1), we have $s=\frac{n(2a+\overline{n-1}\,d)}{2}$

Reversing all the operations that must be performed on a to produce this result, we find that $a=\left(\frac{2s}{n}-(n-1)\,d\right)\div 2.$

5. From formulas (2) and (3), how may we find the value of l, when d, n, and s are known?[a]

Ans. $l=\left(\frac{2s}{n}+\overline{n-1}\,d\right)\div 2.$

---

[a] There are two formulas, which the pupil could hardly be expected to obtain, without considerable knowledge of Algebra. They are, therefore, inserted here, in order that he may have all the formulas that are necessary to solve any question in Arithmetical Progression that can possibly occur.

When a, d, and s are given,

$$n=\frac{d-2a+\sqrt{(d-2a)^2+8ds}}{2d}.$$

6. A laborer agreed to dig a well 39 yards deep, for which he was to be paid as follows: 75 cents for the first yard, $1.25 for the second yard, and so on, increasing 50 cents for each subsequent yard. What would the last yard cost, and what would he receive for the whole job?

2d *Ans.* $399.75.

7. The formula for determining the sum of any geometrical series is $s=\frac{l\times r-a}{r-1}$. Determine from this formula, the value of r when s, a, and l are given.

Ans. $r=\frac{s-a}{s-l}$.[a]

8. When one of the extremes, the ratio, and the sum of the terms are given, how would you find the other extreme? Give separate answers for each extreme.

Ans. $a=l\times r-(r-1)\times s$.

$l=\frac{(r-1)\times s+a}{r}$.

9. What is the sum of the series 2, 1, ½, ¼, &c., to infinity? (The last term in any infinite decreasing series is 0.)

Ans. 4.

10. If a man commences at 21 years of age, and annually puts $500 at compound interest, how much will he be worth when he is 50 years old? *Ans.* $36819.90.

11. Insert 2 mean proportionals between 1 and 343. (As

When d, l, and s are given,

$$n=\frac{2l+d\pm\sqrt{(2l+d)^2-8ds}}{2d}.$$

The sign following d in the second formula, is sometimes +, and sometimes —. The proper sign can easily be determined by trial.

[a] If $(l\times r-a)\div(r-1)=s$, $r\times s-s=l\times r-a$. Then $r\times s=l\times r+s-a$. Subtracting $l\times r$ from $r\times s$, we have $r\times s-l\times r=s-a$. But $r\times s-r\times l=r\times(s-l)$; and therefore, dividing by $s-l$, we obtain the answer, $r=(s-a)\div(s-l)$. This analysis will be more difficult to follow than any of those required in arithmetical progression; but the pupil should pass nothing over until he understands it perfectly.

there are to be 2 means, the number of terms is 4, and the extremes 1 and 343.) *Ans.* 7, 49.

12. Insert 5 mean proportionals between 4 and 2916.
Ans. 12, 36, 108, 324, 972.

13. Every oviparous fish deposits annually, at the spawning season, many thousands of ova. If we estimate the average number deposited by each pair of herrings to be only 2000, to what number would the offspring of a single pair amount in the eighth year, supposing that every egg produced a fish?

Ans. 2 septillion, a number which would make a mass larger than the whole globe.

14. According to some experiments, it has been found that one stem of the hyoscyamus sometimes produces more than 50000 seeds. At this rate, if every seed should produce a fertile plant, what number of plants would be contained in the fourth crop from a single seed?

Ans. 6250 quadrillion, a number that the whole surface of the earth would not be sufficient to contain.

15. If the human race, after making a proper deduction for those who died, had doubled every twenty years, how many of the descendants of Adam would have been living when he was 500 years old? *Ans.* 33554430.

90. Harmonical Progression.[a]

When three numbers are such that the first is to the third, as the difference between the first and second is to the difference between the second and third, they are said to be in Harmonical Proportion; and a series of numbers, in continued harmonical proportion, constitutes a Harmonical Progression.

The *reciprocal* of a number, is the quotient of 1 by the

[a] So called, because if a musical string be divided in harmonical proportion, the different parts will vibrate in unison.

number. Thus, $\frac{1}{2}$ is the reciprocal of 2; 4 is the reciprocal of $\frac{1}{4}$; $\frac{3}{2}$ is the reciprocal of $\frac{2}{3}$, &c. *The reciprocals of any equidifferent series form a harmonical proportion.*

I. Two numbers being given, to find a third in harmonical proportion.

Consider the reciprocals of the numbers as two terms of an equidifferent series. The third term will be the reciprocal of the number sought.

Find a third harmonical proportional to 120 and 40.

The reciprocals are $\frac{1}{120}$, and $\frac{1}{40}$, or $\frac{3}{120}$. The third term of the equidifferent series is $\frac{5}{120}$, and its reciprocal 24 is the harmonical proportional sought.

II. To insert any number of harmonical means between two numbers.

Find as many arithmetical means between the reciprocals of the given numbers. These means will be the reciprocals of the harmonical means.

Insert 4 harmonical means between 20 and 120.

The reciprocals are $\frac{1}{24}$ and $\frac{1}{120}$, or $\frac{6}{120}$ and $\frac{1}{120}$. The four arithmetical means are $\frac{5}{120}$, $\frac{4}{120}$, $\frac{3}{120}$, and $\frac{2}{120}$, whose reciprocals are 24, 30, 40, and 60,—the desired harmonical means.

EXAMPLES.

1. The first two terms of a harmonical progression are 60 and 30. Required the ten succeeding terms.

2. The first two terms of a harmonical proportion are 348075 and 69615. Find the six succeeding terms.

3. Insert 6 harmonical means between 630 and 5040.

4. Insert 8 harmonical means between 10 and 60.

5. Insert 2 harmonical means between $\frac{1}{2}$ and $\frac{1}{8}$.

6. Insert 4 harmonical means between $\frac{1}{6}$ and $\frac{1}{10}$.

91. COMPOUND INTEREST.

Compound Interest may be computed by Geometrical Progression; $a =$ the amount of $1 for the time that should elapse between two successive payments of interest; $r = a$; $n =$ the number of payments.

The labor of computing Compound Interest, may be abridged by a table in which the amount of $1 is computed at different rates, and for a number of years. (See Table I., p. 293.)

I. To find the amount of any sum by the Table, *multiply the given sum by the amount of $1 for the given time.*

EXAMPLE. What will be the amount, at 7 per cent. compound interest, of $200 for 15yr.?

$1 in 15yr. at 7 per cent. amounts to 2.759031, and 2.759031 × $200 = $551.8062.

II. To compute compound discount, or to find the present worth at compound interest, of any sum due at a future time, *divide the given sum by the amount of $1 for the given time.*

EXAMPLE.—When money is worth 5 per cent. compound interest, what is the present worth of $5000 due in 19yr. 4mo. 24dy.?

$1 at 5 per cent. would amount in 19yr. 4mo. 24dy. to $2.577489, and $5000 ÷ 2.577489 = $1939.87.

III. To find the time in which any principal will amount to a given sum, *divide the amount by the principal, and look for the quotient in the Table, under the given rate.*

EXAMPLE.—In what time, at 6 per cent. compound interest, will $25 amount to $48?

$\frac{48}{25} = 1.92$. $1 would amount to 1.898299 in 11 years, or to 2.012196 in 12yr.

1.92 exceeds 1.898299 by .021701, and 2.012196 exceeds 1.898299 by .113897. Then if the gain in 12 months is .113897, in what time would there be a gain of .021701?

.113897 : .021701 : : 12mo. : 2mo. 8dy. *very nearly.*

IV. To find the rate at which any principal will amount to a given sum in a given time, *divide the amount by the principal, and look for the quotient in the Table, opposite the given time.*

EXAMPLE.—At what rate of compound interest, will $250 amount to $550 in 18 years?

$\frac{550}{250} = 2.2$. In the line of 18 years, we find 2.2 under $4\frac{1}{2}$ per cent.

EXAMPLES.

1. Find the amount of $637.25, at 5 per cent. compound interest, for 16yr. 3mo. 15dy. *Ans.* $1411.32.

2. Allowing 7 per cent. compound interest, what is the present worth of $1000, due in 35yr. 5mo. 6dy.?
Ans. $90.91.

3. At 6 per cent. compound interest, in what time will $250 amount to $1000? *Ans.* 23yr. 9mo. 13dy.

4. At what rate of compound interest will $127.75 amount to $201.22, in 10yr. 3mo. 24dy.?
Ans. $4\frac{1}{2}$ per cent.

92. ANNUITIES.

Any sum of money to be paid regularly, at stated periods, is called an ANNUITY. The payment may be stipulated for a given number of years, in which case it is called an *annuity certain*, or it may be dependent upon some particular circumstance, as the life of one or more individuals. The latter is called a *contingent annuity*. A *perpetual annuity*, is one which can only be terminated by the grantor, on the payment of a sum whose interest will be equivalent to the annuity. Of this character is the consolidated debt of England.

An *annuity in possession*, is one on which there is a present claim; an *annuity in reversion*, or deferred annuity, is one that does not commence until the lapse of a stated time, or the occurrence of some uncertain event, as the death of an individual.

The *present worth* of an annuity, is the sum which, at compound interest for the time of its duration, would amount to the sum of all the payments, each being placed at compound interest as it became due.

TABLE I.

SHOWING THE AMOUNT OF $1.00, AT COMPOUND INTEREST, FROM 1 YEAR TO 50.

Year.	3 p. cent.	3½ p. cent.	4 p. cent.	4½ p. cent.	5 p. cent.	6 p. cent.	7 p. cent.
1	1.030000	1.035000	1.040000	1.045000	1.050000	1.060000	1.070000
2	1.060900	1.071225	1.081600	1.092025	1.102500	1.123600	1.144900
3	1.092727	1.108718	1.124864	1.141166	1.157625	1.191016	1.225043
4	1.125509	1.147523	1.169859	1.192519	1.215506	1.262477	1.310796
5	1.159274	1.187686	1.216653	1.246182	1.276282	1.338226	1.402552
6	1.194052	1.229255	1.265319	1.302260	1.340096	1.418519	1.500730
7	1.229874	1.272279	1.315932	1.360862	1.407100	1.503630	1.605781
8	1.266770	1.316809	1.368569	1.422101	1.477455	1.593848	1.718186
9	1.304773	1.362897	1.423312	1.486095	1.551328	1.689479	1.838459
10	1.343916	1.410599	1.480244	1.552969	1.628895	1.790848	1.967151
11	1.384234	1.459970	1.539454	1.622853	1.710339	1.898299	2.104852
12	1.425761	1.511069	1.601032	1.695881	1.795856	2.012196	2.252192
13	1.468534	1.563956	1.665073	1.772196	1.885649	2.132928	2.409845
14	1.512590	1.618694	1.731676	1.851945	1.979932	2.260904	2.578534
15	1.557967	1.675349	1.800943	1.935282	2.078928	2.396558	2.759031
16	1.604706	1.733986	1.872981	2.022370	2.182875	2.540352	2.952164
17	1.652848	1.794675	1.947900	2.113377	2.292018	2.692773	3.158815
18	1.702433	1.857489	2.025816	2.208479	2.406619	2.854339	3.379931
19	1.753506	1.922501	2.106849	2.307860	2.526950	3.025599	3.616526
20	1.806111	1.989789	2.191123	2.411714	2.653298	3.207135	3.869683
21	1.860295	2.059431	2.278768	2.520241	2.785963	3.399564	4.140561
22	1.916103	2.131512	2.369919	2.633652	2.925261	3.603537	4.430400
23	1.973586	2.206114	2.464715	2.752166	3.071524	3.819750	4.740528
24	2.032794	2.283328	2.563304	2.876014	3.225100	4.048935	5.072365
25	2.093778	2.363245	2.665836	3.005434	3.386355	4.291871	5.427431
26	2.156591	2.445959	2.772470	3.140679	3.555673	4.549383	5.807351
27	2.221289	2.531567	2.883369	3.282009	3.733456	4.822346	6.213866
28	2.287928	2.620177	2.998703	3.429700	3.920129	5.111687	6.648836
29	2.356565	2.711878	3.118651	3.584036	4.116136	5.418388	7.114255
30	2.427262	2.806794	3.243397	3.745318	4.321942	5.743491	7.612253
31	2.500080	2.905031	3.373133	3.913857	4.538039	6.088101	8.145110
32	2.575083	3.006708	3.508059	4.089981	4.764941	6.453387	8.715268
33	2.652335	3.111942	3.648381	4.274030	5.003188	6.840590	9.325337
34	2.731905	3.220860	3.794316	4.466361	5.253348	7.251025	9.978110
35	2.813862	3.333590	3.946089	4.667348	5.516015	7.686087	10.676578
36	2.898278	3.450266	4.103932	4.877378	5.791816	8.147252	11.423939
37	2.985227	3.571025	4.268090	5.096860	6.081407	8.636087	12.223614
38	3.074783	3.696011	4.438813	5.326219	6.385477	9.154252	13.079277
39	3.167027	3.825372	4.616366	5.565899	6.704751	9.703507	13.994827
40	3.262038	3.959260	4.801021	5.816364	7.039989	10.285718	14.974465
41	3.359899	4.097834	4.993061	6.078101	7.391988	10.902861	16.022677
42	3.460696	4.241258	5.192784	6.351615	7.761587	11.557033	17.144265
43	3.564517	4.389702	5.400495	6.637438	8.149667	12.250455	18.344363
44	3.671452	4.543342	5.616515	6.936123	8.557150	12.985482	19.628469
45	3.781596	4.702358	5.841176	7.248248	8.985008	13.764611	21.002461
46	3.895044	4.860941	6.074823	7.574420	9.434258	14.590487	22.472634
47	4.011895	5.037284	6.317816	7.915268	9.905971	15.465917	24.045718
48	4.132252	5.213589	6.570528	8.271455	10.401267	16.393872	25.728918
49	4.256219	5.396065	6.833349	8.643671	10.921333	17.377504	27.529943
50	4.383906	5.584927	7.106683	9.032636	11.467400	18.420154	29.457039

TABLE II.

THE AMOUNT OF AN ANNUITY OF $1.00, FROM 1 YEAR TO 50.

Year.	3 p. cent.	3½ p. cent.	4 p. cent.	4½ p. cent.	5 p. cent.	5½ p. cent.	6 p. cent.
1	1.000000	1.000000	1.000000	1.000000	1.000000	1.000000	1.000000
2	2.030000	2.035000	2.040000	2.045000	2.050000	2.055000	2.060000
3	3.090900	3.106225	3.121600	3.137025	3.152500	3.168025	3.183600
4	4.183627	4.214943	4.246464	4.278191	4.310125	4.342266	4.374616
5	5.309136	5.362466	5.416322	5.470710	5.525631	5.581091	5.637093
6	6.468410	6.550152	6.632975	6.716892	6.801913	6.888051	6.975319
7	7.662462	7.779408	7.898294	8.019152	8.142008	8.266894	8.393838
8	8.892336	9.051687	9.214226	9.380014	9.549109	9.721573	9.897468
9	10.159106	10.368496	10.582795	10.802114	11.026564	11.256259	11.491316
10	11.463879	11.731393	12.006107	12.288210	12.577893	12.875354	13.180795
11	12.807796	13.141992	13.486351	13.841179	14.206787	14.583498	14.971643
12	14.192029	14.601962	15.025805	15.464032	15.917127	16.385590	16.869941
13	15.617790	16.113030	16.626838	17.159913	17.712983	18.286798	18.882138
14	17.086324	17.676986	18.291911	18.932109	19.598632	20.292572	21.015066
15	18.598914	19.295681	20.023588	20.784054	21.578564	22.408663	23.275971
16	20.156881	20.971030	21.824531	22.719337	23.657492	24.641139	25.672528
17	21.761588	22.705016	23.697512	24.741707	25.840366	26.996402	28.212880
18	23.414436	24.499691	25.645413	26.855084	28.132385	29.481205	30.905653
19	25.116868	26.357180	27.671229	29.063562	30.539004	32.102671	33.759992
20	26.870374	28.279682	29.778078	31.371423	33.065954	34.868318	36.785592
21	28.676486	30.269471	31.969202	33.783137	35.719252	37.786075	39.992727
22	30.536780	32.328902	34.247970	36.303378	38.505214	40.864309	43.392290
23	32.452884	34.460414	36.617888	38.937030	41.430475	44.111846	46.995828
24	34.426470	36.666528	39.082604	41.689196	44.501999	47.537998	50.815577
25	36.459264	38.949857	41.645908	44.565210	47.727099	51.152588	54.864512
26	38.553042	41.313102	44.311745	47.570645	51.113454	54.965979	59.156383
27	40.709634	43.759060	47.084214	50.711324	54.669126	58.989109	63.705766
28	42.930923	46.290627	49.967583	53.993333	58.402583	63.233510	68.528112
29	45.218850	48.910799	52.966286	57.423033	62.322712	67.711353	73.639798
30	47.575416	51.622677	56.084938	61.007070	66.438847	72.435478	79.058186
31	50.002678	54.429471	59.328335	64.752388	70.760790	77.419429	84.801677
32	52.502759	57.334502	62.701469	68.666245	75.298829	82.677498	90.889778
33	55.077841	60.341210	66.209527	72.756226	80.063770	88.224760	97.343165
34	57.730177	63.453152	69.857909	77.030256	85.066959	94.077122	104.183755
35	60.462082	66.674013	73.652225	81.496618	90.320307	100.251363	111.434780
36	63.275944	70.007603	77.598314	86.163966	95.836323	106.765188	119.120867
37	66.174223	73.457869	81.702246	91.041344	101.628139	113.637274	127.268119
38	69.159449	77.028895	85.970336	96.138205	107.709546	120.887324	135.904206
39	72.234233	80.724906	90.409150	101.464424	114.095023	128.536127	145.058458
40	75.401260	84.550278	95.025516	107.030323	120.799774	136.605614	154.761966
41	78.663296	88.509537	99.826536	112.846688	127.839763	145.118923	165.047684
42	82.023196	92.607371	104.819598	118.924789	135.231751	154.100464	175.950545
43	85.483892	96.848629	110.012382	125.276404	142.993339	163.575989	187.507577
44	89.048409	101.238331	115.412877	131.913842	151.143006	173.572669	199.758032
45	92.719861	105.781673	121.029392	138.849965	159.700156	184.119165	212.743514
46	96.501457	110.484031	126.870568	146.098214	168.685164	195.245720	226.508125
47	100.396501	115.350973	132.945390	153.672633	178.119422	206.984234	241.098612
48	104.408396	120.388257	139.263206	161.587902	188.025393	219.368367	256.564529
49	108.540648	125.601846	145.833734	169.859357	198.426663	232.433627	272.958401
50	112.796867	130.997910	152.667084	178.503028	209.347996	246.217477	290.335905

TABLE III.

THE PRESENT WORTH OF AN ANNUITY OF $1.00, FROM 1 YEAR TO 50.

Year.	3 p. cent.	3½ p. cent.	4 p. cent.	4½ p. cent.	5 p. cent.	5½ p. cent.	6 p. cent.	Year.
1	0.97087	0.96618	0.96154	0.95694	0.95238	0.94786	0.94339	1
2	1.91347	1.89969	1.88609	1.87267	1.85941	1.84631	1.83339	2
3	2.82861	2.80164	2.77509	2.74896	2.72325	2.69793	2.67301	3
4	3.71710	3.67308	3.62990	3.58753	3.54595	3.50514	3.46511	4
5	4.57971	4.51505	4.45182	4.38998	4.32948	4.27028	4.21236	5
6	5.41719	5.32855	5.24214	5.15787	5.07569	4.99553	4.91732	6
7	6.23028	6.11454	6.00205	5.89270	5.78637	5.68297	5.58238	7
8	7.01969	6.87396	6.73274	6.59589	6.46321	6.33457	6.20979	8
9	7.78611	7.60769	7.43533	7.26879	7.10782	6.95220	6.80169	9
10	8.53020	8.31661	8.11090	7.91272	7.72173	7.53762	7.36009	10
11	9.25262	9.00155	8.76048	8.52892	8.30641	8.09254	7.88687	11
12	9.95400	9.66333	9.38507	9.11858	8.86325	8.61852	8.38384	12
13	10.63495	10.30274	9.98565	9.68285	9.39357	9.11708	8.85268	13
14	11.29607	10.92052	10.56312	10.22283	9.89864	9.58965	9.29498	14
15	11.93794	11.51741	11.11839	10.73955	10.37966	10.03759	9.71225	15
16	12.56110	12.09412	11.65230	11.23401	10.83777	10.46216	10.10589	16
17	13.16612	12.65132	12.16567	11.70719	11.27407	10.86461	10.47726	17
18	13.75351	13.18968	12.65930	12.15999	11.68959	11.24607	10.82760	18
19	14.32380	13.70984	13.13394	12.59329	12.08532	11.60765	11.15812	19
20	14.87747	14.21240	13.59033	13.00794	12.46221	11.95034	11.46992	20
21	15.41502	14.69797	14.02916	13.40472	12.82115	12.27524	11.76408	21
22	15.93692	15.16712	14.45112	13.78442	13.16300	12.58317	12.04158	22
23	16.44361	15.62041	14.85684	14.14777	13.48857	12.87504	12.30338	23
24	16.93554	16.05837	15.24696	14.49548	13.79864	13.15170	12.55036	24
25	17.41315	16.48151	15.62208	14.82821	14.09394	13.41391	12.78336	25
26	17.87684	16.89035	15.98277	15.14661	14.37518	13.66250	13.00317	26
27	18.32703	17.28536	16.32959	15.45130	14.64303	13.89810	13.21053	27
28	18.76411	17.66702	16.66306	15.74287	14.89813	14.12142	13.40616	28
29	19.18845	18.03577	16.98371	16.02189	15.14107	14.33310	13.59072	29
30	19.60044	18.39205	17.29203	16.28889	15.37245	14.53375	13.76483	30
31	20.00043	18.73628	17.58849	16.54439	15.59281	14.72393	13.92909	31
32	20.38877	19.06887	17.87355	16.78889	15.80268	14.90420	14.08404	32
33	20.76579	19.39021	18.14765	17.02286	16.00255	15.07507	14.23023	33
34	21.13184	19.70068	18.41120	17.24676	16.19290	15.23703	14.36814	34
35	21.48722	20.00066	18.66461	17.46101	16.37419	15.39055	14.49825	35
36	21.83225	20.29049	18.90828	17.66604	16.54685	15.53607	14.62099	36
37	22.16724	20.57053	19.14258	17.86224	16.71129	15.67400	14.73678	37
38	22.49246	20.84109	19.36786	18.04999	16.86789	15.80474	14.84602	38
39	22.80822	21.10250	19.58448	18.22965	17.01704	15.92866	14.94907	39
40	23.11477	21.35507	19.79277	18.40158	17.15909	16.04612	15.04630	40
41	23.41240	21.59910	19.99305	18.56611	17.29437	16.15746	15.13802	41
42	23.70136	21.83488	20.18563	18.72355	17.42321	16.26299	15.22454	42
43	23.98190	22.06269	20.37079	18.87421	17.54591	16.36303	15.30617	43
44	24.25427	22.28279	20.54884	19.01838	17.66277	16.45785	15.38318	44
45	24.51871	22.49545	20.72004	19.15635	17.77407	16.54772	15.45583	45
46	24.77545	22.70092	20.88465	19.28837	17.88007	16.63283	15.52437	46
47	25.02471	22.89943	21.04294	19.41471	17.98102	16.71357	15.58903	47
48	25.26671	23.09124	21.19513	19.53561	18.07716	19.79011	15.65003	48
49	25.50166	23.27656	21.34147	19.65130	18.16872	16.86266	15.70757	49
50	25.72976	23.45562	21.48218	19.76201	18.25593	16.93143	15.76186	50

Nearly all the questions that occur in annuities, can be easily solved by the first method of Analysis, given in Section 57, p. 190. This will be readily seen, by comparing the following

RULES.

1. *To find the amount due on an annuity that has remained unpaid for a given time.*

Multiply the amount of an annuity of ONE dollar for the given time,[a] by the NUMBER of dollars in the given annuity.

2. *To find the present worth of an annuity certain.*

Multiply the present worth of an annuity of ONE dollar for the given time,[b] by the NUMBER of dollars in the given annuity.

3. *To find the present worth of a perpetual annuity.*

Divide the annuity by the interest that ONE dollar would yield, in the time that elapses between the several payments of the annuity.

4. *To find the annuity, when the present worth, or the amount is given.*

If the present worth is given, divide by the present worth of an annuity of ONE dollar for the time the annuity is to continue. If the amount is given, divide by the amount of an annuity of ONE dollar.

5. *To find the present worth of an annuity in reversion.*

Multiply the present worth in reversion, of an annuity

---

[a] This amount may be taken from Table II., p. 294, or it may be determined by finding the sum of a geometrical progression, (Ex. 7, sect. 89,) in which $a =$ the annuity, $r =$ the amount of $1 for the time that should elapse between two successive payments, and $n =$ the number of payments due.

[b] The present worth may be taken from Table III. p. 295, or it may be found by dividing the amount of an annuity of $1 for the given time, by the amount of $1 at compound interest, for the same time.

of ONE dollar,[a] by the NUMBER of dollars in the given annuity.

EXAMPLES.

1. An estate that yields an annual income of $2000, is offered for sale for the amount of 10 years' income at 6 per cent. compound interest. What is the price of the estate?
Ans. $26361.59.

2. A gentleman wishes to present his estate to his children, reserving enough to yield $700 per annum for 15 years. How much must he reserve, allowing 5 per cent. compound interest? *Ans.* $7265.76.

3. For how much should an estate that rents for $175 per year, be sold, to allow the purchaser 6 per cent. interest on his investment? *Ans.* $2916⅔.

4. What sum of money must a man lay up annually, to amount to $10000 in 20 years, the investments being all made at 6 per cent. compound interest? *Ans.* $271.85.

5. A father leaves an annual rent of $400 to his eldest child for 5 years, and the reversion of it for the 8 succeeding years to his youngest child. What is the present worth of each legacy, at 6 per cent.?
Ans. $1684.94; $1856.13.

6. If a person saves $250 per annum, and invests it at 6 per cent. compound interest, how much will he be worth at the end of 25 years? *Ans.* $13716.13.

7. What sum invested at 6 per cent. compound interest, will yield me an income of $1600 per annum for 25 years?
Ans. $20453.38.

---

[a] This value may be found from Table III., p. 295, by subtracting the present worth of an annuity continuing until the reversion *commences*, from the present worth of an annuity, continuing until the reversion *terminates*. Thus, the present worth of an annuity of $1, to commence in 3 years, and continue 8 years, computing interest at 6 per cent., is $7.88687 — $2.67301 = $5.21386.

8. What sum will build a wall worth $1000, and renew it every 15 years, at 5 per cent. compound interest?

N. B. At 5 per cent. compound interest, $1 in 15 years will amount to $2.078928. The *interest* is therefore $1.078928. The amount necessary to renew the wall, is $1000 ÷ 1.078928, to which must be added the $1000 expended for building the wall at first.

Ans. $1926.85.

9. A builder takes a lease of a lot of ground for 25 years, and erects buildings on it which cost him $20000. Allowing money to be worth 6 per cent. compound interest, what *clear*[a] annual rent must he receive from the buildings to reimburse his expenditure, at the termination of the lease,—the rent commencing one year after the lease is given?

Ans. $1593.58.

10. What sum must be paid, allowing 6 per cent. compound interest, to extend a lease 7 years,—the clear annual rent being $500, and the lease having 4 years to run?

Ans. $2210.88.

11. What is the amount of a pension of $400 a year, payable semi-annually, for 3 years and 6 months, at 7 per cent. per annum? *Ans.* $1555.88.

12. What is the par value of an annual income of £500 in the 4 per cent. consols?[b] *Ans.* £12500.

13. A railroad has been constructed through a farm, in consequence of which, the owner of the estate is obliged to expend $400 in fencing, that must be renewed at the expiration of every 12 years. What sum should he now receive, to compensate him for the required expenditure, money being worth 6 per cent. compound interest? *Ans.* $795.18.

---

[a] The *clear* annual rent, is the amount received after deducting ground-rent, taxes, and other expenses.

[b] CONSOLS is an abbreviation for the consolidated annuities of the British National Debt.

14. The executors of an estate wish to dispose of an unexpired lease that has 8 years to run, for a premium of $1500. What amount must be added to the annual rent, for that purpose? *Ans.* $241.55.

15. What is the present worth of a reversion of $700 per annum, to commence in 20 years, and continue 30 years thereafter, allowing 6 per cent. compound interest?

Ans. $3004.36.

16. The British National Debt is about £800000000. If £8000000 were applied annually to the reduction of this debt,[a] in what time would it be paid off, calculating compound interest at the rate of 5 per cent.?

Ans. 36yr. 10mo. 13dy.

17. There are two adjoining farms, each renting for $400 per annum, but the rent of one is payable semi-annually, the rent of the other quarterly. What will be the difference in the income from the two farms, at the expiration of 20 years, provided all the rent is invested as fast as it becomes due, at 6 per cent. compound interest?[b]

Ans. $190.67.

18. An estate is sold for $50000, of which $5000 is to be paid in cash, and the rest in semi-annual instalments of $2250. But the purchaser proposing to discharge the whole debt at once, he wishes to know what sum of money will be required, allowing discount at the rate of 7 per cent. compound interest. *Ans.* $36977.90.

19. A gentleman takes a lease for ten years, at $450 per annum. At the expiration of two years, he wishes to give up the lease, but the landlord will not consent, unless the tenant will either pay down a year's rent in advance, or $60

---

[a] The interest is supposed to be regularly paid, in addition to the sinking fund for the reduction of the debt.

[b] The amount of 80 quarterly payments of $1 each, would be $152.7092.

per annum, during the whole term of the contract. Which proposal is the more favorable, and how much will he save by accepting it, money being worth 6 per cent. per annum?

2d *Ans.* He will save $77.41.

20. What is the difference in value, between the present worth of a lease, for 100 years, of an estate that rents for $1500 per annum, and the perpetuity of the same estate, computing interest at 7 per cent.?

Ans. $24.69, or less than one week's rent.

N. B. The present worth of an annuity of $1, to continue 100 years, at 7 per cent., is $14.269251.

XX. POSITION.

THE answers to many difficult analytical questions, can be obtained by assuming one or more numbers, and working with them as if they were the true numbers sought. The method of obtaining the correct result in such cases, is called POSITION.

SINGLE POSITION requires only one assumed number. It is used in solving questions in which the required number is increased or diminished by any of its parts or multiples, either by addition, subtraction, multiplication, or division.

DOUBLE POSITION requires two assumed numbers. It is applicable to all questions that can be solved by Single Position, and to nearly all questions that can be solved by algebraical equations.

93. SINGLE POSITION.

Questions in Single Position may be solved by the following rule:

Assume any convenient number, and proceed with it

according to the conditions of the question. Multiply the assumed number by the number given in the question, and divide the product by the result obtained with the assumed number.

Example for Illustration.

Divide $1584 among three persons, in such manner that ½ of the first share, ⅓ of the second share, and ¼ of the third share, shall all be equal.

Suppose ½ the first share 250.

Then the first share will be	500	250 × 1584 ÷ 2250 = 176,
second share	750	is ½ the first share.
third share	1000	*Ans.* 1st share $352
		2d " $528
Result	2250	3d " $704.

Solve by the above rule, Examples 8, 15, 22, 23, 25, 28, 29, 30, 32, 33, 42, 46, 47, 58, 59, 64, 71, 73, 74, 79, 81, Section 60.

94. Double Position.

RULE I.

Assume two convenient numbers, proceed with them separately, according to the conditions of the question, and note the result obtained from each operation.

Multiply the error of either result by the difference of the assumed numbers, and divide the product by the difference of the results. The quotient will be a correction to be added to the assumed number, if it gives a result too small, or to be subtracted from it, if the result is too large.

In many cases the correct result is obtained at the first trial, but if greater accuracy is required, take the number obtained by the first trial, and the nearer of the two numbers first assumed, or any other that appears more nearly correct, as new assumed numbers. Repeat the operation as before, and you will obtain a new answer, more accurate than the former. This process may be repeated until you obtain the true answer, or a number sufficiently correct for your purpose.

This Rule fails in those questions in which the result of the operations to be performed is not a known number, but the required number, or one depending upon it, such as some multiple, or some part of it. Such cases may usually be solved by the following rule.

RULE II.

Assume two convenient numbers, and perform on each of them the operations indicated by the question. Note the errors of the results, and mark each of them with the sign + or —, according as it is in excess or defect.

Then form the following proportion:—

The difference of the errors, when they are *alike*,[a] or their sum, when they are *unlike* : the difference of the assumed numbers : : either error : the correction of the assumed number which produced that error.

EXAMPLES FOR ILLUSTRATION.

1. "A hunter wishes to measure the width of a ravine, and having no other means of doing it, he fires a rifle ball at a small knot on a tree, which stands on the opposite bank. On going over, he finds that the ball struck 6 inches below the knot. By previous experiments, he knows that the ball drops 2 inches in going 50 rods, and its deflection is proportional to the square of the distance added to $\frac{1}{3}$ of its cube. What is the width of the ravine?"

Suppose 100

$$\left(\tfrac{100}{50}\right)^2 + \tfrac{1}{3} \text{ of } \left(\tfrac{100}{50}\right)^3 \times 2 = 13\tfrac{1}{3} \text{ *first result.*}$$

Subtract 6

Error in excess $7\frac{1}{3}$

Suppose 75

$$\left(\tfrac{75}{50}\right)^2 + \tfrac{1}{3} \text{ of } \left(\tfrac{75}{50}\right)^3 \times 2 = 6\tfrac{3}{4} \text{ *second result.*}$$

Subtract 6

Error in excess $\frac{3}{4}$

Then by Rule I, $7\frac{1}{3} \times 25 \div 6\frac{7}{12} = 27\frac{67}{79}$, correction for first supposition. $100 - 27\frac{67}{79} = 72\frac{12}{79}$ rods, approximate width.

Working with this result, we obtain for the deflection 6.168 inches, which is so near the true deflection that 72 rods may be assumed as the true width.

---

[a] The errors are said to be *alike*, when both are too great, or both too small; and *unlike*, when one is too great, and the other too small.

2. A person being asked the time, replied: "The sun now rises at 5 and sets at 7. Now if you add $\frac{1}{7}$ of the hours that have passed since sunrise, to $\frac{16}{21}$ of those which must elapse before sunset, you will have the exact time of the day."

To find how many hours have elapsed since sunrise, by Rule II.

Suppose 6.	Suppose 3.
$\frac{6}{7} + \frac{16}{21}$ of $8 = 6\frac{20}{21}$.	$\frac{3}{7} + \frac{16}{21}$ of $11 = 8\frac{17}{21}$.
The result should be 11.	The result should be 8.
1st Error $-4\frac{1}{21}$.	2d Error $+\frac{17}{21}$.

$4\frac{18}{21} : 3 :: 4\frac{1}{21} : 2\frac{1}{2}$ correction to be subtracted from 6.

$6 - 2\frac{1}{2} = 3\frac{1}{2}$ $\quad 5 + 3\frac{1}{2} = 8\frac{1}{2}$ the true time.

Proof. $\frac{1}{7}$ of $3\frac{1}{2} + \frac{16}{21}$ of $10\frac{1}{2} = 8\frac{1}{2}$.

Examples for the Pupil.

1. A farmer engaged a servant, agreeing to pay him $1.50 for every day he should work, and to charge him $.50 per day for his board, every day he should be idle. At the end of 13 weeks, the man received $86.50. How many days did he work? *Ans.* 66 days.

2. Two men have the same income. A. saves $\frac{1}{5}$ of his, but B. spends $325 per year more than A., and at the end of 5 years finds himself $625 in debt. What is the annual income of each? *Ans.* $1000.

3. A person distributed in charity 2d. apiece among several poor children, and had 4d. left. He would have given them 3d. apiece, but had not enough money by 10d. What was the number of children? *Ans.* 14.

4. A man has a chaise worth $130, and two horses. If the first horse be harnessed to the chaise, their joint value will be 3 times that of the second horse; but if the second horse be harnessed to the chaise, their joint value will be twice that of the first horse. Required the value of each horse. *Ans.* First horse $104; second $78.

5. Find two numbers, whose difference is 29, and their product 546. *Ans.* 13 and 42.

6. A man sold a horse for $144, and thereby gained as much per cent. as was equivalent to the number of dollars that the horse cost him. How much did he give for the horse? *Ans.* $80.

7. The fourth power of a certain number, diminished by 7 times the number, and increased by 3 times its cube, equals 3381. What is the number? *Ans.* 7.

8. Find a number to which if 7 times its square be added, the sum will be 500. *Ans.* 8.3804+.

9. John is now three times as old as Charles, but five years ago he was four times as old. Required the age of each. *Ans.* John 45; Charles 15.

10. Five times a certain number, increased by 12, is equivalent to 7 times the number, diminished by 20. What is the number? *Ans.* 16.

11. What number is that whose half is as much less than 75, as its double is greater than 94? *Ans.* $67\frac{3}{5}$.

12. A farmer purchased a number of geese for £6 5s. He retained 5, and sold the remainder for 1s. 3d. apiece more than he paid, thus receiving what he paid for the whole. How many did he buy? *Ans.* 25.

13. The area of a certain field is 187 square rods, and the length exceeds the breadth by 6 rods. What are the dimensions?

14. There is a fish, whose head weighs 9 pounds; his tail weighs as much as his head and half his body; and his body weighs as much as his head and tail both. What is the weight of the fish? *Ans.* 72 lb.

15. What would have been the width of the ravine, in the example given on p. 302, if the ball had struck 8 inches below the knot? *Ans.* $80\frac{3}{4}$ rods, nearly.

XXI. APPROXIMATIONS.

95. Multiplication.

In Multiplication, if only a certain degree of accuracy is desired, the product may be obtained by writing the units' figure of the multiplier under that figure of the multiplicand whose place we would reserve in the product, and inverting the order of the remaining figures. In multiplying, we commence, for each partial product, with the figure of the multiplicand immediately above the multiplying figure, carrying the tens which would arise from the multiplication of the two rejected figures at the right.

EXAMPLE.

Required the product of 287.613952 by 15.98421, correct to the fourth decimal place.

287.613952	287.613952	
12489.51	15.98421	
2876.1395	28	7613952
1438.0698	575	227904
258.8525	11504	55808
23.0091	230091	1616
1.1505	2588525	568
575	14380697	60
29	28761395	2
4597.2818	4597.28180769792	

The units' figure of the multiplier being placed under the 4th decimal of the multiplicand, and the whole multiplier reversed, the product of each figure by the one above it will be ten-thousandths. Therefore, the right-hand figure of each partial product will fall in the column of ten-thousandths. In the second product, multiplying 52 by 5, we obtain 260, which, being nearer 300 than 200, we carry 3 to the product of 9 by 5.

The multiplication has also been performed in the usual way, the vertical line showing the figures that are rejected.

If the multiplicand does not contain enough decimal figures to correspond with the inverted multiplier, the deficiency should be supplied by annexing zeros. The same contraction may be applied to integers, if we wish only to obtain the thousands, millions, &c., of the product.

96. DIVISION.

In DIVISION, a similar contraction may be made when the divisor is large, a contraction which is also applicable in the extraction of roots.

The first quotient figure is of the same numerical value as the figure of the dividend which stands immediately over the units of the divisor, at the first step of the division.

After the first remainder has been obtained, instead of bringing down the remaining figures of the dividend, we may cut off the right-hand figure of the divisor at each step, as in the following example.

```
342.15 ) 28417.95255 ( 83.057        342.15 ) 28417.9 | 5255 ( 83.057
 .....    27372 0                              27372 0 |
          -------                             ---------------
           10459                               10459 | 5
           10264                               10264 | 5
           -----                               -------------
             195                                 195 | 025
             171                                 171 | 075
             ---                                 -----------
              24                                  23 | 9505
              24                                  23 | 9505
              --                                  -----------
```

In the complete division, the contraction is indicated by the vertical line. In each multiplication, the tens arising from the product of the quotient figure by the suppressed figure of the divisor, must always be carried as in contracted multiplication.

The right-hand figure of the quotient thus obtained, cannot always be relied upon. If greater accuracy is desired, the division may be extended further before commencing the contraction.

IN DIVISION OF CIRCULATING DECIMALS, we may adopt the following rule.

Make the repetends of the divisor and dividend similar and conterminous, and from the result, considered as whole numbers, subtract the finite part of each. Perform the division with the remainders as with whole numbers, and the true quotients will be obtained.

EXAMPLE.

Divide $36.\dot{9}\dot{1}$ by $5.\dot{2}7\dot{3}$.

The example is here solved by contracted decimal division. The exact fractional quotient is $7\frac{6279}{5273268}$ or $7\frac{23}{19316}$. The effect of subtracting the finite parts of the divisor and dividend is the same as reducing the two numbers to improper fractions, and dividing the numerators.

```
5.2̇73273̇) 36.9̇19191̇
     5          36
---------  ----------
5273268)   36919155 (7.001191
  ......   36912876
           --------
               6279
               5273
               ----
               1006
                527
               ----
                479
                474
                ---
                  5
                  5
```

97. Continued Fractions.

Continued Fractions arise from the approximate valuation of fractions whose terms are large, and prime to each other. If, for example, we desire approximate values for the fraction $\frac{89}{487}$, we may commence by dividing both terms of the fraction by the numerator, which gives us $\frac{1}{5\frac{42}{89}}$. Disregarding the $\frac{42}{89}$, we have $\frac{1}{5}$ for a first approximate value, which is greater than the true value, because the approximate denominator is less than the true denominator. But as the denominator is between 5 and 6, the fraction is between $\frac{1}{5}$ and $\frac{1}{6}$.

If we desire greater accuracy, we may divide $\frac{42}{89}$ in the same manner as the first fraction, which gives us $\frac{1}{2\frac{5}{42}}$ for the value of $\frac{42}{89}$, or

$$\frac{1}{5+\frac{1}{2\frac{5}{42}}},$$

for a second value of the original fraction. Disregarding the $\frac{5}{42}$, the continued fraction becomes,

$$\frac{1}{5\frac{1}{2}},$$

or $\frac{2}{11}$, which is less than the true value, because the supposed denominator is greater than the true denominator. We therefore know that the fraction is between $\frac{1}{5}$ and $\frac{2}{11}$.

Still greater accuracy may be obtained by reducing $\frac{5}{42}$, which gives us

$$\cfrac{1}{5+\cfrac{1}{2+\cfrac{1}{8\frac{2}{5}}}}.$$

Rejecting the $\frac{2}{5}$, we have,

$$\cfrac{1}{5+\cfrac{1}{2\frac{1}{8}}}, \text{ or } \frac{1}{5\frac{8}{17}}, \text{ or } \frac{17}{93},$$

for a third approximate value, greater than the true value. The fraction is, therefore, between $\frac{17}{93}$ and $\frac{2}{11}$.

After one farther approximation, we should obtain the original fraction. In fractions whose terms are very large, as in the ratio of the diameter to the circumference of a circle, these approximate values are often very useful. They, moreover, have the advantage of admitting any required degree of accuracy, for the error in adopting any approximation, is always less than the difference between the fraction taken and the one following. Thus, in the present example, if we had adopted $\frac{1}{5}$ as the true value of $\frac{89}{487}$, the error would have been less than $\frac{1}{5}-\frac{1}{6}$, or $\frac{1}{30}$.

For forming the successive approximations, we have the following

RULE.

Divide the greater term by the less, and the divisor by the remainder, &c., as in finding the greatest common measure.

Assume 1 for the numerator, and the first quotient for the denominator of the first approximate value.

Multiply the terms of this fraction by the second quotient, and add 1 to the product of the denominator, for the second approximate value.

For each succeeding approximation, multiply the terms of the last approximate fraction by the following quotient, and add the corresponding terms of the preceding fraction.

If the fraction given is improper, the reciprocals of the fractions thus obtained, will be the approximations desired.

EXAMPLE.

Required less approximate values for the ratio of the circumference to the diameter of a circle, one approximate ratio being $\frac{314159}{100000}$.

```
100000)314159(3
       300000
       ------
        14159)100000(7
               99113
               ------
                  887)14159(15
                        887
                        ----
                        5289
                        4435
                        ----
                         854)887(1
                             854
                             ---
                              33)854(25
                                  66
                                  ---
                                  194
                                  165
                                  ---
                                   29)33(1
                                      29
                                      --
                                       4)29(7
                                         28
                                         --
                                          1)4(4
                                            4
```

Equivalent Continued Fraction.

$$3+\cfrac{1}{7+\cfrac{1}{15+\cfrac{1}{1+\cfrac{1}{25+\cfrac{1}{1+\cfrac{1}{7\frac{1}{4}}}}}}}$$

For convenience, the fraction may be written,

$$3+\frac{1}{7+}\,\frac{1}{15+}\,\frac{1}{1+}\,\frac{1}{25+}\,\frac{1}{1+}\,\frac{1}{7+}\,\frac{1}{4}.$$

$\frac{1}{3}$ 1st approximate value.

$\frac{1 \times 7}{3 \times 7 + 1} = \frac{7}{22}$ 2d approximate value.

$\frac{7 \times 15 + 1}{22 \times 15 + 3} = \frac{106}{333}$ 3d approximate value.

$\frac{106 \times 1 + 7}{333 \times 1 + 22} = \frac{113}{355}$ 4th approximate value.

&c., &c.

The reciprocals of these values are,

$$3,\ \frac{22}{7},\ \frac{333}{106},\ \text{and}\ \frac{355}{113}.$$

The second ratio is the one given by Archimedes. The fourth is that of Adrian Metius, and is *even more exact* than the ratio 3.14159, from which we have derived it.

An infinite continued fraction may be made equivalent to any given number by the following rule:

Assume any number you please for a denominator; add the assumed number to the given number, and multiply the sum by the given number, for a numerator.

EXAMPLES. $5 = \frac{40}{3+}\frac{40}{3+}\frac{40}{3+}$; $2 = \frac{6}{1+}\frac{6}{1+}\frac{6}{1+}$.

The value of any infinite continued fraction, with but one numerator and denominator, may be found by the following rule:

To 4 times the numerator add the square of the denominator. From the square root of the sum, subtract the denominator, and divide the remainder by 2. The quotient is the value sought.

EXAMPLES. $\frac{40}{3+}\frac{40}{3+} = (\sqrt{160+9} - 3) \div 2 = 5$;

$\frac{3}{4+}\frac{3}{4+} = (\sqrt{12+16} - 4) \div 2 = .6457+$

The SQUARE ROOT of any number may be expressed in the form of a continued fraction, after part of the root is found,—*by making each numerator equal to the remainder, and each denominator equal to twice the root found.* Thus, in extracting the square root of 17, the first root figure is 4, and the remainder 1. Then the true root is $4+$ the continued fraction

$$\frac{1}{8+}\frac{1}{8+}\frac{1}{8+}\frac{1}{8+}\frac{1}{8+}, \text{ \&c.}$$

In like manner, the square root of 14, is

$$3 + \frac{5}{6+}\frac{5}{6+}\frac{5}{6+}\frac{5}{6+}, \text{ \&c.}$$

Reducing the fraction, we have, first,

$$\frac{5}{6\frac{5}{6}} = \tfrac{30}{41}, \text{ or } \tfrac{3}{4},$$

nearly, giving the first approximate root $3\frac{3}{4}$. Second,

$$\frac{5}{6\frac{30}{41}} = \tfrac{205}{276} \text{ or } \tfrac{17}{23},$$

nearly, giving a second approximate root $3\frac{17}{23}$. Third,

$$\frac{5}{6\frac{205}{276}} = \tfrac{1380}{1861} \text{ or } \tfrac{23}{31},$$

nearly, giving a third approximate root $3\frac{23}{31}$. This approximation is of use in affording convenient fractional expressions for those roots which are of most frequent occurrence. Thus, the diagonal of a square is to its side as $\sqrt{2}$ is to 1. By the rule just given, we obtain successively for approximate values of $\sqrt{2}$,

$$1\tfrac{1}{2},\ 1\tfrac{2}{5},\ 1\tfrac{5}{12},\ 1\tfrac{12}{29},\ 1\tfrac{29}{70}.$$

The last of these values, $1\frac{29}{70}$ or $\frac{99}{70}$, is a very convenient one.

98. EVOLUTION.

In the Extraction of Roots, we may commence with any complete divisor, cutting off the right hand figure at each step, as in contracted division. At whatever place this contraction is commenced, as many additional root figures will be obtained as are equal to the number of figures in the divisor *less* 1, but the last figure so obtained cannot be relied upon. To illustrate this mode of contraction, we will extract the 5th root of 69.

0	0	0	0	69.(2.3323285
2	4	8	16	32
2	4	8	16	37
2	8	24	64	32.36343
4	12	32	80	4.63657
2	12	48	27.8781	4.29498
6	24	80	107.8781	34159
2	16	12.927	32.0424	29334
8	40	92.927	139.9205	4825
2	3.09	13.881	3.246·	4407
10.3	43.09	106.808	143.166	418
3	3.18	1.4··	3.288	294
10.6	46.27	108.2	146.454	124
···	···	1.4	22·	117
		109.6	146.67	7
		··	22	7
			146.89	
			····	

After obtaining the third trial divisor, we commence rejecting *one* figure from the trial divisor, *two* from the number at the foot of the preceding column, *three* from the third column, &c., and proceed in a similar way with each subsequent trial divisor, until the figures from the preceding columns are entirely cancelled. But in every instance, allowance must be made for the product of the figures rejected, as in simple contracted division.

The following is a general rule for the approximation of ANY ROOT desired.

RULE.

Call the first two figures of the root found in the usual way, the ASCERTAINED ROOT.

Involve the ascertained root to the given power, and multiply by the index of the root for a dividend.

Subtract the power of the ascertained root from the corresponding periods of the given number, for a divisor. Divide, and reserve the quotient.

To 6 times the reserved quotient, add the index of the root, plus 1, for a second dividend.

To 6 times the reserved quotient, add 4 times the index of the root, subtract 2 from the sum, and multiply by the reserved quotient for a second divisor. Divide, add 1 to the quotient, and multiply by the ascertained root for the true root nearly. If greater accuracy is desired, repeat the process with the root thus found.

By this rule, the number of figures in surd roots, may generally be tripled at each operation.

EXAMPLE.

The following is the application of the rule, in extracting the 5th root of 659901.

Ascertained root 14.

$14^5 =$	537824	given no.	659901
index	5	$14^5 =$	537824
dividend	2689120	1st divisor	122077

2689120 ÷ 122077 = 22.02806, reserved quotient.

reserved quotient	22.02806
	6
	132.16836
index + 1	6.
second dividend	138.16836
6 × reserved quotient =	132.16836
4 × 5 — 2 =	18.
	150.16836
Multiply by	22.02806
2d divisor	3307.91764

138.16836 ÷ 3307.91764 = .041768

1.041768 × 14 = 14.584752, approximate root, correct to the fourth decimal place.

This contraction is of use in extracting the higher roots. Any root below the 10th may be obtained in the usual way, nearly as readily, and with much greater accuracy.

99. Examples in Approximation.

1. Multiply 11817.93642 by 2581.36, and reserve two decimal places.

2. Divide 2704.1583 by 361.8901.

3. Divide $4.30\dot{9}\dot{7}$ by $18.6\dot{1}584\dot{3}$.

4. What are the approximate values of .785398, which is nearly the ratio of the area of a circle, to that of its circumscribing square?

Ans. 1; $\frac{3}{4}$; $\frac{4}{5}$; $\frac{7}{9}$; $\frac{11}{14}$; $\frac{172}{219}$; $\frac{355}{452}$, &c.

5. Form infinite continued fractions, equivalent to 15; to 7; to 20; to 5½.

6. Determine the value of the infinite fraction $\frac{3.75}{7+}\,\frac{3.75}{7+}\,\frac{3.75}{7+}$.

Ans. $\frac{1}{2}$.

7. Express in the form of a continued fraction $\sqrt{19}$; $(18)^{\frac{1}{2}}$; $\sqrt{37}$.

8. Find the 5th root of 729. *Ans.* 3.73719.

9. Extract the 17th root of 1.004. *Ans.* 1.00023.

XXII. PROPERTIES OF NUMBERS.[a]

100. Properties of Squares and Cubes.

1. Every square number terminates in 1, 4, 5, 6, or 9, or in an even number of ciphers preceded by one of these figures. If a square number ends in 1, 4, 5, or 9, the last figure but one will be even, but if it ends in 6, the preceding figure will be odd. If a square ends in 5, it will end in 25, and the figure preceding 25 must be even.

No square number can end in two even digits, except two ciphers, or two fours. No square number can end in three equal digits, except three fours; nor in more than three equal digits, unless they are ciphers.

---

[a] Hutton, Barlow, and private sources.

2. Every square number is divisible by 3, or becomes so when diminished by 1. The same remark may be made of 4. If 1 be deducted from any odd square number, the remainder will be divisible by 8. If a square be either multiplied or divided by a square, the product or the quotient will be a square.

3. Every number is either a square, or is divisible into two, three, or four squares.

4. Every power of 5, or of any number terminating in 5, necessarily ends in 5. A similar remark may be made of 1, and 6.

5. Assume any two numbers whatever; then one of them, or their sum, or their difference, must be divisible by 3.

6. If the sum of two squares forms a square, the product of their square roots will be divisible by 6.

7. If 1 be added to the product of two numbers whose difference is 2, the sum will be the square of the intermediate number.

8. If a cube be divisible by 6, its root will also be divisible by 6. And if a cube, when divided by 6, has any remainder, its root divided by 6 will have the same remainder.

9. All exact cubes are divisible by 4, or can be made so by adding or subtracting 1. The same remark may be made of 7, and 9.

10. The cube root of any exact cube, consisting of not more than six figures, may be determined by inspection. Divide the number into periods, (as in the usual mode of extracting the cube root,) and to the root of the greatest cube contained in the first period, affix the root of the cube that terminates in the right hand figure of the second period.

11. If a cube terminates in ciphers, the number of ciphers must be divisible by 3.

12. Any square may be divided into two other squares, in the following manner:—

Assume any two numbers at pleasure, and by their product multiply double the root of the given square, for a numerator. Take the sum of the squares of the assumed numbers for a denominator. The resulting fraction will be the root of one of the squares sought. Subtract the square of this root from the given square, and the remainder will be the other square required.

13. If we add the cubes of the series, 1, 2, 3, 4, commencing at the beginning, and taking any number of terms whatever, the sum will always be a square. Thus, $1+8=9$; $1+8+27=36$; $1+8+27+64=100$.

14. If we write down the series of squares of the natural numbers, and take the difference between the successive terms, and the difference of these differences, the second differences will always be 2, as may be seen below:

Squares.	1		4		9		16		25		36
1st Diff.		3		5		7		9		11	
2d Diff.			2		2		2		2		

15. If the successive differences of the series of cubes be taken, the third differences are always 6, $=1\times2\times3$, as may be seen below:

Cubes.	1		8		27		64		125		216
1st Diff.		7		19		37		61		91	
2d Diff.			12		18		24		30		
3d Diff.				6		6		6			

16. The fourth differences of the series of fourth powers are always equal to $1\times2\times3\times4=24$; the fifth differences of the series of fifth powers are always equal to $1\times2\times3\times4\times5=120$; and so on.

EXAMPLES.

1. By which of the foregoing rules do you know that neither of the following numbers can be a square? 952; 827; 1814; 2795; 3725; 308; 711; 866; 299; 25000; 334; 779; 426; 47800.

2. Divide 24 into three squares. *Ans.* $16+4+4$.

3. Find four square numbers, whose sum will make 30.

4. If you divide the cube of 8709512863 by 6, what will be the remainder?

5. By what rules do you know that neither of the following numbers is an exact cube? 87042; 14284; 730176; 4080000; 51858.

6. Determine by inspection, the cube root of each of the following exact cubes. 12167; 21952; 103823; 39304; 42875; 97336; 79507; 132651; 24389. *Ans.* 23; 28; 47, &c.

7. Divide 49 into two other squares.

Ans. Assuming 2 and 3; $\left(\frac{6 \times 14}{4+9}\right)^2 = \frac{7056}{169}$, $49 - \frac{7056}{169} = \frac{1225}{169}$;

Assuming 1 and 8; $\left(\frac{8 \times 14}{1+64}\right)^2 = \frac{12544}{4225}$; $49 - \frac{12544}{4225} = \frac{194481}{4225}$.

&c. &c. &c.

101. Prime and Composite Numbers.

Every number that cannot be divided by any other number, (except 1,) without a remainder, is called a PRIME NUMBER.

Two or more numbers that have no common divisor, are said to be *prime to each other.* Every prime number is prime to all other numbers except its own multiples.

There are no known means of determining at once whether a proposed number is a prime; but the following properties and rules will enable us to determine all the divisors of any number.

1. 2 is a factor of all numbers terminated by 0, 2, 4, 6, or 8. For, as 2 will divide 10, it will also divide any number of tens, or any number of tens *plus* 2, 4, 6, or 8. Numbers divisible by 2 are called EVEN; all others, ODD numbers.

2. 5 is a factor of all numbers terminated by 0 or 5. For, as 5 will divide 10, it will also divide any number of tens, or any number of tens *plus* 5.

3. 3, or 9, is a factor of all numbers in which the sum of the figures is exactly divisible by 3, or 9. For, if from any power of 10, as 10, 100, 1000, &c., we subtract 1, the remainder consists entirely of 9's, and is, therefore, divisible by both 3 and 9. Hence, any power of 10 is divisible by 3 and 9 with 1 remainder; therefore, any number of tens, hundreds, thousands, &c., diminished by as many units, will be divisible by 3 and by 9. Let us, then, examine the number 34794. 3 ten thousands — 3; 4 thousands — 4; 7 hundreds — 7; 9 tens — 9; and 4 units — 4; each divided by 3 or 9, give no remainder. Therefore, 34794 — 3 — 4 — 7 — 9 — 4, is divisible by 3 and by 9, and if the sum of the numbers subtracted, or in other words, the sum of the digits, is similarly divisible, the number itself will be so.

4. 11 is a factor of all numbers in which the sum of the odd digits, (the 1st, 3d, 5th, &c.,) and the sum of the even digits, (the 2d, 4th, 6th, &c.,) are equal, or their difference is some multiple of 11. For any number of tens, thousands, hundred thousands,

&c., (which represent the even digits,) increased by as many units, will be divisible by 11. Any number of hundreds, ten thousands, millions, &c., (which represent the odd digits,) diminished by as many units, will also be divisible by 11. Take, then, the number 635173. 6 hundred thousands + 6; 3 ten thousands — 3; 5 thousands + 5; 1 hundred — 1; 70 + 7; and 3 — 3; each divided by 11 give no remainder. Therefore, 635173 — 18 + 7 or 635173 — 11, is divisible by 11, and 635173 itself must be so.

5. 4 is a factor of all numbers, in which the two terminating figures are divisible by 4. For, as 4 will divide 100, it will also divide any number of hundreds, or any number of hundreds *plus* any number of units divisible by 4.

6. 25 is a factor of all numbers terminating in 25, 50, 75, or two zeros. For, as 25 will divide 100, it will also divide any number of hundreds, or any number of hundreds *plus* 25, 50, or 75.

7. Every number that is divisible by two or more numbers prime to each other, is divisible by their product. Take, for example, 105, which is divisible by both 3 and 5. This number may be resolved into the factors 5 × 21; 5 × 21, must therefore be divisible by 3. But as 3 will not divide 5, it must divide the other factor 21, and the number may be resolved into the factors 5 × 3 × 7 or 15 × 7. Hence we deduce the following additional properties.

8. Every even number that is divisible by 3 is also divisible by 6; and every even number that is divisible by 9 is also divisible by 18.

9. Every number divisible by 3 or 9, in which the two terminating figures are divisible by 4, is divisible by 12 or 36.

10. Every number divisible by 3 or 9, whose terminating digit is 0 or 5, is divisible by 15 or 45.

11. Every prime number greater than 2, is one greater or one less than some multiple of 4.

12. Every prime number greater than 3, is one greater or one less than some multiple of 6.

13. Every number that has no prime factor, equal to or less than its square root, is itself a prime number. For the product of any two factors, each greater than the square root of a number, would evidently be greater than the number itself. *Therefore, if we attempt the division of any supposed prime, by all the primes less*

than its square root, and discover no factor, the number is itself a prime.

TO FIND ALL THE DIVISORS OF A NUMBER.

What numbers will divide 5940 without a remainder?

2	5940
2	2970
3	1485
3	495
3	165
5	55
11	11
	1

We first resolve the number into all its prime factors, by commencing with 2 and dividing as often as possible, by each of the prime numbers in succession. We thus find that $5940 = 2^2 \times 3^3 \times 5 \times 11$, or $2 \times 2 \times 3 \times 3 \times 3 \times 5 \times 11$. It may, therefore, have as many composite divisors as we can form distinct products of these prime factors. In order to determine all the possible products, we arrange 1, with the powers of the factor that is employed the greatest number of times, in a horizontal line. We then multiply each of the numbers in the first line, by each of the powers of another factor; each of the numbers of the preceding lines, by each of the powers of a third factor, &c., as in the following table.

1	3	9	$27 = 3^3$
2	6	18	$54 = 3^3 \times 2$
4	12	36	$108 = 3^3 \times 2^2$
5	15	45	$135 = 3^3 \times 5$
10	30	90	$270 = 3^3 \times 2 \times 5$
20	60	180	$540 = 3^3 \times 2^2 \times 5$
11	33	99	$297 = 3^3 \times 11$
22	66	198	$594 = 3^3 \times 2 \times 11$
44	132	396	$1188 = 3^3 \times 2^2 \times 11$
55	165	495	$1485 = 3^3 \times 5 \times 11$
110	330	990	$2970 = 3^3 \times 2 \times 5 \times 11$
220	660	1980	$5940 = 3^3 \times 2^2 \times 5 \times 11$

The numbers of the first line having been arranged as directed, we multiply them separately by 2 and 2^2.

All the numbers of these *three* lines, are multiplied by 5, which gives us three new lines of divisors.

All the numbers of these *six* lines are multiplied by 11, which gives us six new lines of divisors. We thus obtain 48 numbers that will divide 5940 without a remainder, and an examination of the table will show that these are *all* the divisors, since the prime factors are combined in every possible way.

We are able to determine without actual trial, the number of exact divisors of any given number. By the foregoing table we perceive that 3^3 had 4, or $3+1$ divisors. $3^3 \times 2^2$ has 12, or $\overline{3+1} \times \overline{2+1}$. $3^3 \times 2^2 \times 5$ has 24 or $\overline{3+1} \times \overline{2+1} \times \overline{1+1}$. In like manner each new factor can be multiplied by all the preceding divisors, as many times as are equivalent to the exponent of its power, thus forming so many new divisors to be added to the preceding. Hence, for finding the number of divisors of any given number, we have the following

RULE.

Add 1 to the exponent of each of the prime factors of the given number, and multiply together the exponents thus increased. The product thus obtained, is the number of divisors sought.

If any other number than 10 were adopted as the base of a system of numeration, the number preceding the base would have the same properties as the figure 9 in our present system. For example, 1183, expressed by a scale of 8 would be 2237. The sum of the digits $2+2+3+7=14$ being divisible by 7, the number itself is so divisible.

A *perfect number* is one that is equal to the sum of all its aliquot parts. Thus, 6, the aliquot parts of which are 1, 2, and 3, is a perfect number, because $1+2+3=6$. The following are the only perfect numbers known:
6, 28, 496, 8128, 33550336, 8589869056, 137438691328, 2305843008139952128, 2417851639228158837784576, 99035203142829718830448816128. Every perfect number must terminate either in 6 or in 28.

Two numbers are said to be *amicable*, when each is equivalent to the sum of all the aliquot parts of the other. Thus, 220 and 284 are amicable numbers, because $220=1+2+4+71+142$, which are the aliquot parts of 284, and $284=1+2+4+5+10+11+20+22+44+55+110$, which are the aliquot parts of 220. There are very few amicable numbers known.

EXAMPLES.

1. Which of the following numbers are prime? 733; 949; 917; 619; 1009; 1001; 989; 11571.

2. Find all the prime factors of 780; 468; 3944; 6972; 1849; 2899; 883; 15664.

3. Find all the divisors of 94800; 21100; 5922.

4. How many divisors has 20736? 44100? 29930? 5940? 16384? 15309?

5. Show that if 7 were adopted as the base of a numerical scale, 548712 would be expressed in such a manner that the sum of its digits would be divisible by 6.

6. Prove that 137438691328 is a perfect number.

7. Prove that 17296 and 18416 are amicable numbers.

8. Prove that 9363584 and 9437056 are amicable numbers.

9. Show that 120 and 672 are each equal to half the sum of their aliquot parts.

102. Figurate Numbers.

Figurate, or polygonal numbers, are formed by adding the successive terms of an arithmetical series.

Thus, if we add the successive terms of the natural series, 1, 2, 3, 4, 5, 6, 7, &c., we obtain the figurate series 1, 3, 6, 10, 15, 21, 28, &c., which are called *triangular numbers*, because they can always be arranged in the form of an equilateral triangle.

If we take the arithmetical progression 1, 3, 5, 7, 9, &c., in which the common difference is 2, we obtain the figurate series, 1, 4, 9, 16, 25, &c., which are called *square numbers*, because they can always be arranged in the form of a square.

The arithmetical progression 1, 4, 7, 10, &c., in which the common difference is 3, furnishes the figurate series 1, 5, 12, 22, &c., which are called *pentagonal numbers*, because they can always be arranged in the form of a polygon with five sides.

In a similar manner are produced the hexagonal, heptagonal, octagonal, and other figurate series, the number of sides of the polygon in which the numbers can be arranged, being always two greater than the common difference of the arithmetical progression from which they are derived.

EXAMPLES.

1. Find the first 20 triangular numbers.
2. What are the first 20 pentagonal numbers?
3. What are the first 10 hexagonal numbers?

4. Find the first 10 dodecagonal numbers.

5. Find the first five 17-gonal numbers.

103. The Fundamental Rules.

It may readily be perceived that the rules of Arithmetic merely indicate *convenient modes* of obtaining a desired result, and that, in many cases, a variety of processes will suggest themselves, either one of which will serve our purpose.

But it may at first seem incredible, that the sum of any number of quantities can be obtained without addition; the difference of two numbers, without subtraction; the product of two numbers, without multiplication; and the quotient of two numbers, without division. Yet such is the case, and the following rules are not only interesting from their curiosity, but from the connexion which they show between the several operations that may be performed upon numbers.

I. *To obtain the sum of a series of numbers, by subtraction.*

Assume any number larger than the required sum, and from the assumed number subtract in succession each of the given numbers. Subtract the final remainder from the number first assumed, and the result will be the sum required.

Example.—Find the sum of 69, 93, and 237.

```
Assume 1000     1000
         69      601                 PROOF.
       ----     ----
        931      399 Ans.   1000 — (1000 — 69 — 93 — 237) =
         93                          69 + 93 + 237.
       ----
        838
        237
       ----
        601
```

II. *To find the difference of two numbers, by multiplication and division.*

Write nine times the subtrahend under the minuend, and add each figure of the upper number to the figures in the same place and all the inferior places of the lower number, carrying as in ordinary addition. Proceed in this manner, stopping at the figure that falls immediately under the left hand figure of the minuend, and the result will be the difference sought.

Example.—Required the difference between 874 and 10757.

7 + 6 = 13; 1 to carry + 5 + 6 + 6 = 18; 1 to carry + 7 + 8 + 6 + 6 = 28; 2 to carry + 0 + 7 + 8 + 6 + 6 = 29; 2 to carry + 1 + 7 + 8 + 6 + 6 = 30, but as the 0 falls under the left hand figure of the minuend, we stop there, and find the true remainder to be 9883.

$$\begin{array}{r} 10757 \\ \underline{7866} = 9 \times 874 \\ 09883 \text{ Ans.} \end{array}$$

III. *To find the product of two numbers by addition, subtraction and division.*

Resolve either factor into a number of submultiples of some power of ten, which submultiples, when combined, either by addition or subtraction, will reproduce the original factor.

Write 1 as a common numerator, and the submultiples as denominators of a series of fractions, and divide the other factor by the decimal expression for each of the fractions thus formed. The several quotients, combined in the same manner as the original submultiples, will give the product desired.

EXAMPLE. What is the product of 5769 × 2841?

$5769 = 5000 + 500 + 250 + 20 - 1.$

$$\tfrac{1}{5000} + \tfrac{1}{500} + \tfrac{1}{250} + \tfrac{1}{20} - \tfrac{1}{1} = .0002 + .002 + .004 + .05 - 1$$

$$\tfrac{2841}{.0002} + \tfrac{2841}{.002} + \tfrac{2841}{.004} + \tfrac{2841}{.05} - \tfrac{2841}{1} = 16389729 \text{ Ans.}$$

IV. *To find the quotient of two numbers, by addition and multiplication.*

1. Employ as a multiplier, the difference between the given divisor, and the least power of 10 which is larger than the divisor.

2. Write the first figure of the dividend in the quotient, and add the product of the first dividend figure by the employed multiplier, to as many of the succeeding figures as are equivalent to the number of figures in the divisor. If the sum has a greater number of figures than the divisor, write the left hand figure under the figure in the quotient, and proceed as before, until a sum is obtained having the same number of figures as the divisor.

3. To this sum annex the remaining figures of the dividend. Place the left hand figure of the result as the second quotient figure, and proceed as in paragraph (2). Continue this process until all the figures of the dividend have been employed.

4. Add the several figures in the quotient, and cut off from the right hand as many figures as there are in the divisor. The figures so cut off will represent a remainder; the remaining figures are the quotient.

[If the remainder is larger than the divisor, subtract the divisor as often as possible, and increase the quotient by a number equivalent to the number of subtractions.

If the divisor is contained two or more times in the next larger power of 10, the quotient may be obtained more readily, by employing as a multiplier the difference between some power of 10 and the greatest multiple of the divisor contained in it. The quotient thus found, should be multipled in the same manner as the original divisor.]

EXAMPLE 1. Required the quotient of 284175 by 89.

100 — 89 = 11, the employed multiplier.

```
           284175 ( 2182
11 × 2 =   22        1 1
           ---      ----
           106      3192
11 × 1 =   11
           -----
           17175
11 × 1 =    11
           -----
            8275          Ans. 3192 87/89.
11 × 8 =     88
            ---
            115
11 × 1 =     11
            ---
            265
11 × 2 =      22
              --
              87
```

Ans. $3192\frac{87}{89}$.

The several multiplications and additions may be made mentally, and the quotient thus obtained by employing very few figures. If the divisor is very near some power of ten, or a submultiple of any number which is but little smaller than some power of ten, the quotient can be obtained in this way with great facility.

EXAMPLE 2. Divide 573612 by 9.

```
      57361 2
      -------
      53634.6
      1 1
      -------
Ans.  63734 6/9
```

Ans. $63734\frac{6}{9}$

EXAMPLE 3. Divide 47281591 by 997.

Ans. $47423\frac{860}{997}$.

```
47281 591
---------
47423.860
```

When the divisor consists entirely of 9's, the following rule for obtaining the quotient will be found more convenient:

RULE FOR DIVIDING BY 9's.

When the divisor consists of any number of 9's, increase it by 1, for a new divisor. Divide the dividend by this new divisor. By the same divisor, divide the integers of the quotient, and proceed in a similar manner, until a quotient is obtained less than the divisor. Add all the quotients together, observing the number of units carried from decimals to integers. Add this number to the right hand decimal figure, and the integers will represent the quotient, and the decimals the remainder. When all but the units figure of the divisor are 9's, increase the divisor by the difference between the units figure and 10, and divide as above directed, multiplying each quotient after the first, by the number added to the divisor. Multiply the number carried from decimals to integers, by the number added to the divisor, and add the product to the decimals for the true remainder. If this increased remainder exceeds the divisor, increase the quotient by 1, and subtract the divisor from the remainder for the true remainder.

EXAMPLES FOR ILLUSTRATION.

Divide 8905473 by 999. The divisor, increased by 1, is 1000. Dividing by the rule, and adding the quotients, we obtain 8914.386. There being 1 unit to carry from decimals, we add 1 to the right-hand decimal figure, and find the quotient is $8914.\dot{3}8\dot{7}$.

$$\begin{array}{r} 8905.473 \\ 8.905 \\ 8 \\ \hline 8914.386 \\ 1 \\ \hline 8914.\dot{3}8\dot{7} \text{ quotient.} \end{array}$$

This rule is founded on the decimal value of $\frac{1}{999}$, and it will be easily seen that the process is nearly the same as in the multiplication by $.\dot{0}0\dot{1}00\dot{1}$ +. In dividing the numerators of fractions obtained by the multiplication of circulating decimals, the rule will often be of use.

Divide 1549638144 by 9991. The divisor, increased by 9, is 10000. Dividing first by this number, we multiply the integers of the first quotient by 9, writing the first figure of the product under the right hand decimal figure, which is equivalent to multiplying by 9, and dividing by 10000.

$$\begin{array}{r} 154963.8144 \\ 139.4667 \\ 1251 \\ \hline 155103.4062 \\ 1 \times 9 = 9 \\ \hline .4071 \text{ remainder.} \end{array}$$

We multiply the integers of this second number, and write them in the same manner, and add the several numbers together. There being 1 unit to carry from decimals, we multiply it by 9, and add the product to 4062, which gives 4071 for the true remainder.

The foregoing rules furnish us with a general method for determining whether any number is divisible by any other given number, without actually performing the division.

EXAMPLES.

1. Find by subtraction, the amount of 1690, 84, 207, 168, and 4493.

2. Find by multiplication and addition, the difference between 847 and 10082; between 19804 and 2973.

3. Multiply 764 by 1972, by Case III.

4. Divide 21709 by 837, by Case IV.

5. Divide 684291708 by 9; by 97; by 99; by 995; by 990; by 999; by 9999.

104. Curious Problems.

1. *To add a column of numbers at a glance.*

The numbers to be added should be arranged in pairs, each member of the pair being the arithmetical complement of the other. The key line may be written in the middle, and the sum of the whole may be found by prefixing the figure which represents the number of pairs, to the key line. For example, if we first write the three numbers, 4719082, 3604227, 1518729, and reserve the third as a key line, then take each figure of the second number from 9 except the right hand figure, which we take from 10, we obtain a fourth number, 6395773. Proceeding in a similar manner with the first number, we obtain 5280918 for our fifth number, thus completing two pairs besides the key line. Then prefixing 2 to the key line, we obtain 21518729 for the sum of the five numbers.

```
 4719082
 3604227
 1518729
 6395773
 5280918
--------
21518729
```

It will be more difficult to detect the method pursued, if we subtract each figure of one of the numbers from 8, except the right hand figure, which we take from 9. Each figure of the key line must then be diminished by 1. Other variations may be made in the process, and the position of the key line in the column may be altered at pleasure.

2. *To tell two or more numbers which a person has thought of, neither number being greater than* 9.

Direct the person to double the first number thought of, add 1 to the product, multiply the sum by 5, and add to the product the second number. If there be a third, let him double the first sum, and add 1 to it, multiply by 5, and add the third number. Thus proceed for each additional number thought of. Finally, if there were but two numbers thought of, direct him to subtract 5 from the result; if three, 55; if four, 555, and so on. The remainder will be composed of the figures thought of, in their proper order.

For example, suppose the numbers thought of to be 2, 2, 8, 9. $2 \times 2 + 1 = 5$; $5 \times 5 = 25$; $25 + 2 = 27$; $27 \times 2 + 1 = 55$; $55 \times 5 = 275$; $275 + 8 = 283$; $283 \times 2 + 1 = 567$; $567 \times 5 = 2835$; $2835 + 9 = 2844$; $2844 - 555 = 2289$, the figures of which indicate, in their order, the four numbers thought of.

3. *What is the product of* £11 11*s.* 11*d. by* £11 11*s.* 11*d.*?

Questions similar to this are to be found in many of the old arithmetics, and different answers have been given, according to the different views of the proposers. But in reality the problem is absurd, the error consisting in the supposition that applicate, or concrete numbers, are capable of being multiplied together. The thorough arithmetician should never lose sight of the fact, that all the operations of arithmetic are performed on *abstract numbers.* We say, indeed, that the area of any rectangular surface is found by multiplying the length by the breadth; but this is merely a convenient expression, adopted to avoid circumlocution. Our meaning is, that if we multiply the NUMBER of feet in the length, by the NUMBER of feet in the breadth, the product will represent the NUMBER of square feet in the area. So in geometry, when we say $AB \times CD =$ the rectangle AD, we mean that if the line AB were repeated for every point of the line CD, we should have the surface AD.

If it were possible to form a product of concrete numbers, we should be obliged to call the product of 1£ by 1£, 1 *square pound;* 1s. × 1s., would then be 1 *square shilling;* 1d. × 1d. = 1 *square d.;* 1qr. × 1qr., = 1 *square qr.* 1 *sq.* £ would then be equal to 400 *sq. s.;* 1 *sq. s.* = 144 *sq. d.;* 1 *sq. d.* = 16 *sq. qr.;* and the product of £11 11s. 11d. by £11 11s. 11d., would be 134 sq. £

185 sq. s. 49 sq. d. The impossibility of conceiving of a *square pound*, a *square shilling*, or a *square penny*, shows at once the absurdity of the original question. But the investigation is a useful one, both because it furnishes exercise in an intricate kind of multiplication, and because it shows the error of the very prevalent idea, that dollars, cents, and mills, can be multiplied by dollars, cents, and mills.

The number of curious problems might be extended indefinitely, if our limits would allow. The few here given may suffice to excite an interest in such investigations, and lead the pupil tc exercise his own ingenuity, and to consult the works of authors who have treated the subject more fully.

XXIII. MISCELLANEOUS PROBLEMS.

105. CHRONOLOGY.

ACCORDING to the Julian Calendar or OLD STYLE, the solar year was considered as being 365 days and 6 hours. The 6 hours in 4 years amounted to a day, therefore every fourth year was called a Leap Year, and consisted of 366 days.

But the true solar year is about 11 minutes less than the Julian year, and on this account, in 1582, it was found that spring commenced 10 days later than at the establishment of the Julian Calendar. Pope Gregory XIII., therefore, caused ten days to be taken out of the month of October in that year, and to prevent the recurrence of a similar variation, he ordered that the centurial years should not be regarded as leap years, unless the number of centuries were divisible by 4.

This computation, which is called the Gregorian or NEW STYLE, was soon adopted in the greater part of Europe; but in England and America, the change was not made until 1752, when the error had amounted tc eleven days.

It was then ordered that the 3d of September should be called the 14th, and the Gregorian calendar adopted for the future. In Russia, the Old Style was retained until the year 1830.

One of the first seven letters, A, B, C, D, E, F, G, is attached to every day in the year; thus A is applied to Jan. 1st, 8th, 15th, &c.; B, to Jan. 2d, 9th, 16th, &c.; C, to Jan. 3d, 10th, 17th, &c. In this manner all days in any year which have the same letter, fall on the same day of the week. The DOMINICAL LETTER for any year is the letter that falls against all the Sundays. Thus, the 6th of January, 1850, fell on Sunday, and the dominical letter was, therefore, the 6th letter, or F. But in leap year there are two dominical letters, the first for January and February, the second for the remainder of the year.

PROBLEMS.

I. *To find the dominical letter for any year, according to the Julian or* OLD STYLE.

To the given year add one fourth of itself, plus 4, and divide the sum by 7. If there is no remainder, the dominical letter is G; if 1 remainder, F; and so on in inverse order. If the given year be leap year, the letter thus found will be the dominical letter for the last 10 months, and the next following letter, for the remainder of the year.

What was the dominical letter for A. D. 1531?

```
Given year   1531
one-fourth    382
                4
           ------
         7 ) 1917
           ------
              273 + 6 remainder.
```

To the given year, we add one fourth of itself, (rejecting the fraction,) and 4. Dividing this sum by 7, we have a remainder 6, which indicates that the dominical letter sought is the 6th from G, counting in retrograde order, which is A.

What were the dominical letters for A. D. 564?

The remainder 2 indicates that the dominical letter is the 2d from G, or E. But the year being leap year, the dominical letter for January and February, will be the next following, or F. The two letters sought are therefore F, E.

$$\begin{array}{r} 564 \\ 141 \\ 4 \\ \hline 7\,)\,709 \\ \hline 101 + 2 \end{array}$$

If the given year were before the Christian era, the remainder would indicate the direct order of the letters. Thus, 1 denotes A; 2 denotes B; 5, E, &c.

II. *To find the dominical letter for any year, according to the Gregorian or* NEW STYLE.

Divide the centuries by 4, and take the remainder from 3. Add twice this remainder to $\frac{5}{4}$ of the odd years, and divide the sum by 7. If there is no remainder, the dominical letter is G; if 1 remainder, F, &c., as in the preceding rule.

What is the dominical letter for 1895?

Dividing 18 centuries by 4, the remainder is 2. Taking this remainder from 3, we have a remainder of 1. Twice 1 added to 95 years plus $\frac{1}{4}$ of 95, (rejecting the fraction,) gives 120, which, divided by 7, gives a remainder 1, indicating that the dominical letter is the 1st before G, which is F.

$$\begin{array}{r} 4\,)\,18 \text{ cent.} \\ \hline 4 + 2 \\ 3 - 2 = 1 \\ 2 \times 1 = 2 \\ \text{Odd years } 95 \\ 23 \\ \hline 7\,)\,120 \\ \hline 17 + 1 \end{array}$$

III. *To find the day of the week corresponding to any given day of the month.*

The dominical letter found by one of the preceding rules, will indicate the day on which the first Sunday in January will fall. The day of the week for the corresponding day of each succeeding month, may be found by the initials of the following couplet:—

At Dover Dwell George Brown, Esquire,
Good Captain French, And David Friar.

On what day of the week was the Declaration of Independence signed?

The dominical letters for 1776 were G, F. Therefore, the first Sunday in January was the 7th of the month. Then A representing the 7th Jan., D would represent the 7th Feb.; D the 7th March; G the 7th April; B the 7th May; E the 7th June, and G the 7th July. But 1776 being a Leap Year, the dominical letter after February is one day earlier in the month, and a day of the month which would otherwise be represented by G, will be represented by A or Sunday. The 7th July, therefore, came on Sunday, and the 4th on Thursday.

The initials O. S. denote the Old Style. In all cases not thus marked, the New Style is understood.

EXAMPLES.

1. Washington was born on the 22d Feb. 1732. What was the day of the week? *Ans.* Friday.

2. The pilgrims landed at Plymouth, Dec. 11, 1620. O. S. What was the day of the week? *Ans.* Monday.

3. The Battle of Waterloo was fought June 18, 1815. Is it probable that a letter, purporting to have been written at the time, and dated Friday, June 18, is authentic?
Ans. No; because the battle was fought on Sunday.

4. Constantine Paleologus was the last Christian Emperor of Constantinople. On the 29th of May, 1453, the city was taken by the Turks, the Emperor Constantine killed, and Mohammed II. ascended the throne, thus founding the present empire of Turkey in Europe. What was the day of the week? *Ans.* Tuesday.

CONTEMPORARY PRINCES,

FROM EGBERT TO QUEEN VICTORIA.

The following table is compiled principally from Wade's British History. The teacher will find in it the materials for a great variety of Chronological questions.

In order to determine the date of the commencement of each reign, add the number following the name to the number at the head of the column. Ex.: Ethelwolf, 836; Athelstan, 925.

800.	900.	1000.
ENGLAND.—Egbert; Ethelwolf, 36; Ethelbald, 57; Ethelbert, 60; Ethelred I., 66; Alfred the Great, 72.	ENG.—Edward the Elder, 1; Athelstán, 25; Edmund, 41; Edred, 46; Edwy, 55; Edgar, 59; Edw. the Martyr, 75; Ethelred II., 78.	ENG.—Edmund Ironside, 16; Canute the Great, 17; Harold Harefoot, 36; Hardicanute, 39; Edward the Confessor, 41; Harold II., 65; Wm. I., the Conqueror, 66; Wm II. Rufus, 87.
SCOTLAND. — Achaius, 787; Congale III., 19; Alpin, 31; Kenneth II., 34; Donald V., 54; Constantine II., 58; Ethus, 74; Gregory the Great, 75; Donald VI., 92.	SCOT.—Constant. III., 1; Malcolm I., 38; Indulphus, 58; Duphus, 68; Cullenus, 72; Kenneth III., 73; Const. IV., 94; Grimus, 97.	SCOT.—Malcolm II., 4; Duncan, 34; Macbeth, 40; Malcolm III., 57; Donald VI., 93; Duncan II., 94; Edgar, 96.
FRANCE.—Charlemagne; Louis I., 14; Charles the Bald, 43; Louis II., the Stammerer, 77; Louis III. and Carloman, 79; Charles the Fat, 84; Hugh, 88; Charles the Simple, 98.	FR.—Robert, 22; Ralph, 23; Louis IV., 36; Lotharius, 54; Louis V., 86; Hugh Capet, 87; Robert the Pious, 97.	FR.—Henry I., 31; Philip I., 60.
GERMANY. — Charlemagne; Louis I., 14; Louis II., 43; Carloman, Louis III., the Younger, and Charles the Fat, 76; Arnold, 87; Louis IV., the Infant, 99.	GER. — Conrad I., 11; Henry I., 19; Otho the Great, 36; Otho II., 73; Otho III.. 83.	GER.—Henry II., the Saint, 2; Conrad the Salic, 24; Henry III., 39; Henry IV., 56.
SPAIN.—Garcia I., 58; Fortunio, 80.	SP.—Sancho I., 5; Garcia II., 26; Sancho II., 70; Garcia III., 94.	SP.—Sancho III., the Great; Ferdinand I., in Castile, 33; Garcia IV. in Navarre, Ramirez I. in Arragon, 35; Sancho IV, Nav., 54; Sancho I., Arr., 63; Sancho I., Cast., 65; Alfonzo I., Cast., 72; Sancho V., Nav. and Arr., 76; Peter I., N. and A., 94.
PAPAL STATES. — Leo III., 795; Stephen V., 16; Paschal I., 17; Eugene II., 20; Valentine, 24; Gregory IV., 27; Sergius II., 43; Leo IV., 47; Benedict III., 55; Nicholas I., 58; Adrian II., 66; John VIII., 73; Martin I., 83; Adrian III., 84; Stephen VI., 85; Formosus, 91; Stephen VII., 97.	PA. ST.—Romanus Formosus and John IX., 1; Benedict IV., 5; Leo V. and Christopher, 6; Sergius III., 7; Anastasius, 10; Lando and John X., 12; Leo VI., 28; Stephen VIII., 29; John XI. 31; Leo VII., 36; Stephen IX., 40; Martin II., 43; Agapet II., 46; John XII., 56; Benedict V., 65; John XIII., 66; Donus II. and Benedict VI., 73; Benedict VII., 74; John XIV., 81; John XV., 85; John XVI., 86; Gregory V., 96; Silvester II., 99.	PA. ST.—Jno. XVII. and XVIII., 3; Sergius IV., 9; Benedict VIII., 12; Jno. XIX., 24; Bened. IX., 33; Gregory VI., 44; Clement II., 47; Damasia II., 48; Leo IX., 49; Victor II., 55; Stephen X., 57; Nicholas II., 58; Alexander II., 61; Gregory VII., 73; Victor III., 85; Urban II., 87; Pascal II., 99.
RUSSIA.	RUS.—Ighor I. 13; Swatoslaw I 45; Jaropolk I., 72; Waldimer the Great, 80.	RUS.—Swatopolk I., 15; Jaroslaw I, of Kiew, 18; Isaslaw I., 51; Swatoslaw II., 73; Wsewolod I., 78; Swatopolk II., 93.
CONSTANTINOPLE.—Irene, 797; Nicephorus I., 2; Michael I., 11; Leo V., 13; Michael II., 20; Theophilus, 29; Michael III., 42; Basil, 67; Leo VI., 86.	CON. — Alexander, 11; Romanus, 19; Constantine VII., 45; Romanus II., 59; Nicephorus II., 63; John Zimisces, 69; Basil II. and Con. VIII., 76.	CON.—Const. IX., alone, 25; Romanus III., 28; Michael IV., 34; Michael V., 36; Zoe and Theodora, and Const. X., 42; Theodora, 54; Mich. VI., 56; Isaac I., 57; Const. XI., 59; Eudocia and Romanus III., 67; Mich. VII., 71; Nicephorus III., 78; Alexius I., 81.

1100.	1200.	1300.
ENGLAND.—Henry I.; Stephen, 36; Henry II., 54; Richard I., Cœur de Lion, 89; John, 99.	ENG.—Henry III., 16; Edward I., 72.	ENG.—Edward II., 7; Edw. III., 27; Richard II., 77; Henry IV., 99.
SCOTLAND.—Alexander I., 7; David I., 24; Malcolm IV., 53; William I., 65.	SCOT.—Alexander II., 14; Alex. III., 45; Margaret, 86; John Baliol, 88; Interregnum, 96.	SCOT.—Robert Bruce, 6; David II., 29; Edward Baliol, 32; Robert II., 70; Robert III., 90.
FRANCE.—Louis VI., the Gross, 8; Louis VII., 37; Philip II., 77; Augustus, 80.	FR.—Louis VIII., 23; Louis IX., (St. Louis,) 26; Philip III., the Bold, 70; Philip IV., the Fair, 85.	FR.—Louis X., K. of Navarre, 14; Philip the Tall, K. of Nav., 16; Chas. IV. (the Fair), K. of Nav., 22; Philip VI. (the Fortunate), 28; John I., (the Good,) 50; Chas. V. (the Wise), 64; Chas. VI., 80.
GERMANY.—Henry V., 6; Lotharius II., the Saxon, 25; Conrad III., 38; Frederick I., Barbarossa, 52; Henry VI., 90; Philip and Otho IV., 98.	GER.—Frederic II., 12; Conrad IV., 50; Wm. of Holland, 54; Rich'd, Duke of Cornwall, 57; Rodolph of Hapsburgh, 73; Adolphus of Nassau, 92; Albert of Austria, 98.	GER.—Henry VII., 8; Louis of Bavaria, and Fred. of Austria, 14; Chas. IV., 46; Winceslaus, 78.
SPAIN.—Alphonso I., N. and Arr., 4; Urraca, Cas., 9; Alphonso II., Cas., 26; Garcia V., Nav., Ramirez II., Ar., 34; Petronilla and Raymondo, Arr., 37; Sancho VI., the Wise, N. 50; Sancho II., Cast., 57; Alphonso II., Arr., 62; Sancho VII., N., 94; Peter II., Arr., 96.	SP.—James I., Ar., 13; Henry I., C., 14; Ferdin. III., C., 17; Theobald I., N., 34; Alphonso IV., C., 52; Theob. II., N., 53; Henry I., N., 70; Joanna I., N., 74; Peter III., Ar., 76; Sancho IV., C. 84; Alphonso III., Ar., 85; Jas. II., Ar., 91; Ferd. IV., C., 95.	SP.—Alphonso V., C., 14; Alph. IV., Ar., 27; Joanna II., N., 28; Peter II., Ar. 36; Chas. II., N., 49; Peter I., C., 50; Henry II., C., 69; John I., C., 79; Charles III., N., 86; John I., A., 87; Henry III., C., 90; Martin, A., 95.
PAPAL STATES.—Ielas II., 18; Calixtus II., 19; Honorius II., 25; Innocent II., 30; Celestine II., 43; Lucius II., 44; Eugene III., 45; Anastasius IV., 54; Adrian IV., 55; Alex. III., 59; Lucius III., 81; Urban III., 85; Gregory VIII., 87; Clement III., 88; Celestine III., 91; Innocent III., 98.	PA. ST.—Honorius III., 17; Gregory IX., 27; Celestine IV., 41; Innoc. IV., 43; Alex. IV., 54; Urb. IV., 62; Greg. X., 64; Clem. IV., 65; Innoc. V., Adrian V., and John XX., 76; Nicholas III., 77; Martin IV., 81; Honorius IV., 85; Nich. IV., 88; Celestine V., 94; Boniface VIII., 95.	PA. ST.—Benedict X., 3; Clem. V., 5; Jno XXI., 16; Alex. II., 27; Bened. XI., 34; Clem. VI., 42; Innocent VI., 53; Urban V., 63; Greg. XI., 71; Urban VI., 78; Boniface IX., 90.
RUSSIA.—Waldimir II., 13; Mistislaw, 25; Jaropolk II., 32; Wsewolod II., 38; Isaslaw II., 46; Jurje I., 49; Andrej, 57; Michel I., 75; Wsewolod III., 77.	RUS.—Jurje II., 13; Constantine, 17; Jaroslaw II., 38; Alexander, 45; Newskoi, 50; Jaroslaw III., 62; Wasilej I., 70; Dimitrej, 75; Andrej, 81; Danilo, 94.	RUS.—Michailow, 5; Jurje III., 17; Iwan I., of Moscow, 28; Semen, 40; Iwan II., 53; Dimitrej II., 59; Dimit. III., 63; Wasilej II., 89.
CONSTANTINOPLE.—Jno. Comnenus, 18; Manuel Com., 43; Alex's Com., 80; Andronicus, 83; Isaac II., 85; Alexius Angelus, 95.	CON.—Isaac II. restored, 3; Mourzouffe, 4; Baldwin, 4; Henry, 6; Peter, 17; Robert, 19; John, with Baldwin II., 28; Baldwin II., alone, 37; Michael Paleologus, 61; Andronicus, 82.	CON.—John Canta, 41; John Paleologus, 55; Manuel Pal., 91.

1400.	1500.	1600.
ENGLAND.—Hen. V., 13; Hen. VI., 22; Edward IV., 61; Edw. V. and Richard III., 83; Hen. VII., 85.	ENG.—Henry VIII., 9; Edward VI., 47; Mary, 53; Elizabeth, 58.	GREAT BRITAIN.—Jas. I., 3; Charles I., 25; Cromwell, 53; Charles II., 60; James II., 85; Wm. and Mary, 89.
SCOTLAND.—James I., 5; James II., 37; James III., 60; James IV., 87.	SCOT.—James V., 12; Mary, 42; James VI., 67.	
FRANCE.—Charles VII., the Victor, 22; Louis XI., the Prudent, 61; Charles VIII., the Affable, 83; Louis XII., 98.	FR.—Francis I., 15; Henry II., 47; Francis II., 59; Charles IX., 60; Henry III., 74; Henry IV., the Great, 89.	FR.—Louis XIII., 10; Louis XIV., the Great, 43.
GERMANY.—Robert; Sigismund, 11; Albert II., 37; Frederic III., 40; Maximilian I., 93.	GER.—Charles V., 19; Ferdinand I., 58; Maximilian II., 64; Rodolph II., 76.	GER.—Matthias, 12; Ferdinand II., 19; Ferd. III., 37; Leopold I., 58.
SPAIN.—John II., Cast., 6; Ferdinand I., Arr., 12; Alphonso V., Arr., 16; Blanche, N., and John I., A., 25; Henry IV., C., 54; Ferdin. II. and Isabella, C., 74; Ferd. II., the Catholic, A., 79; Eleanor and Francis Phœbus, Nav., 79; Catharine, N., 83.	SP.—Charles I., 16; Philip II., 56; Philip III., 98.	SP.—Philip IV., 21; Charles II., 65.
PAPAL STATES.—Innocent VII., 4; Gregory XII., 6; Alexander V., 9; John XXII., 10; Martin V., [illegible]7; Eugene IV., 31; Nicholas V., 47; Calixtus III., 55; Pius II., 58; Paul II., 64; Sixtus IV., 71; Innoc. VIII., 84; Alex. VI., 92.	PA. ST.—Pius III. and Julius II., 3; Leo X., 13; Adrian VI., 22; Clement VII., 23; Paul III., 34; Julius III., 50; Marcellinus II., 55; Paul IV., 56; Pius IV., 59; Pius V., 66; Gregory XIII., 72; Sixtus V., 85; Urban VII. and Gregory XIV., 90; Innocent IX., 91; Clement VIII., 92.	PA. ST.—Leo XI. and Paul V., 5; Greg. XV., 21; Urban VIII., 23; Innoc. X., 44; Alexander VII., 55; Clement IX., 67; Clem. X., 70; Innoc. XI., 76; Alex. VIII., 89; Innoc. XII., 91.
RUSSIA.—Wasilej III., 25; Iwan Wasilej III., 62.	RUS.—Wasilej IV., 5; Iwan Wasilej IV., 33; Feodor I., 84; Boris Godunow, 98.	RUS.—Wasilej Schuiskoi, 6; Mich. Fedrowitsch, 13; Alexej, 45; Feodor II., 76; Iwan V., 82; Pet. the Great, 85.
CONSTANTINOPLE.—Jno. Paleologus, 25; Constantine Paleol., 48; Mohammed II., 53; Bajazet, 81.	CON.—Selim I., 12; Solyman, 20; Selim II., 66; Amurath, 74; Mohammed III., 95.	CON.—Achmed I., 4; Mustapha, 17; Osman I., 18; Mustapha restored, 22; Amurath IV., 23; Ibrahim, 40; Mohammed IV., 48; Solyman II., 87; Achmed II., 91; Mustapha II., 95.

1700.	1800.
GREAT BRITAIN.—Anne, 2; George I., 14; Geo. II., 27; Geo. III., 60.	GT. BRIT.—George IV., 20; William IV., 30; Victoria, 37.
FRANCE.—Louis XV., 15; Louis XVI., 74; Republic, 92.	FR.—Napoleon Emperor, 4; Louis XVIII., 14; Charles X., 24; Louis Philippe, 30; Republic, 48.
GERMANY.—Joseph I, 5; Chas. VI., 11; Chas. VII., 42; Francis I. and Maria Theresa, 45; Joseph II., 65; Leopold II., 90; Francis II., 92.	AUSTRIA.—Francis I., 6; Ferdinand I., 35; Francis Joseph I., 48.
SPAIN.—Philip V.; Ferdinand VI., 51; Charles III., 59; Chas. IV., 88.	SP.—Ferd. VII. and Joseph Napoleon, 8; Ferd. VII., 14; Isabella II., 33.
PAPAL STATES.—Clement XI.; Innocent XIII., 21; Benedict XIII., 24; Clem. XII., 30; Bened. XIV., 40; Clem. XIII., 58; Clem. XIV., 69; Pius VI., 75.	PA. ST.—Pius VII., Leo XII., 23; Pius VIII., 29; Gregory XVI., 31; Pius IX., 46.
RUSSIA.—Catharine I., 25; Peter II., 27; Anne, 30; Iwan VI., 40; Elizabeth, 41; Peter III. and Catharine II., 62; Paul I., 96.	RUS.—Alexander I., 1; Nicholas I., 25.
CONSTANTINOPLE.—Achmed III., 3; Mohammed V., 30; Osman II., 54; Mustapha III., 57; Abdul-Hamet, 74; Selim III., 89.	CON.—Mohammed VI., 8; Abdul-Medjid, 39.

THE SMALLER EUROPEAN STATES, FROM 1700.

1700.	1800.
SWEDEN.—Charles XII., 1697; Ulrica Eleanora, 19; Frederic, 20; Adolphus Frederic, 59; Gustavus III., 71; Gustavus IV., 92.	SWE.—Charles XIII., 9; Charles John XIV., 18; Oscar I., 44.
DENMARK.—Fred. IV., 1699; Christian VI., 30; Fred. V., 46; Christian VII., 66.	DEN.—Frederic VI., 8; Frederic VII., 30; Christian VIII., 39; Frederic VIII., 48.
NAPLES.—Charles II., 13; Charles III., 35; Ferdinand IV., 59.	NAP.—Joseph Napoleon, 8; Joachim Murat, 15; Ferdinand I. (of the two Sicilies), 21; Francis, 26; Ferdinand II., 30.
PORTUGAL.—John V., 6; Joseph Emmanuel, 50; Maria, 77; John VI., 99.	PORT.—Pedro IV., 26; Maria da Gloria, 28.
PRUSSIA.—Frederic I., 1; Frederic William I., 13; Frederic II., the Great, 40; Frederic William II., 86; Frederic William III., 97.	PRUS.—Frederic William IV., 40.

BIRTHS AND DEATHS OF CELEBRATED PERSONS.

	Born.	Died.		Born.	Died.
Solon	B.C. 650	B.C. 559	Oliver Cromwell	A.D. 1599	A.D. 1658
Confucius	" —	" 479	Milton	" 1608	" 1674
Socrates	" 470	" 400	Algernon Sidney	" 1617	" 1683
Plato	" 430	" 347	Moliere	" 1622	" 1673
Demosthenes	" 381	" 322	Pascal	" 1623	" 1662
Alexander the Great	" 356	" 323	George Fox	" 1624	" 1690
Hannibal	" 247	" 183	Christopher Wren	" 1632	" 1723
Scipio Africanus	" 235	" 184	Isaac Newton	" 1643	" 1727
Cicero	" 106	" 43	Leibnitz	" 1646	" 1716
Virgil	" 70	" 10	Fenelon	" 1651	" 1715
Constantine	A.D. 274	A.D. 337	Peter the Great	" 1672	" 1725
Mahomet	" 570	" 632	Handel	" 1680	" 1760
Charlemagne	" 742	" 814	Young	" 1681	" 1760
Haroun Alraschid	" 765	" 809	Berkley	" 1684	" 1753
Canute	" —	" 1036	Voltaire	" 1694	" 1778
Jenghis Khan	" 1160	" 1227	John Wesley	" 1705	" 1790
Dante	" 1265	" 1321	Euler	" 1707	" 1783
Petrarch	" 1304	" 1374	Linnæus	" 1707	" 1778
Gutenberg	" 1400	" 1468	Rousseau	" 1712	" 1778
Joan of Arc	" 1410	" 1431	Hume	" 1717	" 1776
Columbus	" 1442	" 1506	Blackstone	" 1723	" 1780
Copernicus	" 1473	" 1543	Adam Smith	" 1723	" 1790
Michael Angelo	" 1474	" 1564	Joshua Reynolds	" 1723	" 1792
Pizarro	" 1475	" 1541	Cowper	" 1731	" 1800
Luther	" 1483	" 1546	Gibbon	" 1737	" 1794
Raphael	" 1483	" 1520	Robespierre	" 1759	" 1794
Loyola	" 1491	" 1556	Napoleon	" 1769	" 1821
Napier	" 1550	" 1617	Fourier	" 1772	" 1837

EXAMPLES.

1. Columbus discovered America Oct. 11, 1492, O. S. What was the day of the week? *Ans.* Thursday.

2. On what day of the week was the commencement of the year in which Henry III. ascended the English throne?[a] *Ans.* Friday.

106. THE MOON'S AGE AND SOUTHING.

I. *The Golden Number.*

The mean time of the moon's revolution around the earth, is 29.53 days. In 19 years, or 228 solar months, there are 235 lunar months. This period of 19 years, is called a lunar cycle, and all the changes and eclipses of the moon, are the same for each cycle.

[a] In England, until the year 1752, the year was considered as beginning on the 25th of March.

The year of the lunar cycle corresponding to any given year, is called the GOLDEN NUMBER for that year. It is found by the following rule:

Add 1 *to the given year, and divide by* 19. *The remainder will be the golden number.* (*If the remainder is* 0, *the golden number is* 19.)

Required the golden number for A. D. 1853. $\frac{1854}{19} = 97\frac{11}{19}$. *Ans.* 11.

II. *The Epact.*

The Epact is the moon's age at the beginning of the year. It generally either increases by 11, or diminishes by 19, every year, and never exceeds 29. It is found by the following rule:

Divide the given year by 19, *and multiply the remainder by* 11. *The product will be the epact, if it does not exceed* 29.

If the product exceeds 29, *divide it by* 30, *and the remainder will be the epact.*

The epact for 1845: $\frac{1845}{19} = 97\frac{2}{19}$; $2 \times 11 = 22$. *Ans.* 22.

The epact for 1857: $\frac{1857}{19} = 97\frac{14}{19}$; $14 \times 11 = 154$; $\frac{154}{30} = 5\frac{4}{30}$. *Ans.* 4.

The epact for 1862: $\frac{1862}{19} = 98$. *Ans.* 0.

III. *The Number or Epact for the Month.*

The Number for any month, shows what the moon's age would be at the beginning of that month, provided it was new moon on the first of January. It is found by the following rule:

Divide the number of days in the preceding months, (reckoning from the beginning of the year,) by 29.53, *and the nearest whole number to the remainder, is the epact or number for the month, required.*

The epact for April. In common years, $\frac{31+28+31}{29.53} = 3\frac{.41}{29.53}$. *Ans.* 0.

The epact for April. In leap years, $\frac{31+29+31}{29.53}=3\frac{1.41}{29.53}$. *Ans.* 1.

IV. *The Moon's Age.*

The Moon's Age, is the number of days that have elapsed since new moon. It never exceeds 30. It is found by the following rule:

To the epact for the year, add the number for the month, and the day of the month, and divide by 30. *The remainder will be the moon's age.*

What was the moon's age, Aug. 18th, 1849? $\frac{6+5+18}{30}=\frac{29}{30}$. *Ans.* 29 days.

V. *The Moon's Southing.*

The SOUTHING is the time when the moon is on the meridian. It is found by the following rule:

Multiply the moon's age by .8, *and the product will be the hours past noon. If the hours exceed* 12, *subtract* 12, *and the remainder will represent the time after midnight.*

Find the time of the moon's southing, for July 4th, 1776.

Epact 9; number for the month 5; moon's age 18. 18 $\times$ .8 = 14.4h. = 2h. 24m. after midnight, moon south.

EXAMPLES.

1. Find the Golden Number for the years of accession[a] of each of the monarchs of Great Britain. *Ans.* Jas. I. 8; Chas. I. 11; Cromwell, 1; Chas. II. 8; &c.

2. Find the epact for the years of accession of William the Conqueror, and of each of the ten following monarchs of England. *Ans.* Wm. I. 22; Wm. II. 14; Hen. I. 7; Stephen 15, &c.

---

[a] See Table of Contemporary Princes.

3. Find the moon's number for each month of the year, both for common years, and for leap years.

Ans. common years; Jan. 0; Feb. 1; Mar. 29; &c.
leap years; " 0; " 1; " 1; &c.

4. What was the moon's age on the day of the battle of Bunker Hill, June 17th, 1775?

5. The French army commenced its retreat from Moscow, Oct. 18th, 1812. At what time on that day was the moon on the meridian?

107. MENSURATION.

In the following formulas, *a* represents the area; *al* the altitude; *b* the base; *br* the breadth; *c* the circumference; *ca* the conjugate axis, (of an ellipse); *cs* the convex surface; *d* the diameter; *fa* the fixed axis, (of a solid of revolution); *h* the hypothenuse; *he* the height; *l* the length; *lb* the lower base, (of a frustum); *p* the perpendicular; *ra* the revolving axis; *s* the solidity; *sh* the slant height, (of a cone, frustum, &c.); *ub* the upper base.

1. THE PARALLELOGRAM. $a = b \times al$.

2. THE TRIANGLE. $a = \frac{b \times al}{2}$.

To find the area of a triangle, when the three sides are given:

I. Add the three sides together, and take half their sum.

II. From this half sum take each side separately.

III. Form the continued product of the half sum and the three remainders, and extract the square root of that product.

3. THE RIGHT ANGLED TRIANGLE. $h = \sqrt{b^2 + p^2}$.

4. THE CIRCLE. $c = 3.1416 \times d$; $a = .7854 \times d^2$; $a = .07958 \times c^2$.

5. THE ELLIPSE. $a = ta \times ca \times .7854$.

6. THE PARALLELOPIPEDON. $s = l \times br \times he$.

7. THE PRISM. $s = al \times a$ of b.

8. THE CYLINDER. $cs = al \times c$ of b; $s = al \times a$ of b.

9. THE CONE AND PYRAMID. $cs = \dfrac{sh \times c \text{ of } b}{2}$; $s = \dfrac{al \times a \text{ of } b}{3}$.

10. THE FRUSTUM. $cs = \dfrac{(c \text{ of } ub + c \text{ of } lb) \times sh}{2}$; $s = \dfrac{(a \text{ of } ub + a \text{ of } lb + \sqrt{a \text{ of } ub \times a \text{ of } lb}\,) \times al}{3}$.

11. THE SPHERE. $cs = c \times d$; $s = .5236 \times d^3$; $s = .01689 \times c^3$.

12. THE ELLIPSOID. $s = fa \times ra^2 \times .5236$.

13. THE POLYGON. *To find the area of a regular polygon when one of its sides is given:* Multiply the square of the side, by the multiplier standing opposite the name of the polygon in the following table. The radius of the inscribed circle, is found by multiplying the side by the proper number in the right hand column.

No. of Sides.	Names.	Multipliers.	Radius of Inscribed Circle.
3	Triangle	.433013	.288675
4	Square	1.000000	.500000
5	Pentagon	1.720477	.688191
6	Hexagon	2.598076	.866025
7	Heptagon	3.633912	1.038262
8	Octagon	4.828427	1.207107
9	Nonagon	6.181824	1.373739
10	Decagon	7.694209	1.538842
11	Hendecagon	9.365640	1.702844
12	Dodecagon	11.196152	1.866025

14. THE POLYEDRON. *To find the surface of a regular polyedron:* Multiply the square of the linear edge, by the tabular number in the column of surfaces.

To find the solidity of a regular polyedron: Multiply the cube of the linear edge, by the tabular number in the column of solidities.

No. of Sides.	Names.	Surfaces.	Solidities.
4	Tetraedron	1.73205	.11785
6	Hexaedron	6.00000	1.00000
8	Octaedron	3.46410	.47140
12	Dodecaedron	20.64578	7.66312
20	Icosaedron	8.66025	2.18169

EXAMPLES.

1. By the second method of Analysis, (§§ 57 and 59,) determine the formulas for finding b, and al, in a parallelogram.
Ans. $b = a \div al$; $al = a \div b$.

2. What are the formulas for b and al, in a triangle?

3. Required the formulas for b and p, in a right-angled triangle. *Ans.* $b = \sqrt{h^2 - p^2}$; $p = \sqrt{h^2 - b^2}$.

4. Find two formulas for d, and an additional formula for c, in a circle. *Ans.* $d = c \div 3.1416$; $d = \sqrt{a \div .7854}$; $c = \sqrt{a \div .07958}$.

5. Obtain formulas for ta and ca in an ellipse.

6. What are the formulas for l, br, and he in a parallelopipedon?

7. Find formulas for al and a of b in a prism.

8. Obtain formulas for c of b and a of b, and two formulas for al in a cylinder.

9. Obtain formulas for sh, al, c of b, and a of b in a cone or pyramid.

10. Find two formulas for c and for d in a sphere.
Partial Ans. $c = \sqrt[3]{s \div .01689}$.

11. Required formulas for c of ub, c of lb, sh, and al, in the frustum of a pyramid, or of a cone.

Ans. c of $ub = \frac{2cs}{sh} - c$ of lb; c of $lb = \frac{2cs}{sh} - c$ of ub; $sh = 2cs \div (c$ of $ub + c$ of $lb)$; $al = 3s \div (a$ of $ub + a$ of $lb + \sqrt{a \text{ of } ub \times a \text{ of } lb})$.

12. What are the formulas for fa and ra in an ellipsoid? *Ans.* $fa = s \div (.5236 \times ra^2)$; $ra = \sqrt{s \div (.5236 \times fa.)}$

13. The solidity of a prism is 516 c. ft., and the altitude $27\frac{1}{2}$ft. What is the area of its base?

Ans. $18\frac{42}{55}$ sq. ft.

14. The convex surface of a pyramid contains 47 sq. ft., and the circumference of its base is $3\frac{1}{3}$ft. Required its slant height? *Ans.* $28\frac{1}{5}$ft.

15. The solidity of a frustum is $1447\frac{1}{2}$ c. ft.; the area of the upper base 49 sq. ft.; the area of the lower base 81 sq. ft. What is its altitude? *Ans.* $22\frac{1}{2}$ft.

16. The solidity of a cylinder is 28 c. ft., and the altitude 14ft. What is the diameter of the base? *Ans.* 1.59ft.

17. Required the radius of a sphere whose solidity is 75 c. ft. *Ans.* 2.616ft.

18. The fixed axis of an ellipsoid measures 10ft.; and its solidity is 80 c. ft. What is the length of its revolving axis? *Ans.* 3.9088ft.

19. The area of a regular dodecagon is 47ft. Required the length of each side, and the radius of the inscribed circle. *Ans.* 2.0488ft.; 3.823ft.

20. What is the area of a triangle, whose sides measure respectively 16 rods, 27 rods, and 19 rods?

Ans. 3R. 29.398r.

21. A regular icosaedron contains 81 c. ft. What is the length of each of its linear edges?[a] *Ans.* 3.336ft.

108. Tonnage of Vessels.

The accurate determination of the tonnage of any vessel, is a very difficult problem. The best method probably, is, to divide the vessel into a number of sections, and to determine by numerous measurements, the average length, breadth, and depth of each section. The solid contents can thus be found in cubic feet, and by dividing the number of cubic feet by 40,[b] the answer will be obtained in tons. The accuracy of the result, will depend on the number of measurements that are taken.

This process is, however, a very tedious one, and numerous rules have been framed, in order to abridge the labor of computation. Of these rules, the following are the most frequently employed.

I. *Carpenters' Rule.*

Measure the length, breadth, and depth, all in feet. Divide the continued product of the three dimensions by 95, and the quotient will be the tonnage.

If the vessel is double-decked, half the breadth is taken as the depth.

II. *United States Government Rule.*

If the vessel is single-decked, take the length from the fore part of the main stem to the after part of the stern post above the upper deck, the breadth at the broadest part above the main wales, and the depth from the under side of the

[a] Questions similar to the foregoing, may be multiplied by the teacher, to any extent that may seem desirable.

[b] Dividing by 40 will give the *actual* tonnage. Each of the following rules gives a result considerably less than the true tonnage. In freighting by the ton, the owner may charge either by the ton of weight, or the ton of measure, whichever will yield the largest tonnage.

deck to the ceiling, in the hold. From the length deduct $\frac{3}{5}$ of the breadth, multiply the remainder by the breadth, and that product by the depth. Divide the last product by 95, and the quotient will be the government tonnage.

If the vessel is double-decked, substitute half of the breadth for the depth, and proceed as directed for single-decked vessels.

III. *British Government Rule.*

Divide the length of the upper deck between the after part of the stem and the fore part of the stern-post, into 6 equal parts. At the foremost, the middle, and the aftermost of these points of division, measure in feet, and decimals of a foot, the depths from the under side of the upper deck to the ceiling at the limber strake. (In the case of a break in the upper deck, the depths are to be measured from a line stretched in a continuation of the deck.) Divide each of the three depths into 5 equal parts, and measure the inside breadths at the following points, viz: at $\frac{1}{5}$ and $\frac{4}{5}$ below the upper deck of each of the extreme depths, and at $\frac{2}{5}$ and $\frac{4}{5}$ below the upper deck of the midship depth. At half the midship depth measure the length from the after part of the stem to the fore part of the stern-post.

To twice the midship depth add the two extreme depths, for the sum of the depths. Add the upper and lower breadths at the foremost division, 3 times the upper breadth and the lower breadth at the midship division, and the upper and twice the lower breadth at the after division, for the sum of the breadths. Multiply the sum of the depths by the sum of the breadths, and the product by the length, and divide the final product by 3500, which will give the number of tons for register. If the vessel has a half-deck, or a break in the upper deck, measure the inside mean length, breadth, and height of such part thereof as may be included within the bulk-head. Divide the continued product of these three measurements by 92.4, and the quotient will be the number of tons to be added to the result as above found.[a]

---

[a] Previous to Jan. 1st, 1846, the British rules for estimating tonnage were similar to our own. But this rule led to the building of ships of forms improper for the purpose of safe navigation, in order that, by measuring less than their real burden, they might evade a part of the

If the vessel is laden, measure the length on the upper deck between the after part of the stem and the fore part of the stern-post, the inside breadth on the under side of the upper deck at the middle point of the length, and the depth from the under side of the upper deck down the pump-well to the skin. Divide the continued product of these three dimensions by 130, and the quotient will be the register tonnage.

In open vessels, the depths are to be measured from the upper edge of the upper strake.

EXAMPLES.

1. Find by rule 1, the tonnage of a double-decked vessel 163 feet long, and 31 feet wide. *Ans.* 824½ tons.

2. Find by rule 2, the government tonnage of a double-decked vessel 180 ft. long, and 32ft. wide. *Ans.* 866.6 tons.

3. Required the British government tonnage of a vessel with the following measurements: length at half the midship depth 175ft.; depths, at the foremost point of division 29ft., at the middle point 30ft., at the aftermost point 28ft.; breadths, at the $\frac{1}{5}$ and $\frac{4}{5}$ depths of the foremost division 33 and 24ft., at the $\frac{2}{5}$ and $\frac{4}{5}$ depths of the midship division 39 and 36ft., at the $\frac{1}{5}$ and $\frac{4}{5}$ depths of the aftermost division, 35 and 29ft.; half-deck, length 75ft., mean breadth 35ft., height 7ft. *Ans.* 1971.41 tons.

109. GAUGING.

The easiest way of finding the contents of casks, is by the diagonal rod. The contents given by the rod are only correct for casks of the most common forms.

The same result as the rod would give, may be found in the following manner:

---

duties. The method now employed, gives the tonnage of all ships, however built, with tolerable accuracy, and therefore removes the temptation to build vessels of an unsuitable form.

Take any rod, and putting it in at the bung, extend it to the opposite corner of either head, and measure the distance in inches. Extend it in a similar manner to the other head, and measure the distance. Half the sum of these two distances, will be the diagonal. Then,

Multiply the cube of the diagonal by $\frac{144}{64000} = \frac{9}{4000}$. The quotient will be the contents in imperial gallons. For wine gallons, substitute 173; for beer gallons, 141.6; for dry gallons, 148.5, in the place of 144.

In order to determine the contents with accuracy, casks are usually divided into four varieties, according to the degree of their curvature. In each of the following formulas for those four varieties, D denotes the bung diameter, d the head diameter, and a the length of the cask.

I. *When the cask is formed like the middle zone of a spheroid.*

$.0009442 \times a \times (2\,D^2 + d^2) =$ contents in imperial gallons. For the contents in wine gallons, substitute .0011333; for beer gallons, .0009284; for dry gallons, .000974, in the place of .0009442.

II. *When the cask is formed like the middle zone of a parabolic spindle.*

$.0001888 \times a \times (8\,D^2 + 4\,D \times d + 3d^2) =$ contents in imperial gallons. For wine, beer, or dry measure, substitute .0002267, .0001856, .000195, in the place of .0001888.

III. *When the cask is formed like two equal frustums of a parabolic conoid.*

$.0014163 \times a \times (D^2 + d^2) =$ contents in imperial gallons. For wine, beer, or dry measure, substitute .0017, .0013926, .001461, in the place of .0014163.

IV. *When the cask is formed like two equal frustums of a cone.*

$.0009442 \times a \times (D^2 + D \times d + d^2)$ = contents in imperial gallons. For the other measures, substitute as in Case I.

The following method is more accurate than either of the foregoing, when the diameter midway between the head and the bung, can be accurately determined :—

Add the squares of the head diameter, of the bung diameter, and of twice the middle diameter. The sum, multiplied by .0004721 times the length, gives the contents in imperial gallons. For wine gallons, substitute .0005667; for beer gallons, .0004642; for dry gallons, .000487, in the place of .0004721.

If the staves are of uniform thickness, the middle diameter may be found by measuring the circumference, dividing by 3.1416, and subtracting twice the thickness of the staves from the result.

EXAMPLES.

1. The diagonal of a cask is 32 inches. Required its contents in wine gallons.

Ans. $(32)^3 \times 173 \div 64000 = 88.576$ gallons.

2. Find the contents in imperial gallons, of casks of each of the four varieties, the length of each cask being 40 inches, and the diameters 30 and 36 inches.

Ans. 1st var. $.0009442 \times 40 \times (2 \times 36^2 + 30^2) =$ 131.886 gallons.

2d var. $.0001888 \times 40 \times (8 \times 36^2 + 4 \times 36 \times 30 + 3 \times 30^2) = 131.314$ gallons.

3d var. $.0014163 \times 40 \times (36^2 + 30^2) = 124.408$ gallons.

4th var. $.0009442 \times 40 \times (36^2 + 36 \times 30 + 30^2) = 123.628$ gallons.

3. The head diameter of a cask is 25 in., the bung diameter 30 in., the middle diameter 28 in., and the length 36 in. Required the contents in beer measure.

Ans. $(25^2 + 30^2 + 56^2) \times .0004721 \times 36 = 79.216$ gallons.

4. The circumference of a cask at the bung is 113 in., at the head 91 in., and at a point midway between the head and the bung 106 in. Required the contents in each measure, the length being 40in., and the thickness of the staves $\frac{1}{2}$ in.[a]

Ans. 119.16 imperial gallons.
143.03 wine gallons.
117.16 beer gallons.
122.92 dry gallons.

110. Miscellaneous Examples.

1. If the multiplicand is 7, and the product 2, what is the multiplier?

2. The dividend is 1, and the quotient 8; what is the divisor?

3. The dividend is $6\frac{1}{2}$, and the quotient $18\frac{2}{3}$; what is the divisor?

4. The sum of two numbers is $7\frac{1}{3}$, and one of the numbers is 4.759; what is the other?

5. The difference of two numbers is $13\frac{3}{7}$, and the greater number is 29.43; what is the less?

6. The sum of two numbers is $7\frac{5}{19}$, and their difference is $6\frac{3}{31}$; what are the numbers?

---

[a] When the circumferences are given, it is not necessary to find each diameter separately. We may proceed with the circumferences precisely as with the diameters, until we obtain the sum of the squares. This sum should be divided by the square of 3.1416 (9.8696), and the quotient will be the sum of the squares of the diameters. A deduction should be made from each outside circumference, of six times the thickness of the staves, and the remainder will be nearly equivalent to the inside circumference.

7. What number must be subtracted from $39\frac{5}{9}$ to leave $17\frac{2}{25}$?

8. What number must be added to $23\frac{5}{11}$ to make 47.432?

9. What number must be multiplied by $3\frac{7}{8}$, and the product divided by 7.365 to give $8\frac{3}{4}$ as a quotient?

10. What is the difference between $\frac{3}{46}$ and $\frac{46}{3}$ of 7T. 15cwt. 3qr.?

11. What is the difference between 28 miles, and 27m. 7fur. 39r. 5yd. 2ft. 11.9in.?

12. Reduce $\frac{1}{7}$ of 9m. 7fur. 39r. 5yd. 2ft. to inches.

13. If from a purse containing £35 7s. 11d., I pay to each of 15 laborers, £1 9s. $8\frac{3}{4}$d., how much will be left?

14. Find the sum, the difference, and the product of 874.91 and $42\frac{7}{13}$.

15. In $128\frac{4}{7}$ lb. of water, there is $14\frac{2}{7}$ lb. of hydrogen. How much hydrogen is there in $273\frac{2}{3}$ lb. of water?

16. A grocer sold 17cwt. 3qr. 17 lb. of sugar, at $6\frac{3}{4}$ cents a pound, receiving in exchange 20 barrels of flour, at $\$4\frac{7}{8}$ per barrel, and the balance in money. How much money did he receive?

17. From $\frac{5}{7}$ of 3T. 17cwt., subtract $\frac{4}{13}$ of 7T. 3cwt. 1qr. 18 lb.

18. What will be the freight of $17\frac{3}{5}$cwt. for $89\frac{5}{7}$ miles, if \$7.63 be paid for carrying $11\frac{3}{13}$ tons $9\frac{1}{3}$ miles?

19. How many raisins, at $8\frac{1}{2}$ cents a pound, must be given in exchange for 163gal. 2qt. 3gi. of molasses, at $\$.37\frac{1}{2}$ per gallon?

20. Bought 16cwt. 3qr. 16 lb. of rice, at \$4.00 per cwt., and 9cwt. 2qr. 5 lb. of pearl barley, at $\$4.37\frac{1}{2}$ per cwt. How much would be gained on the whole, by selling each at $4\frac{3}{4}$ cents a pound?

21. If $63\frac{5}{8}$yd. of broadcloth cost \$255, at what price must it be sold per yd., in order to gain \$25.50?

22. A hogshead of sugar at \$7.00 per cwt. cost \$43.75. What did it weigh?

23. How much shalloon that is $\frac{3}{4}$yd. wide, will line $14\frac{5}{8}$ yards of cloth, that is $1\frac{3}{4}$yd. wide?

24. What is the price of 16 boxes of raisins, each holding $9\frac{2}{5}$ lb., at $9\frac{1}{4}$ cents per lb.?

25. How much money, that is 9 per cent. below par, will pay a debt of \$187.50?

26. If 9 men mow 10 acres of grass in a day, how much will 11 men mow in $2\frac{1}{2}$ days?

27. In what time will \$150 gain $\$6.37\frac{1}{2}$, at 6 per cent., simple interest?

28. When molasses is $31\frac{1}{4}$ cents a gallon, how many hogsheads, each holding 97gal. 3qt., can I buy for $\$366.56\frac{1}{4}$?

29. What will be the price of 7 bales of sheeting, each bale containing 9 pieces, and each piece measuring $30\frac{3}{4}$yd., if 26yd. cost $\$2.92\frac{1}{2}$?

30. Bought $39\frac{1}{2}$ bushels of potatoes for $\$12.87\frac{1}{2}$. At what price per bushel must they be sold, in order to gain 15 per cent.?

31. What is the interest of \$9431 for 3yr. 7mo. 13dy., at 7 per cent.?

32. How much may a man spend per day, whose income is \$500 a year, after deducting $\frac{5}{10}$ per cent. for taxes?

33. The words in Johnson's Dictionary have been classified as follows:[a] Articles, 3; nouns, 20409; adjectives,

[a] Gregory.

9053; pronouns, 41; verbs active 5445, neuter 2425, passive 1, defective 5, auxiliary 1, impersonal 3; verbal noun, 1; participles, 38; participial adjectives, 125; participial nouns, 3; adverbs in *ly*, 2096; other adverbs, 496; prepositions, 69; conjunctions, 19; interjections, 68. What per-centage of the entire number belongs to each part of speech?

34. Owing to the curvature of the earth, the difference of level in 1 mile is 8 inches. What would be the difference in $\frac{1}{2}$ mile, the level varying as the square of the distance? in $5\frac{3}{4}$ miles? *Ans.* 2in.; $22\frac{1}{24}$ft.

35. Suppose a school of 180 boys to breathe 20 times each per minute, requiring for each respiration 30 c. in. of air, what amount of carbonic acid would they produce in two sessions, of 3 hours each, estimating that 5 per cent. of the air inhaled is changed into an equal volume of carbonic acid? *Ans.* 1125 c. ft.

36. A vessel of 400 tons has a keel 48 feet long. What length of keel has a vessel of 750 tons, that is built on the same model? *Ans.* 59.19ft.

37. There are two cannon balls, one weighing 28 pounds, and the other 9 pounds. What is the diameter of the greater, that of the less being 5 inches?

Ans. 7.3in., nearly.

38. In an establishment in Lowell 50 cows were kept, of which the average number giving milk was 35. Each cow consumed annually 4.18 tons of hay at $18.50 per ton, and green vegetables worth $20.36. What was the loss in two years, exclusive of the amount expended for attendance, the whole amount of milk obtained being 99705 quarts, worth five cents a quart?

39. Bought 6 bales of cinnamon, weighing gross 4cwt. 3qr. 2 lb., tare 9 lb., at $11 per cwt., and paid charges for

freight, duties, &c. $11.43¾. What rate per cent. would be gained or lost, by selling the whole at $.15 per lb.?

Ans. 25 per cent. gained.

40. Find the value of 4 cases of gum tragacanth, at £20 8s. per cwt., duty 25 per cent. ad valorem, the cases weighing as follows, viz:

		cwt.	qr.	lb.			
No. 15	gross	2	1	7	tare	41 lb.	Draft 2 lb. per case.
No. 16	"	2	1	11	"	42 "	
No. 17	"	2	1	7	"	40 "	
No. 18	"	2	0	27	"	39 "	

Ans. £196 5s. 2¼d.

41. Required the amount of commission at 2½ per cent., and brokerage at ½ per cent., and the net proceeds of 50 bags of cotton, weighing gross 115cwt. 1qr. 11 lb., draft 1 lb. per bag, tare 4 lb. per cwt., the whole being sold at 7½cts. per lb. *Ans.* Commission and brokerage $27.92.
Net proceeds $902.68.

42. In the British Sanitary Report of 1843, it is stated that the proportionate numbers of the population at different ages, in the United States, and in England and Wales, were as follows:—

	United States.	England & Wales.
Under 5 years	1744	1324
5 and under 10	1417	1197
10 " " 15	1210	1089
15 " " 20	1091	997
20 " " 30	1816	1780
30 " " 40	1160	1289
40 " " 50	732	959
50 " " 60	436	645
60 " " 70	245	440
70 " " 80	113	216
80 " " 90	32	59
90 and upwards	4	5

Required the average age, both in England and in

America, of all the inhabitants;—of all above 15;—above 20;—above 50.

43. The commissioners of a certain county are about building a new court house, which will cost $75000. They hire money for the purpose at an annual interest of 5 per cent., and propose to pay the debt thus incurred by 50 equal annual instalments. What amount must be paid each year?

Ans. $4108.25.

44. Three men bought a grindstone 50 inches in diameter, for which A. paid 75 cents, B. $1.50, and C. $2.00. What part of the diameter ought each to wear away, allowing the diameter of the axle to be 2 inches?

Ans. A. 4.62in.; B. 11.05in.; C. 32.33in.[a]

45. An estate of $20000 is to be divided between two sons in the following manner: the elder is to receive $100 the first month, $300 the second month, &c., in arithmetical progression, and the younger is to receive $1000 per month, until the whole is paid. What is the share of each, and how long will they be in receiving it? *Ans.* $10000; 10mo.

46. A ladder standing upright against a wall reaches the top, but the foot being removed 12 feet from the wall, it reaches to a point 6 feet from the top. Required the length of the ladder, and the height of the wall. *Ans.* 15ft.

47. A block of stone 12ft. long, 3ft. wide, and 2ft. thick, is to be floated on a pine raft, which is 20ft. long, and 6ft. wide. What must be the depth of the raft, in order that it may float 4 inches above the water, the specific gravity of the pine being 575, and the sp. grav. of the stone 2500?

Ans. 4ft. 3.76+in.

48. Wishing to estimate the height of a hill which is 5 miles distant, I hold a foot rule at the distance of 2ft. from my eye, and find that 1 inch on the rule intercepts the rays

[a] This answer is obtained by supposing that A. uses the stone first, B. second, and C. last. The question admits of five other solutions.

from the top of the hill and from the horizon. What is the height of the hill? *Ans.* 1100ft.

49. What is the distance of a thunder cloud, if three seconds elapse between the flash and report?

Ans. 3270ft.

50. If the travel over a hill, by friction and gravity causes 8000 days' work of a horse, at 75 cents per day, which can be partially avoided by a road along the base of the hill, the travel over which would require only 4000 days' work, and if the new road will require an extra annual outlay of $600 for repairs, how much can be saved by expending $12000 in making the improvement, computing interest at 5%?

Ans. $36000.

51. If an engine has sufficient force to draw 100 tons over level ground, what additional power must be exerted on an ascending grade of 40ft. per mile?

Ans. 15cwt. $16\frac{32}{33}$ lb.

52. What is the power of a steam engine with a cylinder 40 inches in diameter, making the usual estimate of the effective force of the steam and the stroke of the piston?

Ans. 64 horse power.

53. What is the amount of pressure on a dam 150ft. by 18ft., the average depth of water being 6ft.?

Ans. 1012500 lb.

54. My expenditure having exceeded my income by 15 per cent., I find that by saving $\frac{1}{6}$ of my income for the succeeding year, I can supply the deficiency with interest, and have $4.60 left. What is my income? *Ans.* $600.

55. Find 11 terms of a harmonical progression, two of the terms being 4 and 8.

Ans. 8, 4, $2\frac{2}{3}$, 2, $1\frac{3}{5}$, $1\frac{1}{3}$, $1\frac{1}{7}$, 1, $\frac{8}{9}$, $\frac{4}{5}$, $\frac{8}{11}$.

56. At the breaking up of the ice in a river, a tree is cut down by a block of ice, which has a surface of 10000

square feet, and is 1 foot thick. How many axes, each weighing 10 lb., and moving with a velocity of 20ft. per second, would have the same momentum, the velocity of the ice being $3\frac{1}{3}$ft. per second, and its specific gravity 930?

Ans. $8687\frac{1}{2}$.

57. Gregory King, in 1695, made the following estimate of the expense of England, France, and Holland, in diet:[a]

	England.	France.	Holland.	Total.
In bread-stuffs . .	£4300000	£10600000	£1400000	
In meats . . .	3300000	5600000	800000	
In butter, cheese, and milk . . .	2300000	4200000	600000	
In malt liquors .	5800000	100000	1200000	
In spirituous drinks	1300000	9000000	400000	
In fish, fowls, and eggs	1700000	3900000	1100000	
In fruits and garden produce . . .	1200000	3600000	400000	
In groceries and sweetmeats . .	1100000	3000000	300000	
Total				

58. Estimating the population of England, in 1695, at $5\frac{1}{2}$ million, France at $13\frac{1}{2}$ million, and Holland at $2\frac{1}{5}$ million, what was the average amount annually expended for each article of diet by each individual, in each nation? What was the average annual expense of each individual in the three nations? 2d *Ans.* £3 3s. 4.7d.

59. A man spends 25 cents a day for wine and cigars; how much will he lose by the expenditure in 48 years, supposing money to be worth 6 per cent.?

Ans. $23427.55.

60. A road has been constructed through a farmer's land, taking $1\frac{1}{4}$ acres, worth $75 per acre. In addition to the loss of his land, he is obliged to expend $100 in

[a] Wade.

building a fence, which must be renewed every 12 years. What damages should he receive, money being worth 6 per cent. compound interest? *Ans.* \$292.54.

61. Find 6 weights with which any number of pounds, from 1 to 364, can be weighed.

62. What is the difference between the area of a circle whose circumference is $157\frac{2}{25}$ft., and the area of the greatest square that can be inscribed in it?

Ans. 713.5ft.

63. What number is that which is divisible by 11, but if divided by any number less than 11, leaves 1 remainder?

Ans. 25201.

64. The average effect of a bushel of coals, weighing 60 lb., when consumed by an engine now working at Wheal Towan, in Cornwall, is sufficient to raise 70 million pounds one foot high.[a] How many pounds of the same coal would furnish sufficient power to raise a man, weighing 150 lb., to the summit of Mont Blanc, an elevation of 15680ft.?

Ans. $2\frac{14}{875}$ lb.

65. The Menai bridge consists of a mass of iron about 4 million pounds in weight, suspended at an average height of 170ft. above the sea.[a] How many bushels of coal would have sufficed to raise it to its present position?

66. Estimating the quantity of granite in the great pyramid at 75614816 c. ft., the specific gravity at 2700, and the average height to which the materials were raised at 125ft., how many chaldrons of coal would have furnished the power necessary for its erection?

Ans. $632\frac{10489}{11200}$ chaldrons, a quantity consumed in some foundries in a week.

67. The entire expense of the Revolutionary War was estimated by the Register of the Treasury, in 1790,

---

[a] Working Man's Friend.

at $135193700, to meet which an issue was made of $359547027\frac{7}{18}$ in continental money.[a] What was the average loss per cent. by the depreciation of the continental currency? *Ans.* 62.398 + per cent.

68. Four men bought a grindstone, 40 inches in diameter, each contributing an equal amount. How much of the diameter ought each to grind away?

Ans. 1st. 5.359in.; 2d. 6.3568in.;
3d. 8.2842in.; 4th. 20in.

69. A. and B. are on opposite sides of a circular field that is 120 rods in diameter, and commence travelling around it in the same direction. How many times will each go round the field before the slower is overtaken, A. going 39 rods in 3 minutes, and B. $66\frac{2}{3}$ rods in 5 minutes?

Ans. A. $19\frac{1}{2}$ times; B. 20 times.

70. A man sold a horse for $65.25, thereby gaining as much per cent. as the horse cost him. What did he give for the horse? *Ans.* $45.

71. Professor Wheatstone endeavored to determine the velocity of electricity by using a revolving mirror, and observing the relative position of the images made by the sparks at the two extremities of a long wire. What interval of time would be indicated by an angular deviation of 1° in the appearance of the two sparks, supposing the mirror to make 800 revolutions in 1 second? *Ans.* $\frac{1}{288000}$ sec.

72. It has been estimated that the average quantity of air contaminated by respiration, insensible perspiration, and lights, is 4 c. ft. per minute for each individual; the average amount cooled by the draught from each door or window, 11 c. ft. per minute; the number of c. ft. cooled by radiation through the windows, $1\frac{1}{2}$ times the number of sq. ft. of the glass exposed to the external air.[b] According to this estimate, how much fresh air should be supplied per

---

[a] Encyclopædia Americana. [b] Tredgold.

minute, in summer, for a hall containing 2000 people? How much heated fresh air, in winter, for a church with a congregation of 600, there being 28 windows and doors, and 1000 sq. ft. of glass? 1st *Ans.* 8000 c. ft. per min.
2d *Ans.* 4208 c. ft. per min.

73. A. and B. own adjoining farms. A. and his family do no labor in winter, except to take the necessary care of his stock and household; but the family of B., by shoemaking, braiding straw, and other similar employments, make $125. To how much will B.'s winter labor amount in 30 years, if the proceeds are all invested at 6 per cent. compound interest? *Ans.* $9882.27.

74. A farmer has a lane leading through his pasture, to the public highway, and instead of fencing in the lane, he has a gate at each extremity. If 1¼ minutes' delay is occasioned by opening and closing each gate, and if he is obliged to pass through the lane 12 times a day, how much will he lose by the gates in a year, supposing his time to be worth 20 cents an hour? *Ans.* $36.50.

75. A pump in which the water is to be raised to the height of 10ft., has a bore 6in. in diameter. What should be the diameter of the bore of a pump, which is to raise the water 25 feet, in order that the two pumps may be worked with equal ease? *Ans.* 3.79in.

76. Suppose a man whose average weight is 160 lbs., to drink three half-pint cups of coffee per day for 40 years, the average specific gravity of the coffee being 1100, to how many times his own weight will the whole amount?
Ans. 24106½ lb., or 150.665625 times his own weight.

77. It has been stated by one of the most careful and successful manufacturers, that on substituting, in one of his cotton mills, a better for a poorer educated class of operatives, he was able to add 12 per cent. to the speed of his machinery, without any increase of damage or danger from the acceleration. What amount would be saved by the

employment of educated labor, from this source alone, in a business of $500000?

78. The number of females engaged in the various manufactures of Massachusetts, has been estimated at 40000, and their average annual wages at $100 each. The superintendent of the Merrimack Mills, in 1841, estimated the average wages of the best educated operatives, at $17\frac{3}{4}$ per cent. above the general average wages of the mills, and the average wages of the least educated, at $18\frac{1}{2}$ per cent. below the general average. According to this estimate, how much would be gained by elevating the whole 40000 to the highest standard, and how much would be lost by degrading them to the lowest standard?

79. The researches and discoveries of M. Meneville, in regard to the fly which was lately so destructive to the olive in the south of France, are said to have increased the value of the annual product of this fruit, $1000000. What amount of profit will France derive in 50 years, from the education which led to such a discovery, supposing the entire increase of value to be invested annually at 5 per cent. compound interest?

80. The aggregate quantity of water annually discharged by the Mississippi river, has been estimated at 14883360636880 c. ft.[a] To how many cubic miles is this equivalent?

Ans. 101.1 c. miles.

81. Estimating the area of the Mississippi valley at 1400000 sq. miles,[a] and the average depth of rain water that falls annually in the valley, at 52 inches,[a] how many cubic feet of water fall in the whole valley during the year, and what part of it passes off by evaporation?

Ans. 169128960000000 c. ft., of which about $\frac{21}{23}$ passes off by evaporation, and $\frac{2}{23}$ is discharged by the river.

---

[a] Proceedings of Amer. Association, 1848.

82. The removal of the forests from the valley of the Mississippi has increased evaporation to such an extent, that the inundations of the river have become much less frequent and less formidable than they were at the first settlement of the country.[a] What amount of water was annually discharged by the Mississippi, 30 years ago, supposing that it has since decreased 25 per cent.?

Ans. 19844480849173 c. ft.

83. If 23232 c. in. of the Mississippi water, deposit 44 c. in. of sediment,[a] how long would it require to deposit the present delta, which is estimated to contain 13600 sq. m. of surface, and to be of the average depth of $\frac{1}{5}$ mile?

Ans. $14203\frac{4}{5}$ years.

84. What is the approximate weight of the earth, estimating its mean diameter at 7912 miles?

Ans. 13527679878424215242145792 lb.

85. If 12 oxen eat $3\frac{1}{3}$ acres of grass in 4 weeks, and 21 oxen eat 10 acres of the like pasture in 9 weeks, how many oxen will eat 24 acres in 18 weeks, the grass being at first equal on every acre, and growing uniformly?

This example is taken from Newton's Universal Arithmetic. It can be solved most readily by making three distinct questions.

(1.) If 12 oxen eat $\frac{10}{3}$ acres of grass, with the growth, in 4 weeks, how many oxen will eat 24 acres, with 4 weeks' growth, in 18 weeks?

Stating the question by analysis, or by proportion, we obtain

$$\frac{12 \times 4 \times 2}{5} = 19\frac{1}{5} \text{ oxen, for the answer.}$$

$\not{3}$	
12	—
4	$\not{1}\not{8}$
2 $\not{2}\not{4}$	$\not{1}\not{0}5$

(2.) If 21 oxen eat 10 acres, with the growth, in 9 weeks, how many oxen will eat 24 acres, with 9 weeks' growth, in 18 weeks?

---

[a] Proc. of Amer. Association, 1848.

The answer, found as before, is

$$\frac{21 \times 6}{5} = 25\tfrac{1}{5}.$$

21	—
$\not{6}$	$\not{1}\not{8}$
6 $\not{2}\not{4}$	$\not{1}\not{0}5$

Now, if $19\frac{1}{5}$ oxen in 18 weeks, eat 24 acres with 4 weeks' growth, and $25\frac{1}{5}$ oxen in the same time, eat the same number of acres with 9 weeks' growth, 5 weeks' growth on 24 acres will support 6 oxen 18 weeks. Then,

(3.) If 6 oxen in 18 weeks eat 5 weeks' growth, how many oxen in the same time will eat 9 weeks' growth?

6	—
9	5

The answer is $10\frac{4}{5}$. We have already found that $25\frac{1}{5}$ oxen in 18 weeks, will eat 24 acres with 9 weeks' growth, and if we add the number which would eat the growth of the remaining 9 weeks, we obtain 36 oxen for the answer sought.

86. If 15 oxen eat $4\frac{1}{2}$ acres of grass in 2 weeks, and 24 oxen eat $14\frac{2}{5}$ acres in 5 weeks, how many oxen will eat 48 acres in 8 weeks, the grass being at first equal on every acre, and growing uniformly? *Ans.* 60 oxen.

87. If 11 oxen eat $24\frac{4}{9}$ acres of grass in 5 weeks, and 10 oxen eat $26\frac{2}{3}$ acres in 4 weeks, how many acres of similar pasture will 42 oxen eat in 7 weeks, the grass growing uniformly? *Ans.* $78\frac{2}{5}$ A.

88. What number is that which is 169 greater than the greatest square number below, and 114 less than the least square number above itself? *Ans.* 20050.

89. A. and B. can do $\frac{1}{10}$ of a piece of work in a day, B. and C. can do $\frac{1}{5}$ of it in $2\frac{1}{4}$ days, and A. and C. can do $\frac{1}{2}$ of it in $4\frac{1}{11}$ days. In what time would each do it alone, and in what time would it be done if they all worked together? *Ans.* A. in 15 days; B. in 30 days; C. in 18 days; all together in $6\frac{3}{7}$ days.

90. There are some monads not exceeding $\frac{1}{24000}$in. in diameter, in which 6 spots have been observed, separated by membranous partitions not thicker than $\frac{1}{20}$ of the diameter

of the spots. If these estimates are correct, what is the thickness of the partitions? *Ans.* $\frac{1}{2880000}$in.

91. What amount can be recovered from the underwriters, upon the following transaction?—£5000 was insured upon goods from New York to London, the goods being valued in the invoice, at £7200. In a storm, a part of this property valued at £1107 5s. was thrown overboard, and the remainder was so much damaged, that it sold for only £5407 15s. 6d., whereas if it had arrived safe, it would have sold for £6723 10s.; besides which, the owner of the goods was obliged to contribute towards a general average, at the rate of 2.225 per cent. on the invoice value of the whole.

Ans. £1025.

92. Exported 45 hogsheads of sugar from New York to Amsterdam, weighing gross 454cwt. 2qr. 18 lb., tare 53 lb. per hhd., which were sold at 12½ groats per lb., subject to a discount of 5 per cent. The original cost and charges in New York, were $2437.50, the amount of charges in Amsterdam, 1149 florins 7 groats, and the agio of the bank of Holland was 4½ per cent. How much was gained or lost by the adventure, the net proceeds being remitted at the exchange of 40cts. per florin? *Ans.* $2638.19 gained.

93. How far can the top of Bunker Hill Monument, which is 282ft. above the level of the sea, be seen from the deck of a vessel, the spectator's eye being 15ft. above the water?[a]

Ans. 25.3 miles.

94. The average power of draft of a horse, moving 5

---

[a] The distance at which bodies may be seen, is found by the following rule:—

To the earth's diameter (41815224 feet), add the height of the eye, and multiply the sum by the height of the eye. The square root of the product is the distance at which an object ON THE SURFACE of the earth or water can be seen.

Work in the same way with the height of the object, and the sum of the two results is the distance at which the object may be seen.

miles per hour for 8 hours a day, being 75 lb., what will be the annual cost of transportation over a road 40 miles long, on which the average friction is $\frac{1}{25}$ of the weight, estimating the amount transported at 90000 tons, and the value of a horse's labor at $62\frac{1}{2}$ cents a day?

Ans. $67200.

95. If the road, in the preceding example, could be improved by macadamizing or otherwise, so that the friction would be reduced to $\frac{1}{60}$ of the weight, how much might profitably be expended in making the improvement, money being worth 6 per cent.? *Ans.* $$653333\frac{1}{3}$.

96. The shadow cast by a Drummond light, at the distance of 80 rods, was observed to be of the same intensity as that cast by the full moon, which was shining at the time. To how many such lights was the moon's light equivalent, her mean distance being 240000 miles?

Ans. 921600000000.

97. Find the least 3 integers, such that $\frac{3}{8}$ of the first, $\frac{5}{14}$ of the second, and $\frac{7}{20}$ of the third, shall be equal.

Ans. 140, 147, 150.

98. For the purchase of a certain estate, A. offers $150 premium, and $300 rent per annum; B. offers $400 premium, and $250 per annum; C. offers $650 premium, and $200 per annum, and D. offers $1800 in ready money. Whose offer is the best, and what is the difference between them, computing 5 per cent. compound interest?

Partial Ans. A.'s offer is the best.

99. Which is of the greater value, the income of an estate of $500 a year for 15 years to come, or the reversion of the same estate for ever, at the expiration of the 15 years, interest at 6 per cent.?

Ans. The income for 15 years.

100. If a ball were put in motion by a force which

would drive it 12 miles the first hour, 10 miles the second, and so on in geometrical progression, what distance would it go in the whole? *Ans.* 72 miles.

101. What is the least number which, if divided by 2, will leave 1 remainder; by 3 will leave 2; by 4 will leave 3; by 5 will leave 4; by 6 will leave 5; but by 7 will leave no remainder? *Ans.* 119.

102. Required the least three numbers, which, divided by 20, will leave 19 remainder; if divided by 19 will leave 18, and so on, (always leaving one less than the divisor), to unity. *Ans.* 232792559; 465585119; 698377679.

103. A trader offers to receive a young man as partner, proposing, if he will advance $500, to allow him $200 per annum; if he will advance $1000, to allow $275 per annum; and if he advances $1500, he will allow $350 per annum. What per cent. is offered for the use of the money, and how much for the young man's time?

Ans. 15 per cent.; $125 per annum.

104. A shepherd sold to one man, half his flock and half a sheep; to a second, half the remainder and half a sheep; and to a third, half the remainder and half a sheep, when he had 20 left. How many had he at first?

Ans. 167.

105. The annual cost of transportation on a road 40 miles long, being estimated at $30000 per mile, what amount of saving can be effected by expending $50000 to shorten the road 3 miles, and $2000000 to reduce the friction to $\frac{1}{2}$ its present amount, the annual cost of repairs being the same in both cases, and the rate of interest being $5\frac{1}{2}$ per cent.? *Ans.* $\$9677272\frac{8}{11}$.

106. A city of 50000 inhabitants is to be supplied with water, from a river 300 feet below the proposed reservoir. Estimating the average daily consumption at 10 ale gallons

for each individual, what must be the power of an engine working 12 hours a day, to raise the requisite supply?

Ans. 48.53 horse power.

107. A commission sale of 60 bags of coffee was effected at Rotterdam, at 22 stivers per lb., with an allowance of 2 per cent., and of 1 per cent. on the remainder; weight gro. 50cwt. 1qr. 11 lb., draft 1 per cent., tare 6 lb. per bag; commission and guarantee, 3 per cent.; other charges, 269 florins 10 stivers. Required the net proceeds of the sale, a bill being remitted at the exchange of 39 cents per florin. *Ans.* \$2003.58.

108. Bought in Cadiz four chests of Peruvian bark, at 25 rials of plate per lb., weighing gro. 6cwt. 3qr. 16 lb., tare 25 lb. per chest. The export duty was 150 rials vellon per quintal of 100 lb., and the amount of other charges was 1210 rials of plate. What was the whole amount of the purchase? *Ans.* \$1851.40.

109. At a time when bills upon the treasury bore at Jamaica a premium of 7½ per cent., and dollars a premium of 4 per cent., 14000 dollars were purchased and consigned to London, to be disposed of there; they weighed 1010 lb. 3oz. troy, and were sold to the Bank of England at 5s. 3d. per oz. The charges were, freight 1½ per cent.; 14 bags, at 6d. each; weighing, 2s. 6d.; brokerage, ⅛ per cent.; commission, ½ per cent.; and insurance on £3000, at 4 per cent. It is required to find the rate per cent. of the amount of the charges, estimated upon the cost of the dollars, exclusive of the premium, and to find what would have been gained or lost, if in preference to dollars, their actual cost had been laid out in government bills.

Ans. Value of the dollars, £3182 5s. 9d.; net proceeds of the sales, £2994 3s. 11d; rate of the charges, 6⅕ per cent.; gain by government bills, £84 9s. 8d.

110. "One evening I chanced with a tinker to sit,
Whose tongue ran a great deal too fast for his wit;
He talked of his art with abundance of mettle,
So I asked him to make me a flat-bottomed kettle.
Let the top and the bottom diameters be
In just such proportion as five is to three:
Twelve inches the depth I proposed, and no more,
And to hold in ale gallons seven less than a score.
He promised to do it, and straight to work went,
But when he had done it he found it too scant.
He altered it then, but too big he had made it;
And when it held right, the diameters failed it.
Thus making it often too big and too little,
The tinker at last had quite spoiled his kettle,
But declared he would bring his said promise to pass,
Or else that he'd spoil every ounce of his brass.
Now to keep him from ruin, I pray find him out
The diameters' length, for he'll ne'er do't without."

Ans. 24.4in.; 14.64in.

111. Two vessels are 30 miles apart, and are sailing, the first with, and the second against a current of $2\frac{1}{2}$ miles per hour. In still water, each would sail 7 miles per hour. In what time will they meet, and what will be the distance of each from its present position?

112. At what time between half past 7 and 8 o'clock, are the hour and minute hands exactly 13 minutes apart?

Ans. $52\frac{4}{11}$m. past 7.

113. The annual loss in the United States, in consequence of intemperance, has been estimated at $108000000.[a] If this amount were saved, how much would be left after purchasing 4000000 sheep at $2.50, 400000 head of cattle at $25, 200000 cows at $20, 40000 horses at $100, 500000 suits of men's clothes at $20, 1000000 suits of boys' clothes

[a] Amer. Temp. Soc. Reports.

at $10, 500000 suits of womens' clothes at $10, 1000000 suits of girls' clothes at $3, 1200000 bbl. flour at $5, 800000 bbl. beef at $10, 800000 bbl. pork at $12.50, 3000000bu. corn at $.50, 2000000bu. potatoes at $.25, 10000000 lb. sugar at $.10, 400000 lb. rice at 5cts., 2000000gal. molasses at 40cts., building 1000 churches at $5000, 8000 school houses at $500, and supporting 50000 families at $300? *Ans.* $180000.

114. What is the difference between 10000 years according to our present calendar, and 10000 solar years, estimating the solar year at 365d. 5h. 48m. 49½sec.?[a]

Ans. 2d. 14h. 30m.

115. In 1840, Prof. Bessel, from the corrected parallax of the star 61 Cygni, estimated its distance from the earth at 592200 times the mean distance of the sun. According to this estimate, how long would the star's light be in reaching us? *Ans.* 3391dy. 9h. 13m. 45s.

116. A micrometer screw is made with threads $\frac{1}{50}$ of an inch apart, and an index is affixed to the head, which points to the degrees on a graduated circle. What thickness would be measured by turning the index 1° 15′?

Ans. $\frac{1}{14400}$in.

117. A. and B. are carrying a weight of 150 lb., which is suspended on a pole, at the distance of 2ft. 6in. from A., and 1ft. 9in. from B. What amount does each one support?

Ans. A. 61 lb. 12$\frac{4}{17}$oz.; B. 88 lb. 3$\frac{13}{17}$oz.

118. Archimedes boasted, that if he had a place to stand, he could move the world. If he weighed 150 lb., how far would he be obliged to move, in order to move the earth 1 inch? [See Ex. 84.]

Ans. 1423366990574938472.448 miles.

119. At Bilin, in Germany, is a bed of tripoli, composed

---

[a] Carpenter.

almost entirely of the sheaths of a kind of animalcule, the length of each sheath being about $\frac{1}{3500}$ of an inch.[a] How many such sheaths would there be in a cubic inch?

Ans. 42875000000.

120. A grain of copper dissolved in nitric acid, and mixed with three pints of water, (wine measure,) gives a blue color to the whole. What is the weight of the quantity contained in $\frac{1}{1000000}$ of a cubic inch, which is sufficiently large to be visible to the naked eye?[a]

Ans. $\frac{1}{86625000}$ of a grain.

121. In the manufacture of embroidery wire, a cylinder of silver, weighing 360oz., is covered with 2oz. of gold. This is drawn into wire, of which 4000ft. weigh 1oz. A foot of this wire may be divided into 1200 parts visible to the naked eye. What would be the weight of a particle of the gold upon this wire, that would be visible under a microscope magnifying 500 times each way?

Ans. $\frac{1}{216000000000000}$oz.

122. The squares of the times of revolutions of the planets around the sun, are proportioned to the cubes of their mean distances. The mean distance of the earth from the sun, is 95 million miles, and it revolves round the sun in 365d. 5h. 48m. 50s. What is the mean distance of Venus, her time of revolution being 224d. 16$\frac{3}{4}$h.?[b]

Ans. 69 million miles, nearly.

123. What do I gain per cent. by purchasing goods at 8 months' credit, and selling them immediately for cash, at the same price, money being worth 6 per cent. per annum?

Ans. 3$\frac{11}{13}$ per cent.

124. A. and B. enter into partnership, A. advancing $4800 to carry on the business. B. has no money, but being thoroughly acquainted with the trade, agrees to be manager. B.'s annual salary, before he engaged in the copart-

---

[a] Carpenter. [b] Nesbit.

nership, was $250. By the terms of the contract, A. is to be allowed 7 per cent. for the interest of his money and the risk of the business, and the net profits, after making this deduction, are to be divided in the proportion which 6 per cent. on A.'s capital bears to B.'s former salary. The profits at the end of the year, amounted to $5000. How should they be divided? *Ans.* A. $2832.71; B. $2167.29.

125. What quantity of each of the following ingredients, would be required to make a ton of flint glass, estimating the waste in melting at 2 per cent.? White sand, 9 parts; red lead, or litharge, 6.5 parts; pearlash with a little nitre, 4.5 parts.[a] *Ans.* Sand 1028.6 lb.; litharge 742.9 lb.; pearlash 514.3 lb.

126. The dust of the puff ball consists of seeds, which vary from $\frac{1}{20000}$ to $\frac{1}{30000}$ of an inch in diameter. How many seeds, of the average diameter of $\frac{1}{25000}$in., would there be in $1\frac{1}{2}$ cubic inches? *Ans.* 234375000000000.

127. Estimating the entire wealth of the world at $1407374883553.28, in what time would 1 cent absorb all the property on the globe, if it were placed at compound interest, so as to double every 11 years?

Ans. 517 years.

128. A merchant failing in business, offers to settle with his creditors by paying them 80 cents on a dollar in 6 months, without interest, or to give security for the payment of the whole in $4\frac{1}{2}$ years, without interest. Which proposal is the more advantageous, supposing that the composition first proposed can be invested at 6 per cent. compound interest?

129. Bought flour for $3.50 per barrel, on 6 months. At what price must it be sold to gain 20 per cent., and allow 3 months' credit, money being worth 8 per cent. per annum?

Ans. 4.11\frac{12}{13}$.

---

[a] Parnell.

130. A. has 70cwt. of sugar, which he would sell for $7 per cwt. for cash, but in barter he values it at $8.50 per cwt. B. has wheat worth in ready money $1.12 per bu., which he wishes to exchange with A. At what price should it be valued in barter? *Ans.* $1.36.

131. Insured 150 hogsheads of sugar, valued at $125 per hogshead, from Jamaica to Boston, at 8 per cent., to return 4 per cent. for convoy and arrival without loss. Of the quantity insured, 30hhds. were shipped on board a vessel that was totally lost, 64hhds. arrived safe by a vessel that sailed with convoy the whole of her voyage, 45hhds. arrived with convoy, but in consequence of a mast being cut away, and some goods being thrown overboard to lighten the vessel in a storm, the merchandise shipped was obliged to contribute to the loss, at the rate of 2 per cent. on its value, and the rest of the sugar was not shipped. Required the amount to be recovered from the underwriters, and the amount of return premium for short interest, and for convoy and arrival.[a] *Ans.* Underwriters to pay $3862.50.
Return premium, $419.00.

132. What is the product of $18.75 by $.06¼?

133. James Davis, of Waverly, Ross county, Ohio, cultivated, in the year 1849, 1800 acres exclusively in Indian corn. With the crop he filled a corn-crib 3 miles long, 10ft. high, and 6ft. wide.[b] What was the average yield per acre, and how many bushels would the whole make, when shelled?
Ans. 422.4bu. in the ear, per acre.
396000bu., when shelled.

---

[a] The custom varies as to the amount of return premium, when part of the risk is avoided. Some companies reserve ½ per cent. on the amount insured; some reserve 10 per cent. of the premium paid in; on "open policies," no premium is usually paid, except for the amount actually at risk. In the present instance, 10 per cent. of the premium paid on the molasses which was not shipped, is supposed to be reserved by the company.

[b] Cincinnati Gazette.

134. Sysla, the reputed inventor of the game of chess, is said to have asked as a reward, one grain of wheat for the first square on the chess-board, two for the second, and so on in geometrical progression. What would have been the amount of his reward, there being 64 squares on the board, and 9200 grains of wheat in a pint? What would be the height of a cubical bin that would contain it, supposing the base to be 10 miles square?

Ans. 18446744073709551615 grains;
31329388712142.58 bushels; height of bin, 2.65 miles.

135. If the cost of a railroad is $40000 per mile, and the annual repairs and expenses are estimated at $2500 per mile, how much may be profitably expended at the outset, in order to shorten the proposed route 2½ miles, provided the stock pays an annual dividend of 8 per cent.?

Ans. $178125.

136. What is the weight of a round iron rod, 10 feet long, and 4 inches in diameter? *Ans.* 538 lb. 10⅜oz.

137. A mule and an ass travelling together, the ass began to complain that her burthen was too heavy. "Lazy animal," said the mule, "you have little reason to complain; for if I take two of your bags, I shall have three times as many as you, but if I give you two of mine, we shall have only an equal number." With how many bags was each loaded?

138. How would you distribute among 3 persons 21 bags, 7 of which are full of corn, 7 half full, and 7 empty, so as to give to each the same quantity of corn and the same number of bags?

This question admits of two solutions.

139. What is the 7th root of 63? *Ans.* 1.80737+.

140. A farmer has a stack of hay, from which he sells a

quantity which is to the quantity remaining, as 4 to 5. He then uses 15 loads, and finds that the quantity left is to the quantity sold, as 1 to 2. Required the number of loads at first in the stack. *Ans.* 45 loads.

141. At what points would the numbering on Fahrenheit's and on Delisle's thermometer-scale be the same, the only difference being in the sign?[a]

Ans. $+96\frac{4}{11}°$ Fah.; $-96\frac{4}{11}°$ Del.

142. If the temperature below 0° on the Russian scale is marked +, and the temperature above 0° is marked —,[b] at what points will the numbering be the same, on the Russian and Centigrade scales? *Ans.* 60°.

143. The prime cost of 109cwt. 3qr. 18 lb. of sugar was \$769.375; the freight, insurance, and other expenses, amounted to \$92.31. What did it cost per cwt., and at what price must it be sold per lb. to gain 20 per cent., supposing 4 per cent. to be lost by overweight in retailing?

Ans. \$7 per cwt.; $8\frac{3}{4}$cts. per lb.

144. A merchant receives an invoice of goods, which will probably sell for \$16000, by which he will realize a profit of $33\frac{1}{3}$ per cent. above the prime cost; but preferring to sacrifice a portion of his profit, rather than to risk the entire loss of the goods, he determines to insure them. The premium is $3\frac{1}{2}$ per cent., the policy \$2.00, the agent's commission $\frac{1}{2}$ per cent., and if there is any loss, the broker will charge 1 per cent. for procuring a settlement with the underwriters. What amount should the merchant insure,

---

[a] The pupil should observe that Reaumur's and the Centigrade scales correspond only at 0°, but all the other scales have the same numbering at two points of equal temperature, differing at one of the points only by the sign.

[b] The signs are sometimes applied in this way on the Russian scale, but the notation given on p. 197, appears more philosophical, on account of its uniformity.

in order to recover the prime cost of the goods, and all expenses attending the insurance, in case of a total loss?

Ans. $12633.68.

145. A trader wishes to sell his merchandise at wholesale, so as to make a profit of 20 per cent., after deducting 30% from the retail price. He therefore adds 50 per cent. to the cost of the articles in cash, and deducts 30 per cent. from the amount, for the wholesale price, allowing a credit of 8 months. Computing interest at 6 per cent. per annum, what is his actual loss on all sales at this rate, supposing that his expenses and bad debts amount to 15 per cent. of his total sales? *Ans.* $14\frac{391}{416}$ per cent.

146. What percentage should have been added to the prime cost of the articles in the foregoing example, to yield a net profit of 10 per cent. after making the discount proposed? *Ans.* $92\frac{32}{119}$ per cent.

147. A merchant received on consignment three bales of sheeting, marked A., B., and C. A. contained 420 yards of a quality 15 per cent. better than B., B. contained 380 yards of a quality 10 per cent. poorer than C., and C. contained 450 yards. The whole were sold together at $12\frac{1}{2}$ cents per yard; how much should be credited to each, after deducting $2\frac{1}{2}$ per cent. commission?

Ans. A. $53.98; B. $42.47; C. $55.89.

148. An estate was offered for sale for $12000, but the price appearing too high, the tenant took a lease for 25 years at $800 per annum. How much did he gain or lose, estimating compound interest at 6 per cent.?

Ans. $1022.67.

149. If 32 oxen in 3 weeks consume all the grass on 23 acres of pasture, and if 22 oxen in 5 weeks consume the grass on 25 acres, in what time will 88 oxen consume 115 acres of similar pasturage, the grass growing uniformly?

Ans. 5.88 weeks.

150. The steam power at present employed in Great Britain and Ireland, is estimated as equivalent to the power of 8000000 men.[a] If the power of a horse is 5 times as great as that of a man, and if a horse requires 8 times the quantity of soil for producing food that a human being does, what population could be supported by the additional food which would be required, if horse power were substituted for steam? *Ans.* 12800000.

151. The hydraulic press of Bramah can, by the exertion of a single man, produce a pressure of 1500 atmospheres.[b] Estimating the mean height of the mercury in the barometer at 29.53 inches, what would be the pressure exerted on a surface 2ft. 3in. long, and 1ft. 9in. wide?
Ans. 5502T. 5cwt. 1qr. 15 lb. 12oz.

152. A species of lace is made by covering an inclined flat surface with a paste made of leaves, and drawing with a camel's hair pencil, the pattern which is to be left open. A number of caterpillars, of a species which spins a strong web, are then placed at the bottom, and they commence eating and spinning their way to the top, carefully avoiding every part touched by the oil, but devouring every other part of the paste.[b] How many square yards of the lace thus made, would there be in 1 lb. avoirdupois, the average weight being $4\frac{1}{3}$ grains troy per sq. yd.?
Ans. $1615\frac{5}{13}$ sq. yd.

153. Estimating the weight of a globe of air 1ft. in diameter, at $\frac{1}{25}$ lb. avoirdupois, and the weight of an equal globe of hydrogen, allowing for impurities, at $\frac{1}{6}$ as much, what weight would be sustained by a balloon 18ft. in diameter, if fully inflated with hydrogen gas?
Ans. 194.4 lb.

154. Having observed that the shadow of a cloud is 15 seconds in passing from one point to another, I measure

[a] Chambers' Mechanics. [b] Babbage.

the distance between the two points, and find that it is $31\frac{1}{2}$ rods. What is the velocity of the wind per hour?

Ans. 23m. 200r.

155. A., B. and C. have a loaf of sugar, weighing 48 lb., which they wish to divide equally between them, but having only a 4 lb. weight and a 7 lb. weight, it is required to find how the division can be made.

156. A hundred hurdles may be so placed, as to enclose 200 sheep, and with 2 more the fold may be so made as to hold 400. How can this be done?

157. The animalcules of iron ochre, are about $\frac{1}{12000}$ of an inch in diameter. How many such animalcules would occupy 1 c. ft.? *Ans.* 2985984000000000.

158. An ounce of gold forms a cube about $\frac{5}{12}$ of an inch thick, but by hammering, it may be extended so as to cover a surface of 146 sq. ft. How many leaves formed in this manner, would equal in thickness a leaf of writing paper $\frac{1}{150}$ of an inch thick? *Ans.* 1938.

159. Bought wheat for cash, at $.90, at $.95, and at $1.10 per bushel. In what proportions may the three kinds be mixed, so as to gain 20 per cent. by selling at $1.25 per bushel, on 6 months' credit, money being worth 7 per cent. per annum? *Ans.* 11.62bu. at $.90;
11.62bu. at $.95; 20.13bu. at $1.10.

160. A farmer has paid in cash $4000 for the lease of a farm for 8 years. If money is worth 5 per cent. compound interest, what income ought he to receive from the farm each year, in order to recover his outlay, and lay up $200 a year, his family expenses being $350? *Ans.* $1168.89.

161. A family of 10 persons hired a house for 6 months, at a rent of $780 per annum. At the expiration of 14 weeks they received 4 new boarders, and at the expiration of every 3 weeks, during the remainder of the term, they

received 4 more. How much of the rent should be paid by one of each class? *Ans.* 1st class \$30.49$\frac{51}{1001}$; 2d class \$9.49$\frac{51}{1001}$; 3d class \$6.27$\frac{89}{143}$; 4th class \$3.77$\frac{89}{143}$; 5th class \$1.73$\frac{1}{13}$.

162. Estimate of the cost of manufacturing pins, "Elevens," of which 5546 weigh 1 lb.:[a]—

NAME OF THE PROCESS.	Hands.	Time of making 1 lb of pins.	Cost of making 1 lb. of pins.	Hand earns per day.		Price of making each part of a single pin, in millionths of a penny.
		Hours.	*Pence.*	*s.*	*d.*	
1. Drawing wire	Man	.3636	1.2500	3	3	225
2. Straightening wire	Woman	.3000	.2840	1	0	51
	Girl	.3000	.1420	0	6	26
3. Pointing	Man	.3000	1.7750	5	3	319
4. Twisting and cutting the heads	Boy	.0400	.0147	0	4½	3
	Man	.0400	.2103	5	4½	38
5. Heading	Woman	4.0000	5.0000	1	3	901
6. Tinning, or Whitening	Man	.1071	.6666	6	0	121
	Woman	.1071	.3333	3	0	60
7. Papering	Woman	2.1314	3.1973	1	6	576

Estimating the premium on sterling money at 9⅛ per cent., what is the cost of each pin, in our currency? At what price should the pins be sold per ounce, after paying 15 per cent. profit to the manufacturer, 30 per cent. ad valorem for duties, 18 per cent. to the American importer, and 25 per cent. to the retailer, allowing 10 per cent. of the retail price for loss of interest?

Ans. $\frac{3121751}{6655200000}$ of a cent; 3.9837+ cents.

163. If a powerful Drummond light, at the distance of 30 feet, casts a shadow of the same intensity as that cast by the sun, to how many such lights is the sun's light equivalent, estimating the distance of the sun at 95 million miles? *Ans.* 279558400000000000000.

164. It has been supposed that 10 turns of Babbage's calculating machine may be made in a minute.[a] At this rate, how many places of figures would the machine reach

[a] Babbage.

in a million centuries, supposing it to be so regulated as to commence with 1, and give all the following numbers in their natural order? *Ans.* 15 places.

165. It has been estimated[a] that a man in a properly ventilated room can work 12 hours a day, with no greater inconvenience than would be occasioned by 10 hours' work in a room badly ventilated, and that where there is proper ventilation, a man may gain 10 years' good labor on account of unimpaired health. According to this estimate, what is the loss, in 30 years, to each individual in a badly ventilated workshop, valuing the labor at 10 cents per hour? *Ans.* $5008.

166. In the town of Bury, England, with an estimated population of 25000, the expenditure for beer and spirits, in the year 1836, was estimated at £54190.[a] If this sum was 24 per cent. of the entire loss, resulting from the waste of money, ill health, loss of labor, and the other evils attendant upon intoxication, what was the average loss from intemperance, for each man, woman, and child, in the place, estimating the pound sterling at $4.80? *Ans.* $43.352.

111. Table of Prime and Composite Numbers.

The following table contains all the prime numbers, and the factors of all *odd* composite numbers, below 12700, the prime numbers being indicated by a dash.

For numbers below 1000, all the factors are given. For the odd numbers above 1000, one or more factors will be found in the table, which will reduce each number to a prime, or to some number less than 1000, and under the latter number the remaining factors may be found.

The hundreds are placed at the head of the table, and the tens and units at the left hand.

---

[a] British Sanitary Reports.

	0	1	2	3	4	5	6	7	8	9
00	—	$2^2.5^2$	$2^3.5^2$	$2^2.3.5^2$	$2^4.5^2$	$2^2.5^3$	$2^3.3.5^2$	$2^2.5^2.7$	$2^5.5^2$	$2^2.3^2.5^2$
01	—	—	3.67	7.43	—	3.167	—	—	$3^2.89$	17.53
02	—	2.3.17	2.101	2.151	2.3.67	2.251	2.7.43	$2.3^3.13$	2.401	2.11.41
03	—	—	7.29	3.101	13.31	—	$3^2.67$	19.37	11.73	3.7.43
04	2^2	$2^3.13$	$2^2.3.17$	$2^4.19$	$2^2.101$	$2^3.7.9$	$2^2.151$	$2^6.11$	$2^2.3.67$	$2^3.113$
05	—	3.5.7	5.41	5.61	$3^4.5$	5.101	5.11^2	3.5.47	5.7.23	5.181
06	2.3	2.53	2.103	$2.3^2.17$	2.7.29	2.11.23	2.3.101	2.353	2.13.31	2.3.151
07	—	—	$3^2.23$	—	11.37	3.13^2	—	7.101	3.269	—
08	2^3	$2^2.3^3$	$2^4.13$	$2^2.7.11$	$2^3.3.17$	$2^2.127$	$2^5.19$	$2^2.3.59$	$2^3.101$	$2^2.227$
09	3^2	—	11.19	3.103	—	—	3.7.29	—	—	$3^2.101$
10	2.5	2.5.11	2.3.5.7	2.5.31	2.5.41	2.3.5.17	2.5.61	2.5.71	$2.3^4.5$	2.5.7.13
11	—	3.37	—	—	3.137	7.73	13.47	$3^2.79$	—	—
12	$2^2.3$	$2^4.7$	$2^2.53$	$2^3.3.13$	$2^2.103$	2^9	$2^2.3^2.17$	$2^3.89$	$2^2.7.29$	$2^4.3$ 19
13	—	—	3.71	—	7.59	$3^3.19$	—	23.31	3.271	11.83
14	2.7	2.3.19	2.107	2.157	$2.3^2.23$	2.257	2.307	2.3.7.17	2.11.37	2.457
15	3.5	5.23	5.43	$3^2.5.7$	5.83	5.103	3.5.41	5.11.13	5.163	3.5.61
16	2^4	$2^2.29$	$2^3.3^3$	$2^2.79$	$2^5.13$	$2^2.3.43$	$2^3.7.11$	$2^2.179$	$2^4.3.17$	$2^2.229$
17	—	$3^2.13$	7.31	—	3.139	11.47	—	3.239	19.43	7.131
18	2.3^2	2.59	2.109	2.3.53	2.11.19	2.7.37	2.3.103	2.359	2.409	$2.3^3.17$
19	—	7.17	3.73	11.29	—	3.173	—	—	$3^2.7.13$	—
20	$2^2.5$	$2^3.3.5$	$2^2.5.11$	$2^6.5$	$2^2.3.5.7$	$2^3.5.13$	$2^2.5.31$	$2^4.3^2.5$	$2^2.5.41$	$2^3.5.23$
21	3.7	11^2	13.17	3.107	—	—	$3^3.23$	7.103	—	3.307
22	2.11	2.61	2.3.37	2.7.23	2.211	$2.3^2.29$	2.311	2.19^2	2.3.137	2.461
23	—	3.41	—	17.19	$3^2.47$	—	7.89	3.241	—	13.71
24	$2^3.3$	$2^2.31$	$2^5.7$	$2^2.3^4$	$2^3.53$	$2^2.131$	$2^4.3.13$	$2^2.181$	$2^3.103$	$2^2.3.7.11$
25	5^2	5^3	$3^2.5^2$	$5^2.13$	$5^2.17$	$3.5^2.7$	5^4	$5^2.29$	$3.5^2.11$	$5^2.37$
26	2.13	$2.3^2.7$	2.113	2.163	2.3.71	2.263	2.313	$2.3.11^2$	2.7.59	2.463
27	3^3	—	—	3.109	7.61	17.31	3.11.19	—	—	$3^2.103$
28	$2^2.7$	2^7	$2^2.3.19$	$2^3.41$	$2^2.107$	$2^4.3.11$	$2^2.157$	$2^3.7.13$	$2^2.3^2.23$	$2^5.29$
29	—	3.43	—	7.47	3.11.13	23^2	17.37	3^6	—	—
30	2.3.5	2.5.13	2.5.23	2.3.5.11	2.5.43	2.5.53	$2.3^2.5.7$	2.5.73	2.5.83	2.3.5.31
31	—	—	3.7.11	—	—	$3^2.59$	—	17.43	3.277	$7^2.19$
32	2^5	$2^2.3.11$	$2^3.29$	$2^2.83$	$2^4.3^3$	$2^2.7.19$	$2^3.79$	$2^2.3.61$	$2^6.13$	$2^2.233$
33	3.11	7.19	—	$3^2.37$	—	13.41	3.211	—	$7^2.17$	3.311
34	2.17	2.67	$2.3^2.13$	2.167	2.7.31	2.3.89	2.317	2.367	2.3.139	2.467
35	5.7	$3^3.5$	5.47	5.67	3.5.29	5.107	5.127	$3.5.7^2$	5.167	5.11.17
36	$2^2.3^2$	$2^3.17$	$2^2.59$	$2^4.3.7$	$2^2.109$	$2^3.67$	$2^2.3.53$	$2^5.23$	$2^2.11.19$	$2^3.3^2.13$
37	—	—	3.79	—	19.23	3.179	$7^2.13$	11.67	$3^3.31$	—
38	2.19	2.3.23	2.7.17	2.13^2	2.3.73	2.269	2.11.29	$2.3^2.41$	2.419	2.7.67
39	3.13	—	—	3.113	—	$7^2.11$	$3^2.71$	—	—	3.313
40	$2^3.5$	$2^2.5.7$	$2^4.3.5$	$2^2.5.17$	$2^3.5.11$	$2^2.3^3.5$	$2^7.5$	$2^2.5.37$	$2^3.3.5.7$	$2^2.5.47$
41	—	3.47	—	11.31	$3^2.7^2$	—	—	3.13.19	29^2	—
42	2.3.7	2.71	2.11^2	$2.3^2.19$	2.13.17	2.271	2.3.107	2.7.53	2.421	2.3.157
43	—	11.13	3^5	7^3	—	3.181	—	—	3.281	23.41
44	$2^2.11$	$2^4.3^2$	$2^2.61$	$2^3.43$	$2^2.3.37$	$2^5.17$	$2^2.7.23$	$2^3.3.31$	$2^2.211$	$2^4.59$
45	$3^2.5$	5.29	5.7^2	3.5.23	5.89	5.109	3.5.43	5.149	5.13^2	$3^3.5.7$
46	2.23	2.73	2.3.41	2.173	2.223	2.3.7.13	2.17.19	2.373	$2.3^2.47$	2.11.43
47	—	3.7^2	13.19	—	3.149	—	—	$3^2.83$	7.11^2	—
48	$2^4.3$	$2^2.37$	$2^3.31$	$2^2.3.29$	$2^6.7$	$2^2.137$	$2^3.3^4$	$2^2.11.17$	$2^4.53$	$2^2.3.79$
49	7^2	—	3.83	—	—	$3^2.61$	11.59	7.107	3.283	13.73

	0	1	2	3	4	5	6	7	8	9
50	2.5^2	$2.3.5^2$	2.5^3	$2.5^2.7$	$2.3^2.5^2$	$2.5^2.11$	$2.5^2.13$	$2.3.5^3$	$2.5^2.17$	$2.5^2.19$
51	3.17	—	—	$3^3.13$	11.41	19.29	3.7.31	—	23.37	3.317
52	$2^2.13$	$2^3.19$	$2^2.7.9$	$2^5.11$	$2^2.113$	$2^3.3.23$	$2^2.163$	$2^4.47$	$2^2.3.71$	$2^3.7.17$
53	—	$3^2.17$	11.23	—	3.151	7.79	—	3 251	—	—
54	2.3^3	2.7.11	2.127	2.3.59	2.227	2.277	2.3.109	2.13.29	2.7.61	$2.3^2.53$
55	5.11	5.31	3.5.17	5.71	5.7.13	3.5.37	5.131	5.151	$3^2.5.19$	5.191
56	$2^3.7$	$2^2.3.13$	2^8	$2^2.89$	$2^3.3.19$	$2^2.139$	$2^4.41$	$2^2.3^3.7$	$2^3.107$	$2^2.239$
57	3.19	—	—	3.7.17	—	—	$3^2.73$	—	—	3.11.29
58	2.29	2.79	2.3.43	2.179	2.229	$2.3^2.31$	2.7.47	2.379	2.3.11.13	2.479
59	—	3.53	7.37	—	$3^3.17$	13.43	—	3.11.23	—	7.137
60	$2^2.3.5$	$2^5.5$	$2^2.5.13$	$2^3.3^2.5$	$2^2.5.23$	$2^4.5.7$	$2^2.3.5.11$	$2^3.5.19$	$2^2.5.43$	$2^6.3.5$
61	—	7.23	$3^2.29$	19^2	—	3.11.17	—	—	3.7.41	31^2
62	2.31	2.3^4	2.131	2.181	2.3.7.11	2.281	2.331	2.3.127	2.431	2.13.37
63	$3^2.7$	—	—	3.11^2	—	—	3.13.17	7.109	—	$3^2.107$
64	2^6	$2^2.41$	$2^3.3.11$	$2^2.7.13$	$2^4.29$	$2^2.3.47$	$2^3.83$	$2^2.191$	$2^5.3^3$	$2^2.241$
65	5.13	3.5.11	5.53	5.73	3.5.31	5.113	5.7.19	$3^2.5.17$	5.173	5.193
66	2.3.11	2.83	2.7.19	2.3.61	2.233	2.283	$2.3^2.37$	2.383	2.433	2.3.7.23
67	—	—	3.89	—	—	$3^4.7$	23.29	13.59	3.17^2	—
68	$2^2.17$	$2^3.3.7$	$2^2.67$	$2^4.23$	$2^2.3^2.13$	$2^3.71$	$2^2.167$	$2^8.3$	$2^2.7.31$	$2^3.11^2$
69	3.23	13^2	—	$3^2.41$	7.67	—	3.223	—	11.79	3.17.19
70	2.5.7	2.5.17	$2.3^3.5$	2.5.37	2.5.47	2.3.5.19	2.5.67	2.5.7.11	2.3.5.29	2.5.97
71	—	$3^2.19$	—	7.53	3.157	—	11.61	3.257	13.67	—
72	$2^3.3^2$	$2^2.43$	$2^4.17$	$2^2.3.31$	$2^3.59$	$2^2.11.13$	$2^5.3.7$	$2^2.193$	$2^3.109$	$2^2.3^6$
73	—	—	3.7.13	—	11.43	3.191	—	—	$3^2.97$	7.139
74	2.37	2.3.29	2.137	2.11.17	2.3.79	2.7.41	2.337	$2.3^2.43$	2.19.23	2.487
75	3.5^2	$5^2.7$	$5^2.11$	3.5^3	$5^2.19$	$5^2.23$	$3^3.5^2$	$5^2.31$	$5^3.7$	$3.5^2.13$
76	$2^2.19$	$2^4.11$	$2^2.3.23$	$2^3.47$	$2^2.7.17$	$2^6.3^2$	$2^2.13^2$	$2^3.97$	$2^2.3.73$	$2^4.61$
77	7.11	3.59	—	13.29	$3^2.53$	—	—	3.7.37	—	—
78	2.3.13	2.89	2.139	$2.3^3.7$	2.239	2.17^2	2.3.113	2.389	2.439	2.3.163
79	—	—	$3^2.31$	—	—	3.193	7.97	19.41	3.293	11.89
80	$2^4.5$	$2^2.3^2.5$	$2^3.5.7$	$2^2.5.19$	$2^5.3.5$	$2^2.5.29$	$2^3.5.17$	$2^2.3.5.13$	$2^4.5.11$	$2^2.5.7^2$
81	3^4	—	—	3.127	13.37	7.83	3.227	11.71	—	$3^2.109$
82	2.41	2.7.13	2.3.47	2.191	2.241	2.3.97	2.11.31	2.17.23	$2.3^2.7^2$	2.491
83	—	3.61	—	—	3.7.23	11.53	—	$3^3.29$	—	—
84	$2^2.3.7$	$2^3.23$	$2^2.71$	$2^7.3$	$2^2.11^2$	$2^3.73$	$2^2.3^2.19$	$2^4.7^2$	$2^2.13.17$	$2^3.3.41$
85	5.17	5.37	3.5.19	5.7.11	5.97	$3^2.5.13$	5.137	5.157	3.5.59	5.197
86	2.43	2.3.31	2.11.13	2.193	2.3^5	2.293	2.7^3	2.3.131	2.443	2.17.29
87	3.29	11.17	7.41	$3^2.43$	—	—	3.229	—	—	3.7.47
88	$2^3.11$	$2^2.47$	$2^5.3^2$	$2^2.97$	$2^3.61$	$2^2.3.7^2$	$2^4.43$	$2^2.197$	$2^3.3.37$	$2^2.13.19$
89	—	$3^3.7$	17^2	—	3.163	19.31	13.53	3.263	7.127	23.43
90	$2.3^2.5$	2.5.19	2.5.29	2.3.5.13	$2.5.7^2$	2.5.59	2.3.5.23	2.5.79	2.5.89	$2.3^2.5.11$
91	7.13	—	3.97	17.23	—	3.197	—	7.113	$3^4.11$	—
92	$2^2.23$	$2^6.3$	$2^2.73$	$2^3.7^2$	$2^2.3.41$	$2^4.37$	$2^2.173$	$2^3.3^2.11$	$2^2.223$	$2^5.31$
93	3.31	—	—	3.131	17.29	—	$3^2.7.11$	13.61	19.47	3.331
94	2.47	2.97	$2.3.7^2$	2.197	2.13.19	$2.3^3.11$	2.347	2.397	2.3.149	2.7.71
95	5.19	3.5.13	5.59	5.79	$3^2.5.11$	5.7.17	5.139	3.5.53	5.179	5.199
96	$2^5.3$	$2^2.7^2$	$2^3.37$	$2^2.3^2.11$	$2^4.31$	$2^2.149$	$2^3.3.29$	$2^2.199$	$2^7.7$	$2^2.3.83$
97	—	—	$3^3.11$	—	7.71	3.199	17.41	—	3.13.23	—
98	2.7^2	$2.3^2.11$	2.149	2.199	2.3.83	2.13.23	2.349	2.3.7.19	2.449	2.499
99	$3^2.11$	—	13.23	3.7.19	—	—	3.233	17.47	29.31	$3^3.37$

	10	11	12	13	14	15	16	17	18	19	20	21	22	23	24	25	26	27	28	29
01	7	3	–	–	3	19	–	3	–	–	3	11	31	3	7^4	41	3	37	–	2
03	17	–	3	–	23	3	7	13	3	11	–	3	–	7	3	–	19	3	–	
05	3	5	5	3	5	5	3	5	5	3	5	5	3	5	5	3	5	5	3	5
07	19	3	17	–	3	11	–	3	13	–	3	7	–	3	29	23	3	–	7	3
09	–	–	3	7	–	3	–	–	3	23	7	3	47^2	–	3	13	–	3	53^2	–
11	3	11	7	3	17	–	3	29	–	3	–	–	3	–	–	3	7	–	3	41
13	–	3	–	13	3	17	–	3	7	–	3	–	–	3	19	7	3	–	29	3
15	5	5	3	5	5	3	5	5	3	5	5	3	5	5	3	5	5	3	5	5
17	3	–	–	3	13	37	3	17	23	3	–	29	3	7	–	3	–	11	3	–
19	–	3	23	–	3	7	–	3	17	19	3	13	7	3	41	11	3	–	–	3
21	–	19	3	–	7	3	–	–	3	17	43	3	–	11	3	–	–	3	7	23
23	3	–	–	3	–	–	3	–	–	3	7	11	3	23	–	3	43	7	3	37
25	5	3	5	5	3	5	5	3	5	5	3	5	5	3	5	5	3	5	5	3
27	13	7	3	–	–	3	–	11	3	41	–	3	17	13	3	7	37	3	11	–
29	3	–	–	3	–	11	3	7	31	3	–	–	3	17	7	3	11	–	3	29
31	–	3	–	11^3	3	–	7	3	–	–	3	–	23	3	11	–	3	–	19	3
33	–	11	3	31	–	3	23	–	3	–	19	3	7	–	3	17	–	3	–	7
35	3	5	5	3	5	5	3	5	5	3	5	5	3	5	5	3	5	5	3	5
37	17	3	–	7	3	29	–	3	11	13	3	–	–	3	–	43	3	7	–	3
39	–	17	3	13	–	3	11	37	3	7	–	3	–	–	3	–	7	3	17	–
41	3	7	17	3	11	23	3	–	7	3	13	–	3	–	–	3	19	–	3	17
43	7	3	11	17	3	–	31	3	19	29	3	–	–	3	7	–	3	13	–	3
45	5	5	3	5	5	3	5	5	3	5	5	3	5	5	3	5	5	3	5	5
47	3	31	29	3	–	7	3	–	–	3	23	19	3	–	–	3	–	41	3	7
49	–	3	–	19	3	–	17	3	43^2	–	3	7	13	3	31	–	3	–	7	3
51	–	–	3	7	–	3	13	17	3	–	7	3	–	–	3	–	11	3	–	13
53	3	–	7	3	–	–	3	–	17	3	–	–	3	13	11	3	7	–	3	–
55	5	3	5	5	3	5	5	3	5	5	3	5	5	3	5	5	3	5	5	3
57	7	13	3	23	31	3	–	7	3	19	11	3	37	–	3	–	–	3	–	–
59	3	19	–	3	–	–	3	–	11	3	29	17	3	7	–	3	–	31	3	11
61	–	3	13	–	3	7	11	3	–	37	3	–	7	3	23	13	3	11	–	3
63	–	–	3	29	7	3	–	41	3	13	–	3	31	17	3	11	–	3	7	–
65	3	5	5	3	5	5	3	5	5	3	5	5	3	5	5	3	5	5	3	5
67	11	3	7	–	3	–	–	3	–	7	3	11	–	3	–	17	3	–	47	3
69	–	7	3	37^2	13	3	–	29	3	11	–	3	–	23	3	7	17	3	19	–
71	3	–	31	3	–	–	3	7	–	3	19	13	3	–	7	3	–	17	3	–
73	29	3	19	–	3	11	7	3	–	–	3	41	–	3	–	31	3	47	13	3
75	5	5	3	5	5	3	5	5	3	5	5	3	5	5	3	5	5	3	5	5
77	3	11	–	3	7	19	3	–	–	3	31	7	3	–	–	3	–		3	13
79	13	3	–	7	3	–	23	3	–	—	3	–	43	3	37	–	3	7	–	3
81	23	–	3	–	–	3	41^2	13	3	7	–	3	–	–	3	29	7	3	43	11
83	3	7	–	3	–	–	3	–	7	3	–	37	3	–	13	3	–	11	3	19
85	5	3	5	5	3	5	5	3	5	5	3	5	5	3	5	5	3	5	5	3
87	–	–	3	19	–	3	7	–	3	–	–	3^7	–	7	3	13	–	3	–	29
89	3	29	–	3	–	7	3	–	–	3	–	11	3	–	19	3	–	–	3	7
91	–	3	–	13	3	37	19	3	31	11	3	7	29	3	47	–	3	–	7	3
93	–	–	3	7	–	3	–	11	3	–	7	3	–	–	3	–	–	3	11	41
95	3	5	5	3	5	5	3	5	5	3	5	5	3	5	5	3	5	5	3	5
97	–	3	–	11	3	–	–	3	7	–	3	13^3	–	3	11	7	3	–	–	3
99	7	11	3	–	–	3	–	7	3	–	–	3	11	–	3	23	–	3	13	–

	30	31	32	33	34	35	36	37	38	39	40	41	42	43	44	45	46	47	48	49
01	-	7	11	-	19	3^2	13	-	7	47	-	3	-	11	3^2	7	43	3	-	13
03	7	29	-	3^2	41	31	3	7	-	3	-	11	3^2	13	7	19	-	-	3	-
05	5	5	5	5	5	5	5	5	5	5	5	5	5	5	5	5	5	5	5	5
07	31	13	3	-	-	7	-	11	3^2	-	-	3	7	59	13	-	17	3^2	11	7
09	17	-	-	3	7	11	3^2	-	13	3	19	7	23	31	-	3^2	11	17	7	-
11	-	17	13	7	3^2	-	23	3	37	-	7	-	-	3^2	11	13	29	7	17	3
13	23	11	7	—	-	3	-	47	31	7	-	3^2	11	19	3	-	7	3	-	17^3
15	5	5	5	5	5	5	5	5	5	5	5	5	5	5	5	5	5	5	5	5
17	7	3	-	31	17	-	-	7	11	-	13	23	-	3	7	-	3^2	53	-	11
19	-	-	29	-	13	3^2	7	-	19	-	-	3	-	7	3^2	-	31	11	61	-
21	19	-	-	3^2	11	7	17	61^2	-	3	-	13	7	29	-	11	-	-	3	7
23	-	3^2	11	-	7	13	-	17	-	-	3^2	7	41	11	-	-	23	-	7	3^2
25	5	5^5	5	5	5	5	5	5	5	5	5	5	5	5	5	5	5	5	5	5
27	3	53	7	3	23	-	13	-	43	7	-	-	3	-	19	3^2	7	29	3	13
29	13	7	-	-	3^2	-	19	11	7	-	17	-	-	3^2	43	7	3	-	11	31
31	7	31	3^2	-	47	11	-	7	3	-	29	17	-	61	7	23	11	19	-	-
33	3^2	13	53	11	-	-	7	-	-	19	37	-	17	7	11	3	41	-	3^2	-
35	5	5	5	5	5	5	5	5	5	5	5	5	5	5	5	5	5	5	5	5
37	-	-	13	47	7	3^2	-	37	3	31	11	7	19	-	17	13	-	3	7	-
39	3	43	41	7	19	-	3	-	11	13	7	-	3^2	-	23	17	-	7	3	11
41	-	3^2	7	13	31	-	11	29	23	7	3^2	41	-	3	-	19	7	11	47	3^2
43	17	7	23	-	11	3	-	19	7	-	13	3	-	43	3	7	-	3^2	29	-
45	5	5	5	5	5	5	5	5	5	5	5	5	5	5	5	5	5	5	5	5
47	11	3	17	-	3^2	-	7	3	-	-	19	11	31	7	-	-	3	47	37	17
49	-	47	19	17	-	7	41	23	3	11	-	3^2	7	-	3	-	-	3	13	7
51	3^2	23	-	3	7	53	3	11	-	3^2	-	7	13	19	-	37	-	-	7	-
53	43	3	-	7	3	11	13	3^2	-	59	7	-	-	3	61	29	11	7	23	13
55	5	5	5	5	5	5	5	5	5	5	5	5	5	5	5	5	5	5	5	5
57	3	7	-	3^2	-	-	23	13	7	3	-	-	11	-	-	7	-	67	3	-
59	7	13	-	-	3	-	-	7	17	37	11	-	-	3	7	47	3	-	43	3^2
61	-	29	3	-	-	3	7	-	11	17	31	19	-	7	3	-	59	3^2	-	11
63	3	-	13	19	-	7	11	53	-	3	17	23	7	-	-	3^2	-	11	3	7
65	5	5	5	5	5	5	5	5	5	5	5	5	5	5	5	5	5	5	5	5
67	-	-	11	7	-	29	19	-	3	-	7	3^2	17	11	3	-	13	7	31	-
69	11	-	7	3	-	43	3	-	53	7	13	11	3	17	41	3	7	19	3^2	-
71	37	7	-	-	13	-	-	3^2	7	11	23	43	-	31	17	7	3^2	13	-	3
73	7	19	3	-	23	3^2	-	7	3	29	-	13	-	-	7	17	-	37	11	-
75	5	5	5	5	5	5	5	5	5	5	5	5	5	5	5	5	5	5	5	5
77	17	3^2	29	11	19	7	-	3	-	41	3^2	-	7	3	11	23	3	17	-	7
79	-	11	3	31	7	3	13	-	3^2	23	-	7	11	29	3	19	-	3^2	7	13
81	13	-	17	7	59^2	-	3^2	19	-	3	7	37	3	13	-	3^2	31	7	3	17
83	-	3	7	17	3^2	-	29	13	11	7	3	47	-	3^2	-	-	7	-	19	11
85	5	5	5	5	5	5	5	5	5	5	5	5	5	5	5	5	5	5	5	5
87	7	-	19	3	11	17	3	7	13	3^2	61	53	3	41	7	11	43	-	3^2	-
89	-	3	11	-	3	37	7	3^2	-	-	29	59	-	7	67^2	13	3^2	-	-	3
91	11	-	3	-	-	7	-	17	3	13	-	11	7	-	3^2	-	-	3	67	7
93	3	31	37	13	7	-	3	-	17	11	-	7	3^2	23	-	3	13	-	7	-
95	5	5	5	5	5	5	5	5	5	5	5	5	5	5	5	5	5	5	5	5
97	19	23	7	43	13	11	-	-	3^2	7	17	3	-	-	3	-	7	3^2	59	19
99	3	7	-	11	-	59	3^2	29	7	31	-	13	3	53	11	7	37	-	23	-

	50	51	52	53	54	55	56	57	58	59	60	61	62	63	64	65	66	67	68	69
01	3	-	7	3^2	11	-	3	-	-	7	17	-	3^2	-	37	11	7	-	3	67
03	-	3^2	11	-	3	-	13	3	7	-	3^2	17	-	11	19	7	31	-	-	3^2
05	7	5	3.5	5	23	3.5	19	7	3^2	5	5	3.5	17	13	3.5	5	5	3^2	5	5
07	3	-	41	29	-	-	3^2	13	-	11	-	31	3	7	43	3^2	-	19	3	-
09	-	13	-	-	3^2	7	71	11	37	19	3	41	7	3^2	13	23	3	-	11	7
11	-	19	3^2	47	7	11	31	-	13	23	-	3^2	-	-	3	17	11	3	7	-
13	3^2	-	13	7	-	37	3	29	-	3^2	7	-	19	59	11	13	17	7	3^2	31
15	17	11	7	5	3.5	5	5	3^2	5	7	3.5	5	11	3.5	5	5	3^2	17	29	3.5
17	29	7	37	13	-	3^2	41	-	7	61	11	3	-	-	3^2	7	13	3	17	-
19	7	-	17	3^2	-	-	3	7	11	3	13	29	3^2	71	7	41	-	-	3	11
21	-	3^2	23	17	13	-	7	3	-	31	3^2	-	-	7	-	-	3	11	19	3^2
23	-	47	3	-	11	7	-	59	3^2	-	19	13	7	-	3	11	37	3^2	-	7
25	5^2	5^2	5^2	5^2	5^2	5^2	3^2	5^2	5^2	5^2	5^2	5^2	5^2	5^2	5^2	5^2	5^2	5^2	5^2	5^2
27	11	3	-	7	3^2	-	17	23	-	-	7	11	13	3^2	-	61	3	7	-	3
29	47	23	3^2	73^2	61	19	13	17	29	7	-	3^2	-	-	3	-	7	3	-	13
31	3^2	7	-	3	-	-	3	11	7	3^2	37	-	31	13	59	7	19	53	3^2	29
33	7	29	-	-	3	11	43	3^2	19	17	3	-	23	3	7	47	3^2	-	-	3
35	19	13	3.5	11	5	3^2	7	31	3.5	5	17	3.5	29	7	3^2	5	5	3.5	5	19
37	23	11	-	3^2	-	7	3	-	13	3	-	17	3^2	-	41	3	-	-	43	7
39	-	3^2	13	19	7	29	-	3	-	-	3^2	7	17	3	47	13	3	23	7	3^2
41	71^2	53	3	7	-	3	-	-	3^2	13	7	23	79^2	17	19	31	29	3^2	-	11
43	3	37	7	13	-	23	3^2	-	-	7	-	-	3	-	17	3^2	7	11	3	53
45	5	7	5	5	3^2	5	5	3.5	7	29	3.5	5	5	3^2	5	7	3.5	19	5	3.5
47	7	-	3^2	-	13	3	-	7	3	19	-	3^2	-	11	7	-	17	13	41	-
49	3^2	19	29	3	-	31	7	-	-	3^2	23	11	3	7	-	37	61	17	3^2	-
51	-	17	59	-	23	7	-	3^2	-	11	3	-	7	29	-	-	3^2	43	17	7
53	31	-	17	53	7	3^2	-	11	3	-	-	7	13	-	3^2	-	-	3	7	17
55	3.5	5	5	3^2	5	11	3.5	5	5	3.5	7	5	3^2	31	5	3.5	5	7	3.5	13
57	13	3^2	7	11	17	-	-	19	-	7	3^2	47	-	13	11	79	7	29	-	3^2
59	-	7	3	23	53	17	-	13	31	59	73	3	11	-	3	7	-	3^2	19^3	-
61	7	13	-	3	43	67	3^2	7	-	3	11	61	3	-	7	3^8	-	-	3	-
63	61	3	19	31	3^2	-	7	17	11	67	43	-	-	3^2	23	-	3	-	-	11
65	5	5	3^2	29	5	3.5	11	5	3.5	5	5	3^2	7	19	3.5	13	31	3.5	5	7
67	3^2	-	23	3	7	19	3	73	-	3^2	-	7	3	-	29	11	59	67	3^2	-
69	37	3	11	7	3	-	-	3^2	-	47	7	31	-	11	-	-	3^2	7	-	23
71	11	-	7	41	-	3^2	53	29	19	7	13	11	-	23	3^2	-	7	37	-	-
73	19	7	-	3^2	13	-	31	23	7	11	-	-	3^2	-	-	7	-	13	29	19
75	5^2	3^2	5^2	5^2	5^2	5^2	5^2	5^2	5^2	5^2	5^2	5^2	5^2	5^2	5^2	5^2	5^2	5^2	5^2	5^2
77	-	31	3	19	-	11	7	53	3^2	43	59	29	-	7	17	-	11	3^2	23	-
79	3	-	-	11	-	7	3^2	-	-	3	-	37	7	-	11	3^2	-	-	3	7
81	-	11	-	-	3^2	-	13	41	-	-	3	7	11	3^2	-	-	17	-	7	13
83	13	71	3^2	7	-	3	-	-	37	31	7	3^2	61	13	3	29	41	7	-	-
85	3^2	17	7	3.5	5	5	3.5	13	11	3^2	5	5	3.5	5	5	3.5	7	23	3^2	11
87	-	7	17	-	31	37	11	3^2	7	-	3	23	-	3	13	7	3^2	11	71	17
89	7	-	41	17	11	3^2	-	7	13	53	-	3	19	-	3^2	11	-	31	83^2	29
91	3	29	11	3^2	17	-	7	-	43	3	-	41	3^2	7	-	13	-	-	3	-
93	11	3^2	67	-	3	7	-	3	71	13	3^2	11	7	3	43	19	23	-	61	3^2
95	5	5	3.5	13	7	3.5	17	19	3^2	11	23	3.5	5	5	3.5	5	13	3^2	7	5
97	3	-	-	7	23	29	3^2	17	-	3	7	-	3	-	73	3^2	37	7	11	-
99	-	3	7	-	3^2	11	41	3	17	7	19	-	-	3^2	67	-	7	13	-	3

	70	71	72	73	74	75	76	77	78	79	80	81	82	83	84	85	86	87	88	89
01	-	3^2	19	7	3	13	11	17	29	-	3^2	-	59	3	31	-	47	7.11	13	3^2
03	47	-	3	67	11	41	-	-	3^2	7	53	37	13	19	3	11	7	3^2	-	29
05	3.5	5.7	11	3.5	5	19	3^2	23	5.7	3.5	5	5	3.5	11	5	3^2	5	5	3.5	13
07	13	23	-	-	3^2	-	-	3.7	37	-	17	67	29	3^2	7	47	19	-	-	3
09	43	-	3^2	-	31	3	7	13	19	11	-	3^2	-	7	3	67	-	3	23	59
11	3^2	13	-	3	-	29	43	11	73	3^2	-	-	3.7	-	13	3	79	31	3^2	67
13	-	3	-	71	3.7	11	23	3^2	13	41	3	19	43	17	47	-	3^2	-	7	3
15	23	5	3.5	5.7	5	3^2	5	5	3.5	5	5.7	3.5	31	5	3^2	13	5	3.5	41	5
17	3.	11	7	3	-	-	3	-	-	3.7	-	-	3^2	-	19	17	7	23	3	37
19	-	3^2	-	13	3	73	19	31	7	-	3^2	23	-	47	-	7	13	-	-	3^2
21	17	-	29	-	41	23	-	7	3^2	89^2	13	3	-	53	3.7	-	37	3^2	-	11
23	3	17	31	3	13	-	3^2	-	-	19	71	-	3	29	-	3^2	-	13	17	-
25	5^2	5^2	5^2	5^2	5^2	5^2	5^2	5^2	5^2	5^2	5^2	5^2	5^2	5^2	5^2	5^2	5^2	5^2	5^2	5^2
27	.	-	3^2	17	7	13	29	-	3	-	23	3^2	19	11	3	-	-	3	13	79
29	3^2	-	-	3.7	17	-	3	59	-	3^2	31	11	13	-	-	3	-	29	3^2	-
31	79	3	7	-	3	17	13	3^2	41	7.11	3	47	-	3	-	19	3^2	-	-	13
33	13	7	3	-	-	3^2	17	19	3.7	-	29	3	-	13	3^2	23	89	41	73	-
35	3.5	5	5	3^2	5	11	3.5	5.7	5	3.5	5	5	3^2	5	5.7	3.5	11	5	3.5	5
37	31	3^2	-	23	37	-	7	3	17	-	3^2	79	-	3.7	13	-	3	-	-	3^2
39	-	59	19	41	43	3.7	-	71	3^2	17	-	3	7	31	29	-	53	3^2	-	7
41	3	37	13	3	7	-	3^2	-	-	3	17	7	41	19	23	3^2	-	-	3.7	-
43	-	3	-	7	3^2	19	-	29	23	13	3.7	17	-	3^2	-	-	43	7	37	11
45	5	5	3^2	13	5	3.5	11	5	3.5	5.7	5	3^2	17	5	3.5	5	5.7	11	29	5
47	3^2	7	-	31	11	-	3	61	19	3^2	13	-	3	17	-	11	-	-	3^2	23
49	19	3	11	-	13	-	-	3^2	47	-	3	29	73	11	17	83	3^2	13	-	19
51	11	-	3	-	-	3^2	7	23	3	-	83	11	37	7	3^2	17	41	3	53	-
53	3	23	-	3^2	29	13	3	-	-	11	-	31	3^2	-	79	3	17	-	13	7
55	17	3^2	5	5	3.5	5	5	11	5	37	3^2	5.7	13	3.5	19	29	3.5	17	5.7	3^2
57	-	17	41	7	-	11	13	-	3^2	73	7	3	23	61	3	43	11	3^2	17	13
59	13	-	17	11	-	-	3^2	-	29	3.7	-	41	3	13	11	3^2	7	19	3	17
61	23	11	53	17	3^2	-	47	13	7	19	3	-	11	3^2	-	7	3	-	-	29
63	7	13	3^2	37	17	3	79	7	3	-	11	3^2	-	-	3.7	-	-	23	-	-
65	3^2	5	5	3.5	5	17	3.5	5	11	3^2	5	23	3.5	5.7	5	3.5	5	5	3^2	11
67	37	3	13	53	19	23	17	3^2	-	31	3	-	7	3	-	13	3^2	11	-	3.7
69	-	67	3	-	7	3^2	-	17	43	13	-	3.7	-	-	3^2	41	-	37	7	-
71	3	71	11	3^2	31	67	3	19	17	3	7	-	3^2	11	43	3	13	7	3	-
73	11	3^2	7	73	47	-	-	3	-	17	3^2	11	-	3	37	-	3.7	31	19	3^2
75	5^2	5^2	5^2	5^2	5^2	5^2	5^2	5^2	3^2	5^2	5^2	5^2	5^2	5^2	5^2	5^2	5^2	5^2	5^2	5^2
77	3.7	-	19	3	-	-	3^2	7.11	-	3	41	13	31	-	7	3^2	-	67	11	47
79	-	3	29	47	3^2	13	7	3	-	79	3	-	17	3^2	61	23	11	-	13	41
81	73	43	3^2	61	-	3.7	-	31	37	23	-	3^2	13	17	11	-	-	3	83	7
83	3^2	11	-	23	7	-	13	43	-	3^2	59	7	11	83	17	3	19	-	3^2	13
85	13	3.5	31	5.7	3.5	31	29	3^2	19	5	11	5	5	3.5	5	17	3^2	5.7	5	3.5
87	19	-	3.7	83	-	3^2	-	13	11	7	-	3	-	-	3^2	31	17	29	-	43
89	17	13	37	3^2	-	-	11	-	23	3	-	19	3^2	-	13	3.7	-	17	3	89
91	7	3^2	23	19	11	-	-	3.7	13	61	3^2	-	-	3	7	71	3	59	17	3^2
93	41	-	11	-	59	3	7	-	3^2	-	-	3	-	7	19	13	-	3^2	-	17
95	11	5	5	3.5	5	5.7	3^2	5	5	3.5	5	11	3.5	23	5	3^2	37	5	3.5	5.7
97	47	3	-	13	3^2	71	43	23	53	11	3	7	-	3^2	29	-	13	19	31	3
99	31	23	3^2	7	-	17	-	11	3	-	13	3^2	43	37	3	-	-	3.7	11	-

	90	91	92	93	94	95	96	97	98	99	100	101	102	103	104	105	106	107	108	109
01	-	19	3	71	17	3	-	89	11	-	73	3.7	101^2	-	3	-	-	29	7	11
03	3	-	-	3.7	-	13	11	31	-	3	7	-	19	-	101	3^3	23	11	13	-
05	5	3.5	5.7	5	3.5	5	17	3.5	37	5.7	3.5	43	13	3.5	5	11	3.5	5	5	3.5
07	-	7	11	41	23	3	13	17	3.7	-	-	3^2	59	11	3	19	-	43	101	13
09	11	-	-	29	97^2	37	3	19	17	3^3	-	11	41	13	7	31	103^2	-	3^2	-
11	-	3	61	-	3	-	7	13	-	17	47	-	-	3.7	29	23	3^3	-	19	3
13	-	13	37	67	-	3.7	-	11	3	23	17	3	7	-	13	-	-	3	11	7
15	3.5	5	19	3.5	5.7	11	3.5	29	13	3.5	5	5.7	3.5	5	5	3.5	11	5	3.5	37
17	71	3^2	13	7	43	31	59	41	-	47	3.7	67	17	19	11	13	3	7	29	3^2
19	29	11	3.7	-	-	19	-	-	3^2	13	43	3	11	17	23	67	37	3^3	31	61
21	31	7	-	13	-	-	3^2	-	23	3	11	29	3	-	17	3.7	13	71	3	67
23	7	3	23	-	3^3	89	-	3.7	19	-	13	53	-	31	7	17	3	-	79	11
25	5^2	5^2	5^2	5^2	5^2	5^2	5^2	5^2	5^2	5^2	5^2	5^2	5^2	5^2	5^2	5^2	5^2	5^2	5^2	5^2
27	17	-	-	3	11	7	3	71	31	3^2	37	13	3.7	23	-	11	-	17	3^3	7
29	-	17	11	19	3.7	13	-	23	-	-	3	7	53	11	-	-	3^2	-	13	3
31	11	23	17	31	-	3^3	-	37	29	-	7	11	13	-	19	-	-	3.7	-	17
33	3	-	7	17	-	-	13	-	-	3.7	79	-	3^3	-	-	3	31	-	23	13
35	13	3.5	5	5	3.5	5	41	3.5	5.7	5	3.5	5	23	3.5	5	5.7	3.5	19	11	3.5
37	7	-	3	-	-	11	23	13	3^2	19	-	31	29	-	3.7	41	11	3^2	-	-
39	23	19	-	11	-	-	3.7	-	-	3	-	-	3	7	13	3^2	-	-	3	-
41	-	11	-	-	3^2	29	31	17	13	-	3	-	19	3^3	53	83	3	23	37	3.7
43	-	41	13	-	19	3	-	-	17	61	83	3.7	-	-	3	13	29	3	7	31
45	3.5	31	5	3.5	5	23	3.5	5	11	3.5	5.7	5	3.5	5	5	3.5	5	5.7	3.5	11
47	83	3	7	13	47	-	11	19	43	29	17	73	-	3	31	53	3.7	11	-	41
49	-	7	3	-	11	3^2	-	-	3.7	-	13	17	37	79	3^3	11	23	3	19	-
51	3.7	-	29	3^2	13	-	3	7	-	31	19	-	17	11	7	3	-	13	3	47
53	11	3^4	19	47	23	41	7	3	59	37	3^2	13	-	3.7	-	61	53	-	-	3^2
55	5	5	3.5	5	31	3.5	5	5	3.5	11	5	3.5	5.7	19	3.5	5	5	3.5	13	5.7
57	3	-	-	3	7	19	29	11	-	3	89	7	13	-	-	17	-	31	3.7	-
59	-	43	47	7	3^2	79	13	3	-	23	3.7	-	-	3^2	-	-	11	29	-	13
61	13	-	3.7	23	-	3	-	43	19	7	-	3^2	31	13	11	59	7	17	-	97
63	19	17	59	3	-	73	3	13	7	3^3	29	-	11	43	-	3.7	-	47	17	19
65	5.7	3.5	17	5	3.5	5	5	3.5	5	5	11	19	5	3.5	5.7	5	3.5	5	41	17
67	-	89	3	17	-	3^2	7	-	11	-	-	3	-	7	3^2	-	-	37	-	11
69	3	53	13	3^3	17	7	11	-	71	3	-	-	3.7	-	19	13	47	89	3	7
71	47	3^2	73	-	11	17	19	3	-	13	3^8	7	-	3	37	31	3	-	7	23
73	43	-	11	13	-	3	17	29	3^2	-	7	3	-	23	3	97	13	3.7	83	-
75	5^2	5^2	5^2	5^2	5^2	5^2	5^2	5^2	5^2	5^2	5^2	5^2	5^2	5^2	5^2	5^2	5^2	5^2	5^2	5^2
77	29	3.7	-	-	13	61	-	3	17	11	3	-	43	3^2	-	7	3	13	73	3
79	7	67	3^2	83	-	31	-	7.11	37	17	-	13	19	97	3.7	71	59	3	23	-
81	3^2	-	-	53	19	13	3.7	-	41	3^2	17	-	23	7	47	3	11	-	13	79
83	31	3	-	11	29	37	23	3^2	-	67	3	17	13	3	11	19	3^2	41	-	3.7
85	23	11	3.5	5	5.7	3.5	13	19	3.5	5	5	3.5	11	31	3.5	29	5	3.5	5.7	13
87	13	-	37	3.7	53	-	3	-	-	3	11	61	3^3	13	-	3	-	23	19	-
89	61	3^2	7	41	3	43	-	13	29	7	19	23	-	3	17	-	3.7	-	-	11
91	-	13	19	-	-	23	11	-	3.7	97	-	43	41	-	13	17	-	11	-	29
93	3.7	29	-	31	11	53	3^3	7	13	3	-	-	47	19	7	11	17	43	3	-
95	17	3.5	11	5	3.5	19	5.7	3.5	5	5	3.5	5	29	3.5	5	13	3.5	17	5	3.5
97	11	17	3^2	-	-	3.7	-	97	3	13	23	11	7	37	3	-	19	59	17	7
99	3^3	-	17	13	23	29	53	41	19	11	-	31	3	-	-	3	13	-	3.7	17

	110	111	112	113	114	115	116	117	118	119	120	121	122	123	124	125	126
01	19	17	23	3	13	31	3^2	–	–	3	11	–	3.7	–	–	3^3	–
03	–	3	17	89	3.7	–	41	47	29	–	3	13	–	3^2	79	–	3
05	31	5	3.5	5.7	5	3.5	5.11	5	3.5	5	5.7	3.5	5	23	3.5	41	5
07	3^2	29	7	3	17	37	53	23	–	3.7	–	–	13	31	19	3.11	7
09	101	3.7	11	43	3	17	13	3^2	7	–	3	–	29	3.11	–	7	3^3
11	13	41	37	–	–	3^2	17	7	31	43	–	3.11	–	13	3.7	–	–
13	3	–	–	3^3	101	29	3.7	13	–	3.11	41	–	23	7	–	43	–
15	5	3.5	5	31	3.5	5.7	23	3.5	17	5	3.5	5	5.7	3.5	13	5	3.5
17	23	–	3	–	7	349	–	–	13	17	61	3.7	19	109	3	–	31
19	3	–	13	3.7	19	–	3^2	–	53	29	17	–	3	97	11	13	–
21	103	337	7	–	3^3	41	–	3	–	13	3	17	101	3^2	–	19	3.7
23	73	7	29	13	–	23	59	19	3.7	–	11	3^3	17	–	41	7	13
25	5^2	5^2	5^2	5^2	5^2	5^2	5^2	5^2	5^2	5^2	5^2	5^2	5^2	5^2	5^2	5^2	5^2
27	–	3	103	47	13	–	7.11	3^2	–	–	19	67	–	3.7	17	–	23
29	41	31	19	–	11	3.7	29	37	3	79	23	13	7	–	3^2	17	73
31	3	–	11	3^2	23	13	3	–	–	41	53	7	3^3	19	31	3	17
33	17	3^2	47	7	37	19	–	3	–	–	3.7	11	13	3	–	83	3
35	5	17	3.5	5	5	3.5	13	5	3.5	5.7	29	3.5	5	5	3.5	23	5.7
37	13	37	17	3	–	83	3^3	97	19	23	–	53	3	13	–	3.7	–
39	19	47	–	23	31	11	103	3.7	–	–	3	61	–	3^3	7	–	3.11
41	61	13	3^2	11	17	3	7	59	3	–	–	19	–	41	3.11	–	–
43	3^3	11	–	19	–	97	3	–	13	3^2	–	–	3.7	–	23	37	47
45	5	3.5	13	5	3.5	5	17	3.5	23	5	3.5	5.7	31	3.5	19	13	3.5
47	–	71	23	7	–	3^2	19	17	359	13	7	3	37	–	3^3	–	–
49	29	–	7	13	107^2	–	353	31	41	3.7	–	–	3^2	53	59	47	13
51	43	3.7	–	–	3.11	–	61	3	7	19	13	29	–	23	–	7	3
53	7	19	31	–	13	3	43	23	3^3	–	17	3	–	11	3.7	–	–
55	3.5	23	5	3.5	29	5	3.5	5	5	3.5	5	5.11	3.5	5.7	47	3.5	5
57	–	3	–	41	19	13	–	3	71	11	3	–	103	3^2	–	29	3
59	–	–	3^3	37	7	3	89	11	59	–	31	3.7	13	17	3	19	–
61	3^2	–	–	3.7	73	11	13	19	29	3^3	7	–	61	47	17	53	11
63	13	3	7	11	3	31	107	3^2	–	7	3	–	–	13	103	17	3.7
65	5	5.7	3.5	5	5	3.5	5	13	3.5	5	19	3.5	5.11	5	3.5	5.7	17
67	3.7	13	19	3^3	–	43	3	41	–	3	11	23^3	29	83	13	59	53
69	–	17	59	–	3	23	7	3	13	–	3^3	43	–	3.7	37	–	41
71	–	–	13	83	–	3.7	11	79	3^2	–	–	3	7	89	3	13	–
73	3	–	–	17	7.11	71	3^2	61	31	13	–	37	3	–	–	3.11	19
75	5^2	5^2	5^2	5^2	5^2	5^2	5^2	5^2	5^2	5^2	5^2	5^2	5^2	5^2	5^2	5^2	5^2
77	19	–	3.7	31	23	17	–	–	37	29	13	3.11	–	–	3	–	7
79	3^2	7	–	3	13	–	17	–	7	3.11	47	19	3	–	–	3.7	31
81	7	3	29	19	43	37	–	3.7	109^2	–	3	13	–	3	7	23	3^2
83	–	53	3	–	–	13	7	–	17	23	43	31	71	29	19	–	11
85	3.5	5	37	3.5	5	5.7	3.5	5	5	17	5	5	3.5	5	5.11	3.5	43
87	–	113	–	59	3.7	–	13	3	–	–	17	7	11	3	–	41	3
89	13	67	53	7	–	3	–	–	3^2	19	7.11	17	–	13	23	–	–
91	3	19	7	3	–	67	3^3	13	23	3.7	107	73	17	–	–	3^2	37
93	–	3.7	23	–	3^2	–	11	3	7	67	29	89	19	17	13	7	3
95	5.7	5	3.5	43	5.11	3.5	5	5.7	3.5	5	41	3.5	5	37	17	5.11	5
97	3^3	–	13	29	–	–	3.7	47	–	31	–	–	3	23	–	13	–
99	11	3	–	–	3	7	–	19	73	13	37	11	7	3	29	43	17

www.ingramcontent.com/pod-product-compliance
Lightning Source LLC
LaVergne TN
LVHW020122110826
845151LV00001B/243

* 9 7 8 1 4 2 5 5 4 1 0 9 5 *